Advances in Geographic Information Science

The ***Advances in Geographic Information Science*** series aims to: present current and emerging innovations in GIScience; describe new and robust GIScience methods for use in transdisciplinary problem solving and decision making contexts, illustrate GIScience case studies for use in teaching and training situations, analyze GIScience tools at the forefront of scientific research, and examine the future of GIScience in an expanding knowledge-based economy. The scope of the series is broad and will encompass work from all subject disciplines that develop and use an explicitly spatial perspective for analysis and problem-solving.

Shray Pathak • Anoop Kumar Shukla
Shreya Sharma
Editors

Geo-Data Revolution

Advances in Spatial Analysis and Natural Hazard Mapping

Springer

Editors
Shray Pathak
Department of Civil Engineering
Indian Institute of Technology Ropar
Rupnagar, Punjab, India

Shreya Sharma
AWaDH, TIF
Indian Institute of Technology Ropar
Rupnagar, Punjab, India

Anoop Kumar Shukla
Manipal School of Architecture and Planning
Manipal Academy of Higher Education
Manipal, Karnataka, India

ISSN 1867-2434 ISSN 1867-2442 (electronic)
Advances in Geographic Information Science
ISBN 978-3-031-94077-4 ISBN 978-3-031-94078-1 (eBook)
https://doi.org/10.1007/978-3-031-94078-1

This Springer imprint is published by the registered company Springer Nature Switzerland AG
The registered company address is: Gewerbestrasse 11, 6330 Cham, Switzerland

Preface

The rapid advancement of geospatial technologies and analytical methods has ushered in a new era of understanding and managing our planet. Geo-data revolution: Advances in spatial analysis and natural hazard mapping explore how cutting-edge developments in geographic information system (GIS), remote sensing, artificial intelligence (AI), and big data analytics are transforming spatial analysis and hazard assessment. In recent years, the demand for precise and real-time geospatial information has surged, driven by climate change, urban expansion, and increasing frequency of natural disasters. This book brings together innovative research and practical applications to demonstrate how spatial data can be leveraged for risk assessment, disaster preparedness, and environmental planning. Through a multidisciplinary approach, we examine the role of geo-data in mapping hazards such as earthquakes, floods, landslides, and cyclones, providing insights into predictive modeling and early warning systems.

The book is structured to guide readers through fundamental concepts, advanced methodologies, and real-world case studies. It is intended for researchers, practitioners, policymakers, and students who seek to deepen their understanding of spatial analysis and its implications for hazard mitigation and sustainable development. We hope that this book serves as a valuable resource in navigating the evolving landscape of geo-data science and contributes to the broader goal of enhancing resilience in the face of natural hazards.

Rupnagar, Punjab — Shray Pathak
Manipal, Karnataka, India — Anoop Kumar Shukla
Rupnagar, Punjab — Shreya Sharma

Contents

About the Editors

Shray Pathak is working as an Assistant Professor in the Department of Civil Engineering, Indian Institute of Technology Ropar, Rupnagar, Punjab, India. Dr. Pathak has received the bachelor's degree in Civil Engineering from the National Institute of Technology (NIT), Kurukshetra, India, in June 2012 and completed his master's and Ph.D. degrees in civil engineering from the Indian Institute of Technology (IIT), Roorkee, Uttarakhand, India, in June 2014 and February 2019, respectively. His research interests include hydrology, remote sensing, GIS, urban sustainability, and climate change. Earlier, he was employed as an Assistant Professor in the Department of Civil Engineering, Netaji Subhas University of Technology, New Delhi, from December 2021 to September 2023. He has a post-doctoral experience of 2.5 years in the School of Geographic Sciences, East China Normal University (ECNU), Shanghai, China (June 2019 to December 2021), and worked on the risk assessment analysis of urban flooding in coastal and non-coastal regions under extreme climate scenarios. Till date, he has amassed over 45 publications, spanning peer-reviewed journals, conferences, and book chapters. He has received the 2022 Best Practice Oriented Paper for the journal paper published in ASCE Journal of Hazardous, Toxic, and Radioactive Waste awarded by the American Society of Civil Engineers (ASCE).

Anoop Kumar Shukla is an Assistant Professor at Manipal School of Architecture and Planning (MSAP), Manipal Academy of Higher Education, Manipal, Karnataka, India. He holds 8 years of research experience in research and development in geomatics engineering, water and wastewater management, remote sensing, and geographical information system (GIS), as well as in water quality, climate change, crop yield modeling, and related subjects. Dr. Shukla has published 35 high-impact journal papers, 45 international/national conference papers, and 23 book chapters across ASCE, Wiley, and Springer publications. He is a member of various national and international academic, research, and administrative bodies including IAH, INCA, IEEE-GRSS, AGU, and ASCE. He earned his M.Tech in geomatics engineering and Ph.D. in remote sensing and GIS applications in water resource planning and management from the Indian Institute of Technology, Roorkee,

Uttarakhand, India. Following this, he pursued a postdoctoral fellowship (PDF) at the Institute of Geographic Sciences and Natural Resources Research (IGSNRR), Chinese Academy of Sciences (CAS), Beijing, China. Dr. Shukla serves as an Associate Editor for the Journal of Hazardous, Toxic, and Radioactive Waste (ASCE) and as a Review Editor on the Editorial Board of Water and Wastewater Management and Organic Pollutants, Frontiers in Environmental Science journal. Dr. Shukla is a reviewer for more than 90 peer-reviewed international SCI journals.

Shreya Sharma serves as the Research and Development (R&D) and Corporate Manager at the Agriculture and Water Technology Development Hub (AWaDH), a Technology Innovation Foundation (TIF) at the Indian Institute of Technology (IIT) Ropar in Rupnagar, Punjab, India. In this capacity, she plays a pivotal role in advancing research initiatives and fostering corporate collaborations within the hub. Before her current role, Dr. Shreya Sharma served as an Assistant Professor at Chitkara Business School, Chitkara University, and also held a similar position at Amity University. She earned her bachelor's degree in commerce (B.Com) in 2015, followed by a master's degree in business administration (MBA) with a specialization in finance in 2017. In 2023, she completed her Ph.D. in management from Devi Ahilya University, Indore. Her research interests span various domains, including capital market finance; corporate finance; green finance; environmental, social, and governance (ESG) factors; as well as sustainability and Sustainable Development Goals (SDGs). To date, she has published five research articles in reputed journals, contributed three book chapters, edited one book, and presented her work at multiple national and international conferences.

Ecosystem Services and Sustainable Development Goals: A Comprehensive Review

Neha G. Paswan and Shray Pathak

1 Introduction

Ecosystem Services (ES) are essential for the survival of human life [1]. Human activities like land use land cover (LULC) change and resource extraction have direct impacts on ecosystem function. LULC changes exacerbate ecosystem degradation, increasing vulnerability to natural catastrophes and biodiversity loss [2–4]. These activities disturb the biological and chemical cycles [5], disrupt the Earth's energy balance, and contribute to phenomena such as global warming and climate change. The complete knowledge of economic and social ES helpful in understanding the interaction within the ES components and helps us to maintain the human life as well as the ES quality. This approach will also be useful in sustainable practices for achieving sustainable development goals. It allows decision-makers to consider the opportunity costs of using ecosystems, including the benefits foregone from other uses, leading to more informed choices. Urbanization has a significant influence on ecological services [5]. The haphazard development of urbanization promotes habitat destruction, loss of nature reserves, pollution, and change in the climate, leading to the deterioration or loss of important ES Gretchen Daily's 1997 study highlights how human lifestyles, pushed by urbanization and industrialization, might damage biodiversity and jeopardize future generations' well-being. Humans benefit from ES, which functions as a route for the development of nature [6, 7]. Maintaining these services is essential for ecosystem systems and human beings. ES are commonly categorized into four primary groups: provisioning, regulating, cultural, and supporting service (Fig. 1).

N. G. Paswan (✉) · S. Pathak
Department of Civil Engineering, Indian Institute of Technology Ropar, Rupnagar, Punjab, India
e-mail: neha.23cez0014@iitrpr.ac.in; shray.pathak@iitrpr.ac.in

S. Pathak et al. (eds.), *Geo-Data Revolution*, Advances in Geographic Information Science, https://doi.org/10.1007/978-3-031-94078-1_1

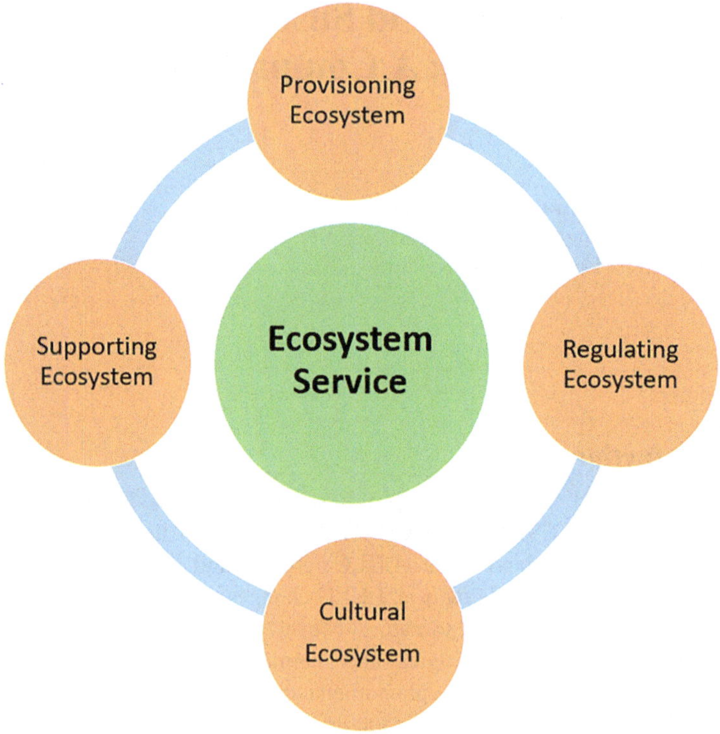

Fig. 1 Types of ecosystem services

Regulating ecosystems performs essential functions by managing and stabilizing environmental processes, ensuring ecosystem balance and performance. These services are crucial for maintaining the health, resilience, and sustainability of natural systems, benefiting both humans and other species. Climate control is a key aspect of ecosystem management. Ecosystems such as forests, oceans, and wetlands play significant roles in regulating the climate. Carbon dioxide is absorbed, gets stored by the process of transpiration, evaporation, and carbon sequestration, and helps to reduce the effects of greenhouse gases. Forest covers lower the carbon dioxide present in the atmosphere and store in wood and soil, which will support the variety of plants and healthy soil. Biodiversity is supported by ecosystem by maintaining the biogeochemical cycle, enhancing soil carbon storage through root systems and litter decomposition. Reforestation initiatives and sustainable forestry practices are essential to preserving their capacity as carbon sinks.

Provision ES covers all the tangible and physical resources from the ecosystem that are basic needs for human for their survival and to maintain the economic system [7]. This service consists of a wide range of resources. Agricultural production facilitated by ecosystem-mediated soil fertility and water availability, as well as the provisions of aquatic resources for fishers [6]. Wooded ecosystems provide essential timber and fiber that are primary products for construction industry. Various manufacturing sectors, including textiles and paper production [8]. The preservation and sustainable management of these ES are imperative to meet current anthropogenic demands while ensuring the resilience and functionality of natural systems for future generations [9].This demands for the implementation of sustainable harvesting practices, conservation strategies, and ecosystem restoration initiatives to maintain the integrity and operational capacity of these vital ecological systems [10].

The essential biogeochemical processes that sustain ecosystem resilience and functionality and enable the provision of extra ecosystem benefits are referred to as supporting ES [11]. To sustain ecosystem productivity and regulate water cycles, rely on soil-related processes like soil formation, organic matter decomposition, rhizosphere interactions, and nutrient cycling, particularly nitrogen fixation and mineralization [12].

Supporting ES are fundamental for the processing of ecosystems, providing essential processes that sustain all other ES. Soil formation involves the gradual breakdown of rocks and minerals through physical, chemical, and biological processes, forming the basis for terrestrial life by providing a medium for plant growth and nutrient cycling. Primary production, driven by photosynthesis, generates organic matter, which serves as the basis of food webs, sustaining various communities of organisms. Nutrient cycling involves the movement of elements like carbon, nitrogen, phosphorus, and other nutrients, ensuring the growth and maintenance of these biological communities.

Cultural ecosystem services (CES) are essential for regulating human-nature relationships, fostering place awareness and belonging, and promoting environmental conservation ethics [13]. CES provide significant cultural, spiritual, and recreational benefits [5]. They enhance quality of life by supporting cultural identity, social cohesiveness, and overall health [14]. Additionally, these spaces often hold spiritual and religious significance, serving as centers of cultural heritage [15].

Ecosystems are valuable for environmental education and scientific research, increasing ecological literacy and fostering a better understanding of natural processes [16]. The diverse nature of these services highlights their importance in the broader context of ecosystem value and sustainable development efforts [17].

Recognizing that the perception and valuation of cultural ES can vary across different sociocultural contexts and temporal scales is crucial. This variability requires adaptive and participatory approaches to their assessment and management [17].

2 ES and SDGs

ES are vital for achieving the Sustainable Development Goals (SDGs) within environmental limits. They provide food and water security, preserve biodiversity, and regulate climate through carbon sequestration. Green infrastructure supports these services and has significant economic impacts beyond urban development. Urban green infrastructure (UGI) is especially crucial for water systems, influencing water quality, managing water flow, and mitigating flood risks.

Climate change increases heat loads and energy consumption in cities, causing extra stress. Cities may use UGI to adapt and create resilience to these energy-related constraints. However, existing human activities have depleted natural capital, and urban development and design paradigms must be fundamentally altered to overcome this deficiency. The bulk of nature's services are in decline worldwide, rendering contemporary wealth and the infrastructure that supports it unsustainable. A cross-cutting, transdisciplinary strategy is required to design and cultivate a more environmentally conscious urban world.

In the context of the SDGs and ES, it is critical to integrate was different knowledge and close gaps, with a special emphasis on societal readiness for urbanization and the use of green infrastructure to achieve urban sustainability goals.

3 Relationships Among Different ES

The interactions between ES (ESs) are complex and interconnected, with different functions frequently supporting or strengthening one another. Supporting services, such as soil formation and nutrient cycling, are essential because they enable the production of food, water, and raw materials. For example, good soil and nutrient-rich ecosystems are critical for sustainable agriculture and agricultural development. Provisioning services, which include items that people may directly consume or utilize (such as food, water, fiber, fuel, and construction materials), rely heavily on supporting services. Regulating services, such as water and air purification, temperature regulation, and disease management, are directly accessible to humans and comprise critical components of ecosystem processes and functions. As a result, human well-being is inextricably linked to providing, cultural, and supporting services. Thus, the notion of ESs is a useful tool for connecting natural systems to human civilization. In Forest Lapland, where livelihoods like forestry, reindeer husbandry, and tourism heavily rely on natural resources and intact ecosystems, these linkages are more apparent compared to industrialized regions of Central Europe [16].

Soil formation, autotrophic plant photosynthesis, and the cycling of energy, matter, and water are all examples of supporting ESs [1]. These services are not just helpful but are necessary for the performance of other ESs. To operate properly, supporting services require specialized ecosystem structures and functions. Ecosystems require the cycling of energy, matter, and water, a variety of essential species, and appropriate abiotic conditions to function [17]. These elements are part of the notions of "ecosystem integrity" and "ecosystem health," which focus on conserving required structures and functions to make the system adaptable (resilient to perturbations) and sustainable in delivering ESs. Supporting services provide critical ecosystem products and services for human well-being [11].

Regulatory services, such as water purification, temperature regulation, and pollination, directly impact provisioning services. Clean water and effective pollination are vital for food production, while climate regulation maintains conditions suitable for agriculture and forestry. Moreover, supporting services facilitate regulating services. Nutrient cycling and primary production help to control climate by sequestering carbon, whereas healthy ecosystems stabilize soils, reduce erosion, and moderate water flow. Cultural services, which provide intangible advantages, rely on the quality and availability of supporting, providing, and regulating services.

The ES framework provides a valuation system that assesses natural capital, extending beyond traditional economic metrics to emphasize ecological functions [6, 7]. This approach integrates ecological considerations into policy formulation and land use planning, enhancing sustainability in decision-making processes [12, 18]. Such collaboration fosters more effective environmental management strategies by incorporating diverse perspectives and areas of expertise [7, 16].

The interaction between different ES involves both synergies and trade-offs that are difficult to predict or manage [13, 19]. Predicting and managing these dynamics can be challenging, often leading to unintended consequences in ecosystem management efforts. Additionally, the distribution of benefits from ES and the associated management costs can vary widely across different populations and regions, raising important issues of environmental justice and equity [20]. Variability in access to and benefits from ES can exacerbate societal inequalities.

The difference between the scales at which ES are produced or function (such as local ecosystems) and the scales at which they are managed or consumed (such as regional or global governance) can create significant governance challenges [13]. Effective management requires aligning the spatial and temporal scales of ecosystem processes with the scales of human activities and decision-making. Despite its benefits, the ES framework risks oversimplifying complex ecological processes [16, 21]. This simplification may limit management techniques and fail to fully appreciate the complexity of ecosystems, potentially undermining conservation and management efforts. While the ES framework enhances our understanding of natural capital and promotes informed decision-making, considering these challenges is crucial for effective and equitable environmental management and conservation. The ES concept offers a valuable heuristic for understanding and valuing nature's contributions to human well-being, but it also presents significant challenges in implementation and interpretation (Fig. 2). Researchers and policymakers must

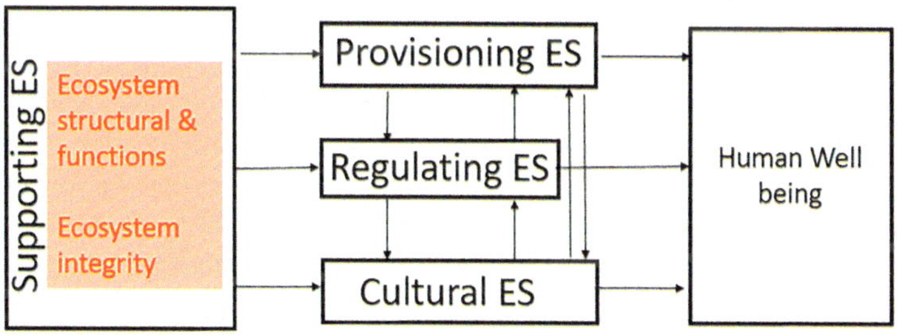

Fig. 2 Interaction of ecosystem services with human well-being

carefully consider these advantages and disadvantages when applying the ES approach to environmental management and decision-making processes.

4 Role of RS–GIS in ES

Remote sensing (RS) and Geographic Information Systems (GIS) play significant roles in mapping, analyzing, and monitoring ES. They enable thorough evaluations, spatial analysis, and access to essential data sources to assess changes in ecosystems and their services. This helps estimate specific ES such as water production and land cover changes, and identify spatial ecosystem dynamics. They also evaluate ecological processes and services such as predicting water yield and monitoring biodiversity, which aid in effective land use planning and policy formulation. These tools are used in both developing and developed nations. RS and GIS are open-source software and satellite-based optical sensors to collect data, enabling the assessment and identification of ecosystem properties, hazards, vulnerabilities, and their relations with ES. They support urban planning, management, and policy-making by improving access to spatial data. RS and GIS techniques empower local communities and decision-makers in managing ecosystems, improving their knowledge and understanding of ecological processes for effective ecosystem management.

ES mapping and monitoring allow for identification of interactions between distinct ES. Ecosystems often provide multiple services, which interact and are influenced by human resource management decisions, leading to various relationships among ES. ES trade-offs exist when one ES increases and another decreases, whereas synergies reflect an improvement in both ES. Recognizing ES interactions is essential for making long-term management decisions that enable the provision of different services. For example, converting forest into agricultural land to boost food supply represents a trade-off, causing declines in services such as water flow and quality regulation, disease control, carbon sequestration, and carbon storage. Conversely, agroforestry systems, which combine agriculture with forestry, create synergies by supporting multiple ES and enabling multifunctional landscapes.

ES trade-offs exist when one ES increases and another decreases, whereas synergies reflect an improvement in both ES [22, 23]. Recognizing ES interactions proves essential for making long-term management decisions that enable the provision of different services [24]. For example, converting forest into agricultural land to boost food supply represents a trade-off, causing declines in services such as water flow and quality regulation, disease control, carbon sequestration, and carbon storage. Also, the agroforestry systems when combine agriculture with forestry create synergies by supporting multiple ES [14] and enabling multifunctional landscapes [8].

Most of the researchers have focused on understanding interactions between provisioning and non-provisioning ES [25]. As a result, trade-offs between ES have been identified nearly three times more frequently than synergies [19]. A common trade-offs observed is between "biomass production" (e.g., timber and fodder) and "climate regulation," while synergies often occur between "habitat and gene pool protection services" and regulating services such as "soil formation" [19].

The dynamics of ES interactions can vary depending on the spatial scale [9, 15, 19]. The couple of ES may exhibit a trade-off at a local level but can be seen as synergy at the national level. Much of the existing research has primarily focused on ecosystem service (ES) interactions at local and plot scales, with less emphasis on regional and global scales [13].

5 Spatial Mapping and Quantification of ES

RS and GIS technologies can spatially map and quantify ecosystem services (ES) across various scales. These technologies produce precise cartographic representations that reveal the spatial distribution and quantities of ES, supporting evidence-based decision-making. For example, multispectral and hyperspectral satellite imagery accurately delineate and classify land cover types. These classifications serve as indicators for ES provision, establishing relationships between landscape characteristics and specific ES. Furthermore, they develop ES indicators and quantify service flows. This integration allows the creation of spatially explicit models that capture the diversity of ES provision across landscapes. Multi-temporal satellite imagery serves as a foundation for analyzing ES dynamics over extended temporal scales. It is crucial for monitoring temporal variations in ES provision, assessing impacts of land use and land cover changes, and evaluating conservation efforts.

The use of time-series satellite data enables researchers to identify trends in ecosystem processes and structures, closely tied to landscapes' ability to deliver ES. Long-term satellite observations assess changes in primary productivity, a key indicator of multiple ES. Moreover, the temporal dimension of remote sensing data detects changes in ES provision due to human activities or natural disasters. This understanding is critical for assessing resilience and vulnerability of ES to environmental change. Furthermore, temporal analysis of ES through remote sensing supports the development and validation of predictive models, enabling scenario-based assessments of future ES under different environmental and management scenarios. This predictive capability informs long-term conservation planning and climate change adaptation strategies.

6 Integration of Heterogeneous Data

GIS serves as a robust platform for the integration and synthesis of disparate datasets, including biophysical, socioeconomic, and cultural variables [6, 11]. This integrative capability facilitates a more holistic and detailed assessment of ES, recognizing intricate interactions between ecological processes and human influences [26, 27]. Integration of diverse data sources within a GIS framework allows for clarification of the multifaceted relationships between ecosystem service provision and human well-being [28, 29]. Moreover, the integration of biophysical data with sociocultural variables allows for the incorporation of stakeholder perspectives and local knowledge into ecosystem service assessments [30]. This participatory approach enhances the relevance and legitimacy of ecosystem service valuations, particularly in complex socio-ecological systems [31, 32]. The GIS-enabled synthesis of heterogeneous data also supports the development of sophisticated ecosystem service models that can capture the non-linear relationships and feedback mechanisms inherent in socio-ecological systems [33]. These integrated models provide a more comprehensive understanding of the trade-offs and synergies among multiple ES across spatial and temporal scales [34].

Furthermore, the integration of remote sensing data with ground-based measurements and socioeconomic statistics within a GIS environment facilitates the validation and calibration of ecosystem service assessments, enhancing their reliability and applicability for decision-making processes [35]. This multi-source data integration approach is particularly valuable for addressing the inherent complexity and uncertainty in ecosystem service quantification and valuation [36]. RS and GIS are pivotal for developing, parameterizing, and applying spatially explicit ecosystem service models [37]. These geospatial technologies integrate diverse data streams to create sophisticated models that simulate the complex dynamics of ecosystem service provision across multiple spatial and temporal scales [38]. The predictive capabilities of these models, combined with scenario analysis techniques, provide a powerful tool for forecasting future trajectories of ecosystem service provision under different environmental, socioeconomic, and policy scenarios [39, 40]. This approach enables researchers and policymakers to assess the potential impacts of alternative land use decisions, climate change effects, and management strategies on ecosystem service delivery [41].

Integrating RS-derived data into ecosystem service models enhances their spatial resolution and temporal relevance, improving the accuracy of model outputs [42]. This integration is particularly valuable for capturing the spatial and temporal variability of ecosystem processes and human influences that affect ecosystem service provision [34, 36]. Scenario-based modeling supported by RS and GIS technologies provides a robust framework for evaluating trade-offs and synergies among multiple ES under alternative future conditions [43]. This analytical capability is essential for guiding adaptive management strategies and evidence-based policy development aimed at optimizing ecosystem service delivery while minimizing potential negative impacts [44]. Furthermore, employing ensemble-modeling techniques facilitated by GIS platforms allows for quantifying uncertainties associated with ecosystem service projections [45]. This approach enhances decision-makers' understanding of the range of possible outcomes, supporting more robust and adaptive long-term planning and policy development [46]. RS technologies provide advanced capabilities for quantitatively assessing and spatially mapping regulating ES, offering crucial insights into the complex biophysical processes that underlie these services [12]. Satellite-based observations offer a synoptic view and repetitive coverage that facilitate monitoring of regulating services across extensive geographic areas and various temporal scales, enabling both local and global assessment [47].

In the realm of carbon sequestration, RS-derived vegetation indices such as the Normalized Difference Vegetation Index (NDVI) and Enhanced Vegetation Index (EVI) serve as proxies for net primary productivity and biomass accumulation. These indices, when combined with allometric models and ground-based measurements, allow for estimating above-ground carbon stocks and fluxes with high spatial and temporal resolution. Advanced techniques like LiDAR (Light Detection and Ranging) enhance forest carbon stock assessments by providing detailed three-dimensional structural information [48].

For water regulation services, RS offers diverse techniques to monitor hydrological processes at landscape to global scales. Microwave remote sensing, particularly through instruments like SMAP (Soil Moisture Active Passive), enables accurate measurement of surface soil moisture. Thermal infrared sensors help estimate evapotranspiration rates, a critical aspect of the water cycle. Integrating these RS-derived hydrological variables with topographic and land cover data in GIS environments allows comprehensive assessments of water regulation services, encompassing flood mitigation and groundwater recharge [49]. RS technologies also play a pivotal role in assessing other regulating services such as climate regulation, air quality management, and erosion control. Aerosol optical depth derived from satellite observations aids in air quality assessments, while multi-temporal analysis of high-resolution imagery enables detection and quantification of soil erosion processes [50].

The integration of RS-derived indicators with process-based models and machine learning algorithms is advancing our ability to quantify and map regulating ES with greater accuracy and detail. This integration enhances understanding of the spatial and temporal dynamics of regulating services, supporting the development of more effective conservation strategies and ecosystem management.

7 Monitoring and Assessment of Biodiversity and Habitat Quality Using Remote Sensing and GIS

Remote Sensing (RS) and Geographic Information Systems (GIS) technologies have become indispensable tools for monitoring and assessing biodiversity and habitat quality, critical for the provision of various ES [16]. These geospatial technologies offer unprecedented capabilities to quantify and map biodiversity patterns and habitat characteristics across different spatial and temporal scales, from local to global extents. High-resolution multispectral and hyperspectral RS data enable detailed characterization of vegetation structure, composition, and phenology, serving as indicators of habitat quality and biodiversity [51]. Advanced techniques like

LiDAR provide three-dimensional information, enhancing assessments of habitat complexity and suitability for various species [52].

Integration of RS data with GIS allows assessment of landscape-level ecological processes crucial for biodiversity conservation. For example, spatial metrics derived from land cover maps quantify habitat fragmentation, evaluating patch size, shape, and isolation [53]. Models using graph theory and circuit theory (e.g., Circuitscape) incorporate RS-derived resistance surfaces to assess habitat connectivity, which is crucial for species movement. Species distribution modeling, vital for biodiversity assessment, also benefits significantly from RS-derived environmental variables. These models, integrating climate data and species records, predict species ranges and assess climate change impacts on biodiversity [54].Time-series analysis of satellite imagery detects changes in biodiversity and habitat quality over time, offering insights into ecosystem dynamics and conservation effectiveness. This temporal perspective is crucial for understanding ecosystem resilience to human activities and natural disturbances.

Machine learning and data fusion techniques applied to RS and GIS data enhance biodiversity mapping accuracy, detecting fine-scale habitat features and individual tree species in complex ecosystems. Integrating RS and GIS with emerging technologies like environmental DNA (eDNA) sampling and acoustic monitoring opens new avenues for biodiversity assessment, especially for elusive species. This comprehensive approach to biodiversity monitoring improves understanding of ecosystem functioning and resilience, guiding conservation strategies and ecosystem management practices [55].

8 Methodologies for Mapping Ecosystem Service Values

One-dimensional proxies are a method for identifying the ES using simple indicators like land cover, meteorological data, and soil datasets. This approach assumes that specific types of land cover or land use directly respond to the provision of particular ES. For instance, forest cover might indicate carbon sequestration potential, while agricultural fields could represent food production services. Researchers such as Costanza et al. [6] have used this technique to map and quantify ES based on the distribution and quantity of different land cover types. One-dimensional proxies leverage easily accessible and understandable data, making them a simple approach to estimating the availability of various ES across regions.

Non-validated models for mapping ES depend on assumptions and expert knowledge rather than real-world data. These models estimate ES by combining variables such as soil type, climate, and plant cover to determine the level of service provided. However, accuracy and reliability may not be correct because these estimates are not based on real-world data. Holzkämper and Seppelt [56] provide instances of this method, in which models estimate ES through theoretical interconnection and expert opinion rather than direct empirical validation. This technique allows for preliminary evaluations and hypothesis development, but the outcomes should be interpreted carefully with caution due to a lack of validation. Validated models for mapping ES are similar to non-validated models in that they predict ecosystem service availability using ecological production functions and other explanatory factors. The only difference is that validated models are calibrated and tested on real data, whether primary (collected directly through experiments or observations) or secondary (obtained from existing datasets). This calibration method increases the model's accuracy and reliability by updating and improving predictions based on actual measurements and observations. Schägner et al. [57] and Wang et al. [58] employed this technique to evaluate models against real-world data on ecosystem service supply, yielding reliable estimates of ES. This method improves the model's reliability, making the results more useful for practical applications and decision-making.

Incorporating the provision of ES into the valuation process is necessary for implicit modeling in value transfer functions. This approach determines the value of ES by taking into account several supply-side aspects, rather than only assessing their provision. Several elements, including socioeconomic conditions, land use patterns, and environmental features, may be taken into account. This approach looks for linkage between these influencing elements and the ES value using statistical or economic techniques. They create transfer functions, or models, that represent the intricate relationships between different variables by utilizing sophisticated modeling approaches and analyzing enormous datasets. Thus, even in situations where direct data is not available, these models can be used to estimate the value of ES in various locations or circumstances.

9 Drivers of Ecosystem Services

The reasons and methods affecting the supply of ES must be understood through a detailed evaluation and comprehension of the fundamental processes in ecosystems. This is important because changes in ecosystems are directly related to the change in the availability of the ES. Drivers can be natural or manmade entity, which can alter the ecosystem directly or indirectly.

The extensive effects of global climate change on ecosystems and their services to human society are profound. One of the most recognize impacts is the change in natural ecosystems and species distribution patterns. As temperatures rise and weather patterns become increasingly unpredictable, many plant and animal species are forced to relocate in search of more favorable and stable living conditions. This leads to instability in the ecosystem and the loss of vital ES essential to human well-being. Freshwater resources, a crucial ecosystem function, are also significantly affected by climate change. Shifting precipitation patterns and the melting of glaciers and ice caps changes the river and lake water flows, affecting the drinking water quality, agriculture, and other critical sectors.

Climate change in a global level can impact the ecosystem as well as human society. With the increasing temperature, the environmental species are forced to relocate in favorable and stable living conditions. Rising carbon dioxide levels result in ocean acidification, which damages aquatic life by preventing it from building protective layers such as coral reefs and shellfish. These considerations make it difficult for forests to properly store carbon while also limiting the amount of wood available. Similarly, overfishing and environmental changes cause fisheries to perish, affecting livelihoods and food security. These all cause a reduce in the resilience of ecosystems that are important for carbon sequestration, lumber production, and fisheries, threatening biodiversity and human well-being.

Moreover, the loss and fragmentation of habitats due to human activities can disturb the basic ES like clean water, fertile soil, and pollination. This may further damage the quality of the remaining habitat. Intact ecosystems, such as forests, help to regulate the climate by sequestering carbon. Habitat loss lowers this capability, which contributes to increased atmospheric carbon dioxide levels and exacerbates climate change. Invasive species have also direct influence on ES. Non-native species, whether purposefully or unintentionally brought into a habitat, have the potential to disturb the ecosystem's functioning. These frequently outcompete native species for resources, causing native population losses or extinctions, and this disturbance can have far-reaching consequences for ecosystem functions critical to human being. Invasive species can alter the structure and function of ecosystems by exceeding, feeding on, or introducing illnesses to native species, lowering ecosystem resilience and functions like pollination, water purification, and disease control. They will influence the basic ecological processes such as nitrogen cycling, water movement, and fire behavior.

Furthermore, invasive species cause considerable economic consequences by damaging the crops, fisheries, forests, and infrastructure, compromising food security and livelihoods, particularly in rural and indigenous populations. Some invasive species have a direct impact on human health because they function as disease vectors or produce poisons. They reduce the quality and quantity of ES, such as when invasive plants clog waterways, lowering water quality and clean water availability, or when invasive animals disrupt food webs, causing the extinction of important species that contribute to services such as pest control and pollination. Effective management and proper preventive techniques are required to control invasive species. Thus, ecosystems and their functions are being harmed by the pollution, resource exploitation, and changes in LULC pattern by the increasing human population, making more sustainable planning and management necessary.

10 Conclusion

The Millennium Ecosystem Assessment (MEA) of 2005 outlined outcomes from direct and indirect human activities as well as natural processes on ecosystems. Human activities like resource extraction and land use changes create widespread degradation and increase the vulnerability of ecosystems to natural disasters and loss of species, which affect ecosystem health and functioning. A sustainable management must be economic and of spirit bases, which reflect the value that society has on the services provided by ecosystems. Urbanization and ES Urbanization has significant adverse effects on the provision of ES due to habitat destruction, pollution, and climate change. There are four classes of ES: provisioning, regulating, maintaining, and cultural. The types of ES are provisioning, regulating, supporting and cultural services. These services are important for human lives as well as the ecological balance of our environment. They provide useful services such as storing carbon and controlling climate, as well as essential goods like food and water; they maintain biodiversity; and they offer cultural and recreational opportunities.

In this context, the achievement of the Sustainable Development Goals (SDGs) needs to preserve the provision of food and water, conserve biodiversity, and store carbon to compensate the climate change, which are some functions of ES. Cities that wish to adapt to climate change can benefit from building an urban green infrastructure (UGI); however, this will only successfully environmentally align the development paradigm of a city if significant changes are made to the current pattern of exploitation and misuse of natural capital caused by human activity. The relationships between the different ES are complex. These connections emphasize how important it is to preserve the integrity and health of ecosystems. To obtain sustainable results, effective management must take into account the trade-offs and synergies among ES.

GIS and RS are important techniques for evaluating and monitoring ecological services. They support policy-making, management, and urban planning by offering data and geographical analysis for measuring ecological change. These technological significantly enhance conservation efforts and improve our understanding of ecological processes and functioning. Various approaches exist for mapping ES, ranging from simple proxies to complex, validated models. Effective management necessitates recognizing the factors influencing ES, such as habitat loss, invasive species, climate change, and human population growth. These variables impacting ecosystem health and service provision requires thorough and sustained management planning.

The study highlights the indispensable role of ecosystem services in achieving sustainable development. The preservation of these services is vital for maintaining environmental sustainability and human well-being. The review emphasizes the importance of using advanced technologies to understand the complex dynamics of ES and integrating these insights into decision-making processes. By adopting comprehensive and scalable management approaches, we can ensure the health of ecosystems and secure long-term benefits for future generations, aligning with the goals of sustainable development.

References

1. MEa, M. E. A. (2005). Ecosystems and human well-being: Wetlands and water synthesis.
2. Deeksha, & Shukla, A. K. (2022). ES: A systematic literature review and future dimension in freshwater ecosystems. *Applied Sciences, 12*(17), 8518.

3. Deeksha, Shukla, A. K., & Rama Devi, N. (2023). Impact of land use and land cover in water resources. In *River conservation and water resource management* (pp. 217–231). Springer.
4. Daily, G. C. (1997). Introduction: What are ES. *Nature's services: Societal dependence on natural ecosystems*. 1(1).
5. Gómez-Baggethun, E., & Barton, D. N. (2013). Classifying and valuing ES for urban planning. *Ecological Economics, 86*, 235–245.
6. Costanza, R., d'Arge, R., De Groot, R., Farber, S., Grasso, M., Hannon, B., Limburg, K., Naeem, S., O'Neill, R. V., Paruelo, J., Raskin, R. G., Sutton, P., & Van Den Belt, M. (1997). The value of the world's ES and natural capital. *Nature, 387*(6630), 253–260.
7. Daily, G. C., Polasky, S., Goldstein, J., Kareiva, P. M., Mooney, H. A., Pejchar, L., Ricketts, T. H., Salzman, J., & Shallenberger, R. (2009). ES in decision making: Time to deliver. *Frontiers in Ecology and the Environment, 7*(1), 21–28.
8. Foley, J. A., DeFries, R., Asner, G. P., Barford, C., Bonan, G., Carpenter, S. R., Chapin, S., Coe, M. T., Daily, G. C., Gibbs, H. K., Helkowski, J. H., Holloway, T., Howard, E. A., Kucharik, C. J., Monfreda, C., Patz, J. A., Prentice, I. C., Ramankutty, N., & Snyder, P. K. (2005). Global consequences of land use. *Science, 309*(5734), 570–574.
9. Hein, L., Van Koppen, K., De Groot, R. S., & Van Ierland, E. C. (2006). Spatial scales, stakeholders and the valuation of ES. *Ecological Economics, 57*(2), 209–228.
10. Bhaduri, A., Bogardi, J., Siddiqi, A., Voigt, H., Vörösmarty, C., Pahl-Wostl, C., & Osuna, V. R. (2016). Achieving sustainable development goals from a water perspective. *Frontiers in Environmental Science*. https://doi.org/10.3389/fenvs.2016.00064
11. Müller, F., & Burkhard, B. (2007). An ecosystem based framework to link landscape structures, functions and services. In *Multifunctional land use: Meeting future demands for landscape goods and services* (pp. 37–63). Springer.
12. Egoh, B., Drakou, E. G., Dunbar, M. B., Maes, J., & Willemen, L. (2012). *Indicators for mapping ES: A review* (p. 111). European Commission, Joint Research Centre (JRC).
13. Raudsepp-Hearne, C., & Peterson, G. D. (2016). Scale and ecosystem services: How do observation, management, and analysis shift with scale—Lessons from Québec. *Ecology and Society, 21*(3).
14. Fagerholm, N., Torralba, M., Burgess, P. J., & Plieninger, T. (2016). A systematic map of ES assessments around European agroforestry. *Ecological Indicators, 62*, 47–65.
15. Willemen, L., Veldkamp, A., Verburg, P. H., Hein, L., & Leemans, R. (2012). A multi-scale modelling approach for analysing landscape service dynamics. *Journal of Environmental Management, 100*, 86–95.
16. Burkhard, B., & Müller, F. (2008). Indicating human-environmental system properties: Case study northern Fenno-Scandinavian reindeer herding. *Ecological Indicators, 8*(6), 828–840.
17. Yuan, M. H., & Lo, S. L. (2020). ES and sustainable development: Perspectives from the food-energy-water Nexus. *ES, 46*, 101217; Adeoti, O., & Fati, B. O. (2020). Barriers to extending piped water distribution networks: The case of Ekiti State, Nigeria. *Utilities Policy*.
18. Maes, J., Egoh, B., Willemen, L., Liquete, C., Vihervaara, P., Schägner, J. P., , Bruna Grizzetti, Evangelia G. Drakou, Alessandra La Notte, Grazia Zulian, Faycal Bourao ui, Maria Luisa Paracchini, Leon Braat, Bidoglio, G. (2012). Mapping ecosystem services for policy support and decision making in the European Union. Ecosystem Services, *1*(1), 31–39.
19. Lee, H., & Lautenbach, S. (2016). A quantitative review of relationships between ecosystem services. *Ecological Indicators, 66*, 340–351.
20. Ernstson, H., & Sörlin, S. (2013). Ecosystem services as technology of globalization: On articulating values in urban nature. *Ecological Economics, 86*, 274–284.
21. Norgaard, R. B. (2010). Ecosystem services: From eye-opening metaphor to complexity blinder. *Ecological Economics, 69*(6), 1219–1227.
22. Bennett, E. M., Peterson, G. D., & Gordon, L. J. (2009). Understanding relationships among multiple ecosystem services. *Ecology Letters, 12*(12), 1394–1404.

23. Rodríguez, J. P., Beard, T. D., Jr., Bennett, E. M., Cumming, G. S., Cork, S. J., Agard, J., Dobson, A. P., & Peterson, G. D. (2006). Trade-offs across space, time, and ecosystem services. *Ecology and Society, 11*(1), 14.
24. Liu, L., Oza, S., Hogan, D., Perin, J., Rudan, I., Lawn, J. E., Cousens, S., Mathers, C., & Black, R. E. (2015). Global, regional, and national causes of child mortality in 2000–13, with projections to inform post-2015 priorities: An updated systematic analysis. *The Lancet, 385*(9966), 430–440.
25. Howe, C., Suich, H., Vira, B., & Mace, G. M. (2014). Creating win-wins from trade-offs? Ecosystem services for human well-being: A meta-analysis of ecosystem service trade-offs and synergies in the real world. *Global Environmental Change, 28*, 263–227.
26. De Groot, R. S., Alkemade, R., Braat, L., Hein, L., & Willemen, L. (2010). Challenges in integrating the concept of ecosystem services and values in landscape planning, management and decision making. *Ecological Complexity, 7*(3), 260–272.
27. Bagstad, K. J., Semmens, D. J., Waage, S., & Winthrop, R. (2013). A comparative assessment of decision-support tools for ecosystem services quantification and valuation. *Ecosystem Services, 5*, 27–39.
28. Sherrouse, B. C., Clement, J. M., & Semmens, D. J. (2011). A GIS application for assessing, mapping, and quantifying the social values of ecosystem services. *Applied Geography, 31*(2), 748–760.
29. Burkhard, B., Kroll, F., Nedkov, S., & Müller, F. (2012). Mapping ecosystem service supply, demand and budgets. *Ecological Indicators, 21*, 17–29.
30. Plieninger, T., Dijks, S., Oteros-Rozas, E., & Bieling, C. (2013). Assessing, mapping, and quantifying cultural ecosystem services at community level. *Land Use Policy, 33*, 118–129.
31. Pascual, U., Balvanera, P., Díaz, S., Pataki, G., Roth, E., Stenseke, M., Watson, R. T., Dessane, E. . B., Islar, M., Kelemen, E., Maris, V., Quaas, M., Subramanian, S. M., Wittmer, H., Adlan, A., Ahn, S. E., Al-Hafedh, Y. S., Amankwah, E., Asah, S. T., Berry, P., & Yagi, N. (2017). Valuing nature's contributions to people: The IPBES approach. *Current Opinion in Environmental Sustainability, 26*, 7–16.
32. Chan, K. M., Guerry, A. D., Balvanera, P., Klain, S., Satterfield, T., Basurto, X., Bostrom, A., Chuenpagdee, R., Gould, R., Halpern, B. S., Hannahs, N., Levine, J., Norton, B., Ruckelshaus, M., Russell, R., Tam, J., & Woodside, U. (2012). Where are cultural and social in ecosystem services? A framework for constructive engagement. *BioScience, 62*(8), 744–756.
33. Bagstad, K. J., Villa, F., Batker, D., Harrison-Cox, J., Voigt, B., & Johnson, G. W. (2014). From theoretical to actual ecosystem services: Mapping beneficiaries and spatial flows in ecosystem service assessments. *Ecology and Society, 19*(2). https://doi.org/10.5751/ES-06523-190264
34. Lavorel, S., Bayer, A., Bondeau, A., Lautenbach, S., Ruiz-Frau, A., Schulp, N., Seppel, R., Verburg, P., van Teeffelen, A., Vannier, C., Arneth, A., Cramer, W., & Marba, N. (2017). Pathways to bridge the biophysical realism gap in ecosystem services mapping approaches. *Ecological Indicators, 74*, 241–260.
35. Eigenbrod, F., Armsworth, P. R., Anderson, B. J., Heinemeyer, A., Gillings, S., Roy, D. B., Thomas, C. D., & Gaston, K. J. (2010). The impact of proxy-based methods on mapping the distribution of ecosystem services. *Journal of Applied Ecology, 47*(2), 377–385.
36. Schulp, C. J., Burkhard, B., Maes, J., Van Vliet, J., & Verburg, P. H. (2014). Uncertainties in ecosystem service maps: A comparison on the European scale. *PLoS One, 9*(10), e109643.
37. Kareiva, P. (2014). New conservation: Setting the record straight and finding common ground. *Conservation Biology, 28*(3), 634–636.
38. Nelson, E., Mendoza, G., Regetz, J., Polasky, S., Tallis, H., Cameron, D., Chan, K. M. A., Daily, G. C., Goldstein, J., Kareiva, P. M., Lonsdorf, E., Naidoo, R., Ricketts, T. H., & Shaw, M. R. (2009). Modeling multiple ecosystem services, biodiversity conservation, commodity production, and tradeoffs at landscape scales. *Frontiers in Ecology and the Environment, 7*(1), 4–11.

39. Goldstein, J. H., Caldarone, G., Duarte, T. K., Ennaanay, D., Hannahs, N., Mendoza, G., Polasky, S., Wolny, S., & Daily, G. C. (2012). Integrating ecosystem-service tradeoffs into land-use decisions. *Proceedings of the National Academy of Sciences, 109*(19), 7565–7570.
40. Arkema, K. K., Verutes, G. M., Wood, S. A., Clarke-Samuels, C., Rosado, S., Canto, M., Rosenthal, A., Ruckelshaus, M., Guannel, G., Toft, J., Faries, J., Silver, J. M., Griffin, R., & Guerry, A. D. (2015). Embedding ecosystem services in coastal planning leads to better outcomes for people and nature. *Proceedings of the National Academy of Sciences, 112*(24), 7390–7395.
41. Schröter, D., Cramer, W., Leemans, R., Prentice, I. C., Araújo, M. B., Arnell, N. W., Bondeau, A., Bugmann, H., Carter, T. R., Gracia, C. A., de la Vega-Leinert, A. C., Erhard, M., Ewert, F., Glendining, M., House, J. I., Kankaanpää, S., Klein, R. J., Lavorel, S., Lindner, M., Metzger, M. J., Meyer, J., Mitchell, T. D., Reginster, I., Rounsevell, M., Sabaté, S., Sitch, S., Smith, B., Smith, J., Smith, P., Sykes, M. T., Thonicke, K., Thuiller, W., Tuck, G., Zaehle, S., & Zierl, B. (2005). Ecosystem service supply and vulnerability to global change in Europe. *Science, 310*(5752), 1333–1337.
42. Cord, A. F., Bartkowski, B., Beckmann, M., Dittrich, A., Hermans-Neumann, K., Kaim, A., Lienhoop, N., Locher-Krause, K., Priess, J., Schröter-Schlaack, C., Schwarz, N., Seppelt, R., Strauch, M., Václavík, T., & Volk, M. (2017). Towards systematic analyses of ecosystem service trade-offs and synergies: Main concepts, methods and the road ahead. *Ecosystem Services, 28*, 264–272.
43. Carpenter, S. R., Mooney, H. A., Agard, J., Capistrano, D., DeFries, R. S., Díaz, S., Dietz, T., Duraiappah, A. K., Oteng-Yeboah, A., Pereira, H. M., Perrings, C., Reid, W. V., Sarukhan, J., Scholes, R. J., & Whyte, A. (2009). Science for managing ecosystem services: Beyond the millennium ecosystem assessment. *Proceedings of the National Academy of Sciences, 106*(5), 1305–1312.
44. Guerry, A. D., Polasky, S., Lubchenco, J., Chaplin-Kramer, R., Daily, G. C., Griffin, R., Ruckelshaus, M., Bateman, I. J., Duraiappah, A., Elmqvist, T., Feldman, M. W., Folke, C., Hoekstra, J., Kareiva, P. M., Keeler, B. L., Li, S., Mckenzie, E., Ouyang, Z., Reyers, B., Ricketts, T. H., Rockström, J., Tallis, H., & Vira, B. (2015). Natural capital and ecosystem services informing decisions: From promise to practice. *Proceedings of the National Academy of Sciences, 112*(24), 7348–7355.
45. Polasky, S., Carpenter, S. R., Folke, C., & Keeler, B. (2011). Decision-making under great uncertainty: Environmental management in an era of global change. *Trends in Ecology & Evolution, 26*(8), 398–404.
46. Martinez-Harms, M. J., Bryan, B. A., Balvanera, P., Law, E. A., Rhodes, J. R., Possingham, H. P., & Wilson, K. A. (2015). Making decisions for managing ecosystem services. *Biological Conservation, 184*, 229–238.
47. Pettorelli, N., Laurance, W. F., O'Brien, T. G., Wegmann, M., Nagendra, H., & Turner, W. (2014). Satellite remote sensing for applied ecologists: Opportunities and challenges. *Journal of Applied Ecology, 51*(4), 839–848.
48. Lefsky, M. A., Cohen, W. B., Parker, G. G., & Harding, D. J. (2002). Lidar remote sensing for ecosystem studies: Lidar, an emerging remote sensing technology that directly measures the three-dimensional distribution of plant canopies, can accurately estimate vegetation structural attributes and should be of particular interest to forest, landscape, and global ecologists. *Bioscience, 52*(1), 19–30.
49. Vigerstol, K. L., & Aukema, J. E. (2011). A comparison of tools for modeling freshwater ecosystem services. *Journal of Environmental Management, 92*(10), 2403–2409.
50. Sepuru, T. K., & Dube, T. (2018). An appraisal on the progress of remote sensing applications in soil erosion mapping and monitoring. *Remote Sensing Applications: Society and Environment, 9*, 1–9.
51. Kerr, J. T., & Ostrovsky, M. (2003). From space to species: Ecological applications for remote sensing. *Trends in Ecology & Evolution, 18*(6), 299–305.

52. Davies, A. B., & Asner, G. P. (2014). Advances in animal ecology from 3D-LiDAR ecosystem mapping. *Trends in Ecology & Evolution, 29*(12), 681–691.
53. McGarigal, K., & Cushman, S. A. (2002). Comparative evaluation of experimental approaches to the study of habitat fragmentation effects. *Ecological Applications, 12*(2), 335–345.
54. Elith, J., & Leathwick, J. R. (2009). Species distribution models: Ecological explanation and prediction across space and time. *Annual Review of Ecology, Evolution, and Systematics, 40*(1), 677–697.
55. Pereira, H. M., Ferrier, S., Walters, M., Geller, G. N., Jongman, R. H., Scholes, R. J., Bruford, M. W., Brummitt, N., Butchart, S. H. M., Cardoso, A. C., Coops, N. C., Dulloo, E., Faith, D. P., Freyhof, J., Gregory, R. D., Heip, C., Höft, R., Hurtt, G., Jetz, W., Karp, D. S., McGeoch, M. A., Obura, D., Onoda, Y., Pettorelli, N., Reyers, B., Sayre, R., Scharlemann, J. P. W., Stuart, S. N., Turak, E., Walpole, M., & Wegmann, M. (2013). Essential biodiversity variables. *Science, 339*(6117), 277–278.
56. Holzkämper, A., & Seppelt, R. (2007). A generic tool for optimising land-use patterns and landscape structures. *Environmental Modelling & Software, 22*(12), 1801–1804.
57. Schägner, J. P., Brander, L., Maes, J., & Hartje, V. (2013). Mapping ecosystem services' values: Current practice and future prospects. *Ecosystem Services, 4*, 33–46.
58. Wang, L. J., Gong, J. W., Ma, S., Wu, S., Zhang, X., & Jiang, J. (2022). Ecosystem service supply–demand and socioecological drivers at different spatial scales in Zhejiang Province. *China. Ecological Indicators, 140*, 109058.

The Role of Carbon Sequestration in Mitigating Climate Change: A Focus on India

Deeksha and Anoop Kumar Shukla

1 Introduction

Effects or threats to the global climate include sea-level rise, droughts and crop failures, inundations, and forest degradation. The solutions for mitigating these calamities should be based on robust environmental policies that steer market conditions toward less carbon-intensive technologies and consequently reduce atmospheric carbon or carbon dioxide concentrations [1]. Such policy interventions would undoubtedly also be reflected in the form of regulations, taxes, and incentives to direct private sector strategies focusing on the implementation of conservation and sequestration initiatives. Great emphasis is being placed on promoting carbon sequestration initiatives in the third world countries such as India, which are facing major challenges on the road to achieving sustainable development. These countries are particular victims of the vagaries associated with global warming and face negative consequences associated with changes in the patterns of extreme weather conditions and desert-like conditions. The proposed sequestration options are becoming additional most popular and sound approaches for local and regional carbon mitigation in sharing the global benefit associated with the Kyoto protocol.

Climate change is a matter of great concern for the entire world, and the scientific community is nearly unanimous in getting across to humanity that climate change is inherently related to mankind's economic activities in the modern age [2–4]. Achieving a global agreement to combat climate change implicates making considerable progress against the potential threats posed to the global economy through a gradual and substantial approach to any mitigation effort. The typical response from the different national governments has been a commitment to

Deeksha · A. K. Shukla (✉)
Manipal School of Architecture and Planning, Manipal Academy of Higher Education, Manipal, Karnataka, India
e-mail: deeksha.msapma2@learner.manipal.edu; anoop.shukla@manipal.edu

S. Pathak et al. (eds.), *Geo-Data Revolution*, Advances in Geographic Information Science, https://doi.org/10.1007/978-3-031-94078-1_2

reducing greenhouse gas emissions primarily through national action plans. These initiatives consist principally of actions directed toward increases in the deployment of advanced technologies that lead to a reduction in emissions, mostly through increased energy efficiency and conservation.

1.1 Definition and Causes of Climate Change

Climate has always shown changes due to natural processes and the periodic rhythms of solar activity and the Earth's orbit which affect the distribution of energy received at the surface. Therefore, the concept of climate change is not new, but the process was rather a response to the natural forces that caused global changes (warming and cooling). Based upon these natural effects, Markusen et al. in his study "Environment and Policy" have classified the climate change into three groups, namely, the ice ages, which are believed to have occurred at great intervals and little changes that occur from 1 day to the next. However, the current understanding of climate change is the rapidly changing pattern in the Earth's climate, mainly caused by anthropogenic processes.

Anthropogenic activities such as deforestation, urbanization, and the overuse of fossil fuels result in an increase in the concentration of anthropogenically produced greenhouse gases [5, 6]. Most of this gas is carbon dioxide, which leads to a disturbance in the balance of natural fluxes of carbon. The overwhelming impacts of such a disturbance have led to global warming, which is leading to climate change. Understandably, the global community needed to arrest the alterations in global temperature. Industrialized countries have already been trying to implement strategies that focus on sustainable development. A tool designed under the Kyoto Protocol is carbon sequestration. Through this process, the Kyoto agreement to limit greenhouse gas concentrations in Earth's atmosphere, specifically carbon dioxide, can be achieved.

1.2 Concept and Importance of Carbon Sequestration

Sequestration can be characterized as the hiding of CO_2 from the carbon cycle of a pollution source, causing the CO_2 not to reach the upper layers, or perhaps causing the CO_2 not to be generated or absorbed through indirect methods. In the latter case, it can be called CO_2 recycling or carbon capture and storage. Forests, organic matter-rich soils, deep oceans, lime, basalt, and peridotite are all natural reservoirs for carbon sequestration. Although heterotrophic structures of organic matter, deep oceans, and lime absorptions require certain conditions, autotrophic structures related to increasing forested areas, and bioenergy with carbon storage applications, which produce or stimulate the capture of CO_2 during functioning, and which are extractive, absorptive, radiative, and synthetic structures, are the principal focus of the sequestration concept.

The Earth has different forms of capacity to act against external disturbances. The ability to resist against the disturbances indicates the structure of the Earth, and the ability to recover after the occurrence of disturbances shows resilience of the structure. One of the most important substances for this resilience is carbon. It is a main component of the structural compounds of living organisms. Its characteristic to bond with each other in long chains makes it the basic structural element of the compounds. Carbon is extensively available on Earth, but carbon dioxide is one of the least concentrated unessential waste products, which does not take part in the reactions.

The importance of carbon sequestration arises from the well-documented increase in the concentration of atmospheric carbon dioxide (CO_2) [7–10]. This increase is due to increased emissions from fossil fuel use and land-use changes. Each year, approximately 6.8 billion metric tons of carbon are emitted from burning of fossil fuels, and 1.6 billion metric tons from deforestation. Over the past few centuries, atmospheric CO_2 concentrations have risen from preindustrial levels of about 280 parts per million to nearly 380 parts per million today. The contemporary concentration of carbon dioxide in the atmosphere as compared with its atmospheric concentration pre-1750 has mainly been affected by anthropogenic sources, fires, and CO_2 absorption by the oceans, vegetation, soil, and dead organic matter in the terrestrial environment. The natural sequestration stock of carbon is much larger than the annual sources or sinks and would balance upward and downward emissions if no human activities were undertaken. Therefore, returning the emission of CO_2 to zero would be expected to reduce the concentration of CO_2 in the atmosphere until its increase was counterbalanced by its natural sequestration.

Carbon is sequestered into three reservoirs: the atmosphere, the terrestrial environment, and the oceans. Carbon sequestration refers to long-term storage of carbon in the terrestrial, aquatic, and oceanic environments in order to ensure that atmospheric carbon dioxide levels are not replenished by human activities, such as deforestation. Various land management techniques can enhance carbon sequestration processes, and consequently sequester carbon dioxide from the atmosphere. These include preserving and creating forests and storing carbon within vegetative matter, including woody biomass, roots, and stable soils. The enhancement of carbon sinks through biological and agricultural measures helps to offset the release of CO_2 from sources. Land management techniques need to be designed so as not to upset the balance between sources and sinks and allow them to have the necessary elasticity to accommodate ongoing human activities. Proposals to increase the amount of carbon sequestered in terrestrial systems, including forests, grasslands, and agricultural soils, will involve, among other measures, policies, and practices that allow incentive and competence working alike to channel effort to where it is most valuable.

2 Carbon Sequestration Techniques

2.1 Natural Vs. Artificial Sequestration Methods

Natural carbon sequestration methods involve the absorption and storage of CO_2 through biological and ecological processes. These methods include afforestation and reforestation, soil carbon sequestration, wetland restoration, and the preservation of blue carbon ecosystems like mangroves, salt marshes, and seagrasses. Natural sequestration is beneficial because it not only captures carbon but also provides numerous co-benefits such as enhancing biodiversity, improving soil health, and protecting coastlines from erosion. However, these approaches have considerable obstacles, including the requirement for huge land expanses, sensitivity to ecological shifts such as global warming and forest destruction, and reduced carbon collection rates when contrasted with technology alternatives. Furthermore, the potential for storing carbon in environments could end up saturated, reducing their long-term usefulness.

Artificial carbon sequestration involves capturing and storing CO_2 using modern technology. Carbon Capture and Storage (CCS) captures emissions from industrial sources and stores them underground, while Direct Air Capture (DAC) extracts CO_2 directly from the atmosphere. Bioenergy with Carbon Capture and Storage (BECCS) combines biomass energy production with carbon capture, and mineral carbonation forms stable carbonates by reacting CO_2 with minerals. These approaches have the potential for significant and quick reductions in atmospheric CO_2 levels. They may be installed in a variety of settings and can run constantly, regardless of seasonal or environmental changes. However, artificial sequestration is often costly, energy-intensive, and necessitates extensive infrastructure. Concerns remain concerning the long-term security of CO_2 storage, as well as the requirement for strong laws and public acceptability to facilitate extensive implementation.

An analytical examination of natural and artificial sequestration strategies demonstrates that both approaches are required for full climate change mitigation. Natural solutions provide significant ecological advantages and are often more sustainable and cost-effective in the long run. However, its scale is constrained by land availability and environmental considerations. Artificial approaches, on the other hand, provide scalable and quick answers while encountering economic and technological obstacles. An integrated strategy that blends natural and artificial sequestration can optimize each approach's strengths while addressing their respective limitations (Table 1). For example, improving natural systems where possible and supplementing them with artificial technology in places requiring quick and large-scale carbon capture can result in a more balanced and successful carbon sequestration framework. Taking a comprehensive strategy is vital for reducing atmospheric CO_2 levels and addressing climate change.

Table 1 Natural vs artificial carbon sequestration

Aspect	Natural sequestration	Artificial sequestration
Mechanism	Biological and ecological processes	Technological and industrial processes
Scalability	Limited by land and ecosystem availability	Potentially scalable with technological advances
Cost	Generally lower, often needs maintenance	Higher due to technology and infrastructure
Timeframe	Long-term, gradual carbon storage	Can be more immediate with large-scale deployment
Co-benefits	Enhances biodiversity, ecosystem services	May provide energy or industrial by-products
Risks and Challenges	Vulnerability to environmental changes	High costs, energy requirements, public acceptance
Monitoring	Complex, requires ecological assessments	Technologically intensive but potentially more precise
Examples	Forests, wetlands, soils, blue carbon ecosystems	CCS, DAC, BECCS, mineral carbonation

2.2 *Biological Carbon Sequestration*

Biological carbon sequestration occurs when living species and ecosystems absorb and store carbon dioxide (CO_2) through processes in forests, soils, and wetlands. It can take place through afforestation and reforestation techniques, soil carbon sequestrations, agroforestry, wetland and peatland restoration, and blue carbon ecosystems. These help in absorption and storage of sequestered carbon resulting in improvement of soil and water conditions.

2.3 *Technological Carbon Sequestration*

Technological carbon sequestration includes capturing and storing carbon dioxide (CO_2) using different modern technologies. These techniques are commonly employed at industrial sizes and include procedures like Carbon Capture and Storage (CCS), Direct Air Capture (DAC), Bioenergy with carbon capture and storage (BECCS), and Mineral Carbonation.

2.4 *Oceanic Carbon Sequestration*

Oceanic carbon sequestration involves capturing and storing carbon dioxide (CO_2) in the ocean, taking use of the ocean's large potential to absorb and store carbon. There are various techniques employed like ocean fertilization technique using iron fertilization and nitrogen and phosphorous addition, then we have biological carbon pump using phytoplankton photosynthesis, carbon transfer and sedimentation, blue carbon ecosystem also plays a major role in carbon sequestration through mangroves, salt marshes, and seagrasses, enhanced weathering is a techniques applied through chemical reaction and application, and last one being artificial upwelling and downwelling technique.

3 Carbon Sequestration Potential in India

India has a considerable potential for carbon sequestration, using both natural and artificial approaches to combat climate change. The country's large and diversified landscapes provide potential for afforestation and restoration, especially on degraded soils and in deforested areas. India's National Afforestation Programme and Green India Mission seek to expand forest cover and improve ecosystem services, hence improving carbon sinks. Furthermore, India's agricultural sector may help to sequester carbon through sustainable farming methods including no-till agriculture, cover cropping, and agroforestry. These activities enhance soil health, boost organic carbon content, and encourage biodiversity. The restoration of mangroves and other coastal ecosystems in India has significant blue carbon sequestration potential, as well as additional advantages like as coastal protection and improved fisheries.

On the technology front, India is looking at CCS to cut emissions from its large coal-fired power production and industrial sectors. Pilot studies and cooperation with foreign partners are now ongoing to develop and implement CCS technology, which may drastically reduce India's carbon footprint. DAC and BECCS are also being investigated; however, they are still in their early phases when compared to natural techniques. Combining technology advancements with regulatory assistance, monetary support, and global collaboration could assist India realize its entire carbon sequestration capacity. By combining natural and artificial sequestration measures, India can play an important role in global attempts to address climate change while also accomplishing sustainable development objectives.

3.1 *Overview of India's Emissions Profile*

India is one of the world's leading producers of greenhouse gases (GHGs), thanks to its fast expanding economic growth, enormous population, and significant dependency on petroleum and coal [10]. The nation's emissions pattern is notable for considerable contributions from a variety of sectors, including energy, industry, agriculture, transportation, and waste management.

Around 74% of GHG in India is emitted due to energy sector, paving its way to the largest GHG emitter. This is primarily due to the urban growth and reliance on fossil fuel. 15% of emissions is due to industrial sector, with an order of cement steel and chemical production, are highly energy-intensive and rely predominantly on fossil fuels. These sectors produce high levels of CO_2, CH_4, and N_2O. Furthermore, factories frequently generate non-energy emissions, which includes the discharge of fluorinated gases; those possess a large climate change capacity. Around 16% of GHG emission is responsible by the agricultural sector, the principal sources of CH_4 in the fermentation processes among cattle and cultivated rice fields. N_2O pollutants from artificial manure and fertilizer treatment also make a substantial contribution. Considering the significance of farming in the nation's economy and food security, decreasing emissions in this sector poses particular difficulties and possibilities. Waste management accounts for around 3% of India's total GHG emissions.

3.2 *Key Sectors for Carbon Sequestration Efforts*

The key sectors in which the efforts has to be made in sequestering carbon is forestry and land use, agriculture, energy, industry, and waste management. Forestry and land use are two of the most important potential for carbon sequestration in India. The country's vast land area, various ecosystems, and massive areas of deteriorated and deforested land present enormous opportunities for reforestation and afforestation initiatives. Initiatives such as the National Afforestation Program and the Green India Mission aim to increase canopy coverage and improve the condition of current forests. Agriculture plays and vital role in carbon sequestration. Implementing sustainable agricultural techniques can greatly improve soil carbon storage. Conservation tillage, cover cropping, crop rotation, and organic farming all boost soil organic matter and prevent soil erosion, resulting in improved carbon storage. The energy industry, particularly the use of BECCS, has enormous potential for carbon sequestration in India. BECCS uses biomass to produce electricity and store CO_2 emissions underground. This technique achieves negative emissions, removing more CO_2 from the atmosphere than it releases.

The industrial sector may play an important role in carbon sequestration by using CCS technology. Industrial sectors may drastically lower their carbon impact by gathering and conserving CO_2 emissions underground. Using collected CO_2 in industrial uses, such as synthetic fuels, chemicals, and construction resources, can promote a circular carbon economy and enhance carbon sequestration initiatives. Optimizing waste management methods will assist with carbon sequestration in India. Organic matter, including crop leftovers, garbage, and sludge from sewers, may be turned into compost or biochar, either of which trap carbon when used on soil. Adopting holistic waste management solutions such as recycling, composting, and anaerobic digestion can improve carbon sequestration and minimize waste-related pollutants.

4 Government Initiatives and Policies

4.1 National Action Plan on Climate Change

The National Action Plan on Climate Change (NAPCC) is India's strategy framework for combating climate change, which was launched by the government in 2008. The NAPCC specifies eight national missions, each addressing a distinct area of climate change mitigation and adaptation. These missions seek to promote sustainable development, increase energy efficiency, and conserve fragile ecosystems while driving economic growth.

4.2 National Solar Mission

- **Objective:** To promote the development and use of solar energy for power generation and other applications.
- **Targets:** Achieve 100 GW of solar power capacity by 2022 and 300 GW by 2030.
- **Initiatives:** Incentives for solar power plants, rooftop solar installations, and solar thermal power projects.

4.3 National Mission for Enhanced Energy Efficiency (NMEEE)

- **Objective:** To enhance energy efficiency across various sectors, thereby reducing emissions.
- **Programs:** Perform, Achieve, and Trade (PAT) scheme, Market Transformation for Energy Efficiency (MTEE), and promoting energy efficiency in small and medium enterprises.

4.4 National Mission on Sustainable Habitat

- **Objective:** To promote energy efficiency in urban planning, transportation, and waste management.
- **Focus Areas:** Energy-efficient buildings, improved urban waste management, and better urban planning to reduce the carbon footprint.

4.5 National Water Mission

- **Objective:** To conserve water, minimize wastage, and ensure more equitable distribution.
- **Goals:** Increase water use efficiency by 20%, promote integrated water resource management, and develop a comprehensive water database.

4.6 National Mission for Sustaining the Himalayan Ecosystem

- **Objective:** To protect and sustain the fragile Himalayan ecosystem.
- **Strategies:** Address the impacts of climate change on the Himalayan glaciers and biodiversity and promote sustainable tourism and livelihoods.

4.7 National Mission for a Green India (Green India Mission)

- **Objective:** To increase forest and tree cover and restore degraded ecosystems.
- **Targets:** Enhance carbon sequestration and ecosystem services, including biodiversity, hydrological services, and livelihoods of forest-dependent communities.

4.8 National Mission for Sustainable Agriculture

- **Objective:** To promote sustainable agriculture practices that are resilient to climate change.
- **Actions:** Develop climate-resilient crops, improve soil health, and optimize water use efficiency through integrated farming systems.

4.9 National Mission on Strategic Knowledge for Climate Change

- **Objective:** To build a scientific knowledge base and promote climate research and data sharing.
- **Focus:** Enhance understanding of climate change impacts, adaptation strategies, and develop innovative technologies and solutions.

5 Challenges and Opportunities

5.1 Technological and Financial Barriers

India confronts severe scientific and financial challenges in its attempts to increase carbon sequestration. Introducing innovations into current industrial processes and the energy sector presents extra obstacles owing to the intricacy and cost of upgrading facilities. Furthermore, limited R&D resources stifle innovation, and there is a scarcity of knowledge in sophisticated carbon sequestration systems. Tracking and verifying sequestered carbon, particularly in natural systems such as forests and soils, remains technically difficult, complicating efforts.

Financially, the high initial and ongoing expenses of implementing carbon sequestration technology are important deterrents. There are little financial incentives, such as subsidies, tax credits, and grants, to encourage investment in these projects. The lack of strong carbon pricing systems limits the financial appeal of carbon sequestration, and the restricted availability of public financing further stifles advancement. Accessing international climate funds and building bilateral and multilateral collaborations also provide extra problems owing to complicated application processes and political obstacles.

Dealing with these impediments necessitates an extensive plan which involves increased investment in R&D to boost technical innovation and lower prices. Improving policy assistance through monetary incentives and implementing a clear carbon pricing system can encourage both public and private investment. It is also vital to strengthen international collaboration and make global climate money more accessible. By overcoming these technological and financial barriers, India may greatly accelerate its carbon sequestration efforts, therefore contributing to global climate change mitigation and fostering sustainable development.

5.2 Role of Public-Private Partnerships

Public-private partnerships (PPPs) play an important role in furthering carbon sequestration initiatives in India by combining the capabilities and resources of two sectors. PPPs enable the public sector to provide regulatory support, money, and policy incentives, while the private sector contributes technical innovation, operational knowledge, and extra capital. Public-private cooperation may boost R&D efforts, resulting in creative solutions that increase the efficiency and cost-effectiveness of carbon sequestration technology.

PPPs can assist to scale up successful pilot initiatives and ensure wider acceptance. PPPs can help to scale up the implementation of sustainable agriculture techniques, renewable energy projects, and forest conservation efforts by integrating public and private sector endeavors. For example, agroforestry efforts funded by both government programs and commercial agribusinesses can increase soil carbon sequestration while also providing farmers with economic benefits. As a result, PPPs are critical for bridging financial and technological barriers, encouraging innovation, and boosting large-scale deployment of carbon sequestration solutions in India.

6 Conclusion and Future Directions

India's commitment to combating climate change through carbon sequestration is critical to global climate mitigation efforts. The country's diversified ecosystems and quickly rising economy provide both possibilities and problems for adopting successful carbon sequestration solutions. These programs not only contribute to global carbon sequestration, but they also promote sustainable development and resistance to climate change. Nonetheless, considerable technological and financial impediments remain. The introduction of sophisticated carbon capture systems such as CCS and DAC is hampered by high costs and technical challenges. The high initial and ongoing expenses, combined with insufficient financial incentives and the absence of efficient carbon pricing systems, limit the scalability of these initiatives. To stimulate innovation and investment in carbon sequestration technology, we need targeted regulations, additional financing, and stronger PPP.

PPP play an important role in resolving these difficulties. By combining the capabilities of both sectors, PPPs can offer the cash, knowledge, and regulatory backing required to scale up carbon sequestration initiatives. Enhancing policy support through subsidies, tax credits, and grants, as well as ensuring a stable carbon market, may assist in bringing in investors and encourage wider use of carbon sequestration methods. Looking ahead, India must continue to enhance its R&D activities in order to innovate and improve carbon sequestration methods. Strengthening international collaboration and making access to global climate funds easier are also critical for funding large-scale initiatives. Integrating natural and technical solutions, involving local people, and encouraging multi-sector collaboration will be critical to optimizing carbon sequestration capability. By tackling current obstacles and harnessing empirical perspectives, India can make a greater contribution to global climate change mitigation, guaranteeing a sustainable and resilient tomorrow.

Further research should focus on integrating carbon sequestration with climate change adaption and resilience-building techniques, as well as investigating nature-based solutions that provide co-benefits for both mitigation and adaptation efforts. Advancing monitoring, reporting, and verification (MRV) methodology is crucial for assuring the credibility and transparency of carbon sequestration programs, necessitating the development of cost-effective and dependable ways for estimating carbon stocks and flows. By tackling these research objectives, India may significantly increase its carbon sequestration initiatives while contributing to global climate change mitigation objectives.

References

1. Valls-Val, K., & Bovea, M. D. (2021, November 1). Carbon footprint in higher education institutions: A literature review and prospects for future research. *Clean Technologies and Environmental Policy, 23*(9), 2523–2542. https://doi.org/10.1007/s10098-021-02180-2
2. Salunkhe, O., Khare, P. K., Kumari, R., & Khan, M. L. (2018, December 1). A systematic review on the aboveground biomass and carbon stocks of Indian forest ecosystems. *Ecological Processes, 7, 1*. https://doi.org/10.1186/s13717-018-0130-z
3. Fang, J., Liu, M., Liu, W., Pathak, S., Li, S., Tang, X., Zhou, L., & Sun, F. (2020). Piloting a capital-based approach for characterizing and evaluating drivers of island sustainability-an application in Chongming Island. *Journal of Cleaner Production, 261*, 121123.
4. Hendriks, K., Gubbay, S., Arets, E., Janssen, J., Dige, G., & Erhard, M. (2020 November). Carbon stocks and sequestration in terrestrial and marine ecosystems: A lever for nature restoration? A quick scan for terrestrial and marine EUNIS habitat types EEA contribution and project management. *Wageningen*. https://doi.org/10.18174/XXXX
5. Basha, U., Pandey, M., Nayak, D., Shukla, S., & Shukla, A. K. (2024, April). Spatial–temporal assessment of annual water yield and impact of land use changes on upper Ganga Basin, India, using InVEST model. *Journal of Hazard Toxic Radioactive Waste, 28*(2). https://doi.org/10.1061/jhtrbp.hzeng-1245
6. Nayak, D., Shukla, A. K., & Devi, N. R. (2024, January). Decadal changes in land use and land cover: Impacts and their influence on urban ecosystem services. *AQUA—Water Infrastructure, Ecosystems and Society, 73*(1), 57–72. https://doi.org/10.2166/aqua.2024.211
7. Nayak, D., Shukla, A. K., & Rama Devi N. (2023, July). Effects of Urbanization on Urban Ecosystem Services (UESS)—A geospatial approach. In *IGARSS 2023–2023 IEEE International Geoscience and Remote Sensing Symposium* (pp. 6216–6219). IEEE. https://doi.org/10.1109/IGARSS52108.2023.10281721
8. Nayak, D., Shukla, A. K., & Devi, N. R. (2024). Applications of remote sensing and GIS techniques in the monitoring of ecosystem services. In *Agri-tech approaches for nutrients and irrigation water management* (pp. 116–126). CRC Press. https://doi.org/10.1201/9781003441175-7
9. Deeksha, Shukla, A. K., Devi, N. R., & Shukla, S. (2023). *Remote sensing and its application on soil: An ecosystem services* (pp. 307–319). Springer. https://doi.org/10.1007/978-981-99-4811-6_17
10. Deeksha, & Shukla, A. K. (2022, September). Ecosystem services: A systematic literature review and future dimension in freshwater ecosystems. *Applied Sciences (Switzerland), 12*(17). https://doi.org/10.3390/app12178518

Assessment and Management of the Gomti River Encroachment Through Geospatial Technique

Sudhakar Shukla, Pallavi Singh, and Mamta Shukla

1 Introduction

'River valleys are the cradle of civilization'. It is a well-established fact, known far and wide, that the world's greatest ancient civilizations had strived, thrived and dwelled on the banks of rivers which continue to sustain dense population and infrastructure on their waters. With the world undergoing the paradigm of 'population explosion', the crisis of land scarcity to support the ever-growing population arises. In order to bridge the widening gap between the availability of land and the population, river banks are being used to cater the needs. The opportunist and desperate section of the society is acquiring the river banks illegally for infrastructure construction, commercialization and agricultural activities to name a few [1] which adversely affect the area covered and the course of the rivers as their catchment areas reduce to hazardously low levels. Such practices have highlighted the long-existing yet recently developed concept of 'River Encroachment'. An unlawful intrusion or invasion in the basin of a navigable river with or without causing an obstruction can be referred to as river encroachment [2]. The multiplying population puts tremendous pressure on the land, creating more demand which results in the infringement of the encroachment laws. River encroachment is significantly visible in the areas along the river with mushrooming settlements, infrastructure, industries, commercial centres and primary activities like agriculture [3]. The river succumbs to biodiversity loss, environmental degradation, aesthetic devaluation, morphological and hydrological retardation resulting in a dead river system [4].

S. Shukla (✉)
School of Geo-Informatics, Remote Sensing Applications Centre—Uttar Pradesh, Lucknow, India

P. Singh · M. Shukla
Remote Sensing Applications Centre—Uttar Pradesh, Lucknow, India

S. Pathak et al. (eds.), *Geo-Data Revolution*, Advances in Geographic Information Science, https://doi.org/10.1007/978-3-031-94078-1_3

The River Gomti, one of the major tributaries of the Ganga River System, passes through the most economically, politically and socio-culturally prominent cities of the state of Uttar Pradesh. With India at the top of the 'Most Populous Countries' in the world and Uttar Pradesh securing the spot as the 'Most Populous State of India', the condition of the water bodies in the country as well as the state can be predicted precisely. The Gomti has been subjected to encroachment at numerous locations throughout its length of 960 km. Lesser inundation, shrunk area, increased occurrence of dry patches, decreased depth due to increased deposition in riverbed and threatening level of pollution are the few challenges being faced as the result of river encroachment. The Gomti which is a groundwater-fed river is under the distress of losing its perennial nature, becoming seasonal, due to mass encroachment at the very origin of the river where the lake is being artificially reduced to facilitate the construction of bridges and embankments which make the region accessible and well-connected. Such an extensive encroachment at its origin has forced the river to turn into a stream, disappearing at short distances in the Pilibhit district. The extraction of groundwater through bore wells and direct pumping at large scale has worsened the groundwater levels in the region which in turn adds to the drying of the river.

When the population started to surge in the country, post-independence, the need for sufficient quantity of food grew simultaneously. The limited land and other natural resources could not compliment the expanding human resource as a result of which any piece of fallow land, irrespective of its past cover pattern, was taken up as agricultural land. The fertile floodplain and wetland of the rivers were the first to be encroached upon with the outcome of increasing food supply nation-wide but decreasing width of the rivers. Uttar Pradesh had become the 'Staple Bowl of India' but its brunt was being borne by its water bodies. The chemical effluents and incessant drilling for groundwater to boost the agricultural productivity further deteriorated the health and size of the rivers. In the early stage of the Gomti, the agricultural fields situated on both the banks are severely harming the river until today.

The Gomti sweeps the banks of some noteworthy cities of Uttar Pradesh. The drainage basin encompasses the State Capital—Lucknow, the Agricultural Trade Centre of the state—Sultanpur and the Communal Hub of the state—Jaunpur. These urban centres have witnessed tremendous demographic, economic and infrastructural growth in the past few decades. The repercussion of drastic urbanization is gravely evident on the floodplain of the river which has been converted into residential colonies or illegal settlements like slums to accommodate the in-migrants from the deficient parts of the state. The Gomti becomes fenced with concrete infrastructure, hindering the natural attribute of free-flow of rivers. The concretizing banks and compounding deposition in the bed of the Gomti in the conurbations has detrimental consequence on its morphology and river pollution. The river in Lucknow, running in its Mature Stage, morphologically, carries several meanders and riverine island-like structures along its course. These landforms are the unsettled and unconsolidated sediments flowing in the river, comprising of sewage, solid, domestic and industrial wastes which are dumped in it from the drains gushing underneath the city. River encroachment as an ally of urban sprawl has oppressed the River Gomti, rolling in the cities of Lucknow, Sultanpur and Jaunpur, to the extremes. The encroachment patches with the 'Built-up' typology hamper the water percolation, groundwater recharge, ecosystem stability and consequently the revival of water in the Gomti the most.

The Gomti is being vandalized by river encroachment in its entire catchment area. The evidence in the form of data and figures from river discharge, flow pattern of river channels, sediment transport, river bank erosion, aquatic biome destruction, ecological devastation and shifting land-use over time elucidates the risks of river encroachment. The reducing width and depth of the river over the span of last 50 years (1972–2022) threatens its existence. River encroachment has become one of the most significant concerns around the world with the availability and quality of freshwater depleting at lightning speed from the surface of the earth. Therefore, the problem of river encroachment is one such man-made hazard which requires proper mitigation steps to tackle it from the grass-root level [5].

2 Objectives of the Study

The objectives have been laid out as follows:

- To mark the points of encroachment along the course of the River Gomti
- To allocate the registered plot number, village and district-wise location of each encroachment patch along the River Gomti
- To demarcate the land-use pattern and types of encroachment along the River Gomti
- To suggest effective measures to tackle the adversities caused by encroachment on the River Gomti

3 Study Area and Data Used

Gushing as the lifeline of Uttar Pradesh, the Gomti carves its path through diverse landscapes. It transcends its geographical significance, weaving itself into the fabric of the region's ecology and society. The river nourishes the land, serving as an essential source of water for agriculture, industry and life. Its vibrant ecosystem provides sanctuaries for countless species while its waters sustain fishery, transportation and other systems of the region. The life-giving Gomti, gliding through the bustling districts of Uttar Pradesh, faces a grim reality, today. Indiscriminate extraction of water, industrial runoff, anthropogenic hindrances and sprawling encroachments are crippling its flow, reflected by the rapidly diminishing groundwater level in the vicinity. The study delves into the pressing issue of uncontrolled river encroachment on the banks and floodplains of the Gomti as it scrutinizes its extent and far-reaching consequences in lieu of the environment and the communities dependent on the river as well as paves the way for effective conservation and management strategies, ensuring the prosperity of the river.

The Gomti, one of the major rivers of North India, is the sixth-order, left-bank tributary of the river Ganga [6]. The total catchment of the river covers an area of 30,437 km^2 with Uttar Pradesh carpeting the entire river system while the river takes up only 12% of the state's area. It constitutes precisely 3.53% of the Ganga river system [7]. The drainage basin of the Gomti impacts the life of millions residing in the lustrous Central Ganga Plain. The course of the river is highly seasonal with peak flows during the monsoons from July to September and stagnant discharge during the dry period in summers [8]. Throughout its journey, the Gomti nurtures numerous urban settlements along its bank. It serves as a crucial source of drinking water for the city of Lucknow, situated on its banks with a population of 2.82 million (*Census of India, 2011*). Villages and towns where agriculture is the primary source of income, rely on its waters for irrigation, domestic and industrial purposes [9].

As the river winds its way through the plains, it receives untreated waste discharged from more than 45 major drainage channels [10]. The Gomti is made up of several small tributaries, namely the Kathana, the Sarayan, the Luni, the Kalyani, the Behta, the Jokhan and the Sai which originate from higher altitude regions within the periphery of the watershed and transport wastewater, loaded with sewage, industrial waste as well as chemical effluents from towns and industrial facilities, into the river. The river Sai is the primary tributary of the Gomti, merging shortly before the Gomti meets the Ganga. As the Gomti approaches its confluence with the mighty Ganga at Kaithi village near the Varanasi-Ghazipur district border, joining India's one of the most culturally revered rivers, it marks the culmination of the journey of Gomti to become a part of the vast network of river systems in a topographically divergent country.

3.1 Geographical Location

The Gomti is a significant groundwater-fed river of India, originating from a natural reservoir named 'Fulhaar Jheel' (also known as the 'Gomat Taal', located on the 28°34′N latitude and 80°07′E longitude) in the Madho Tanda village, approximately 30 km east of the Pilibhit district headquarter in Uttar Pradesh and 50 km south of the majestic Himalayan foothills [11]. It flows through the state capital, Lucknow and marks the dividing boundary between Varanasi and Ghazipur districts just before its confluence with the Ganga. From its origin, the Gomti flows in a south and south-east ward direction and plunges into the river Ganga after traversing a distance of 960 km with an average elevation of 200 m (660 ft) above the mean sea level.

The Gomti river basin is elongated in shape and situated within the central portion of the fertile Ganga plain. Its upper course falls in the districts of Pilibhit, Lakhimpur Kheri and Shahjahanpur, the central section includes Sitapur, Hardoi,

Lucknow, Barabanki, Sultanpur and Jaunpur districts while the lower segment comprises parts of Varanasi and Ghazipur. The underlying map is a representation of the location of the Gomti river basin in Uttar Pradesh (*refer to* Fig. 1).

3.2 *Data Used*

3.2.1 Datasets

High quality data is the cornerstone of a wholesome study, unveiling its scope, potential challenges and resources. It is the guiding light in sound decision-making, fostering operational efficiency and minimizing uncertainties. The absence of reliable data increases the probability of errors, delays and budget overruns, thereby, jeopardizing the credibility of the study. The present study has invoked diverse data sources to achieve its objectives by incorporating high-resolution satellite imagery for detailed visual representations of the study area. The supplementary data has enriched the comprehensiveness of the analysis. The descriptions detailing data sources, employed in the investigation, are as follows.

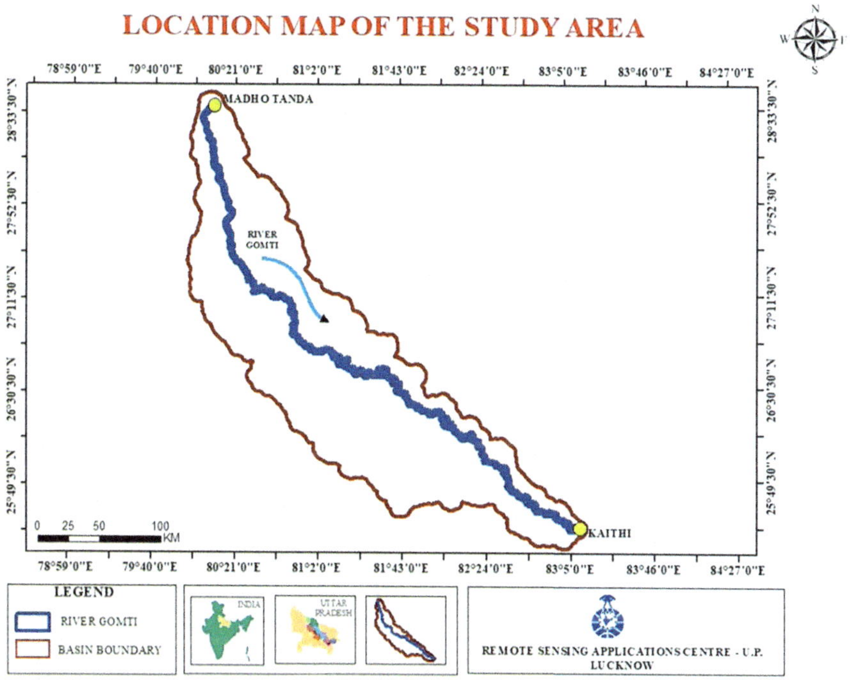

Fig. 1 Map representing the location of the study area

3.2.1.1 Survey of India (SOI)

Toposheet often referred to as 'Topographical Map', 'Open Series Map' or 'OSM' serves as a window into the intricate details of specific geographic regions. Crafted by the Survey of India (SOI), the national custodian of the Indian maps, toposheets are visual representations that empower geospatial endeavours ranging from precisely planned urban landscapes, robust infrastructural development to sustainable management of natural resources and environmental studies. Toposheets are the detailed blueprints of the earth's surface, a comprehensive directory which unfolds a wealth of information to delve deep into the unique features of the area of interest.

The study leveraged the topographical maps drawn to the scale of 1:50,000 which provide a comprehensive overview of the entire area to reveal crucial insights into its geomorphology, landforms and their origin, associated tectonic characteristics and pinpoint accuracy in the latitudinal and longitudinal information allowed by the scale. The topographical map, utilized as the foundation dataset for the study, facilitates seamless integration of the data, findings, analysis and interpretation. The toposheet numbers incorporated in the study are - 62D2/3/4/8, 63A1/5/6/7/11/12/16, 63B13, 63F1/2/5/6/10/11/14/15, 63J3/4/8, 63K5/9/10/14 and 6302/3.

3.2.1.2 Satellite Data

3.2.1.2.1 Satellite Imagery

The Google Earth platform is a multi-faceted integration of satellite images spanning across 37 years which offers the public an extensive insight into the earth's evolving landscape. It is a collaborative effort aligning the European Space Agency (ESA), the European Commission, the National Aeronautics and Space Administration (NASA) and the United States Geological Survey (USGS) to provide a unique and comprehensive depiction of the planet's history (*ESA, 2021*). Landsat-8 was launched on February 11, 2013 from the Vandenberg Air Force Base, California using an Atlas-V 401 rocket. The satellite carries two significant instruments—the Operational Land Imager (OLI) and the Thermal Infrared Sensors (TIRS), which captures seasonal coverage of the earth surface with a spatial resolution of 30 m in the visible, near and shortwave infrared bands (*refer to* Table 1).

Copernicus SENTINEL-2, a part of the European programme, pioneers in capturing high spatial resolution imagery. It is equipped with a set of optical instruments and sensors detecting different wavelengths comprising of 13 spectral bands where 4 bands offer 10 m resolution, 6 band offer 20 m and 3 offer 60 m (*refer to* Table 1) screening atmospheric interference and cloud cover along the swath of

Table 1 Satellites employed in the study with their spatial resolution

S. no.	Sensor/satellite	Resolution
1	SENTINEL-2	10–60 m
2	LANDSAT-8	15–30 m

290 km. It is apt for agricultural monitoring, natural disasters and land-cover classification. Landsat and SENTINEL-2, distinct in their operational characteristics, offer a robust synergy for interactive earth observation. The unparalleled temporal coverage of Landsat coupled with the high-resolution capability of SENTINEL-2 equips the study with versatility for unfolding the intricacies of Gomti river encroachment.

3.3 Ancillary Data and Software

3.3.1 Cadastral Map

The inclusion of the cadastral maps of innumerable villages of Uttar Pradesh from Pilibhit, Shahjahanpur, Lucknow, Barabanki, Sultanpur, Jaunpur, Varanasi and Ghazipur has enhanced the quality and comprehensiveness of the study multifold. The cadastral maps find their source in the web-based application 'Bhunaksha', developed by the National Informatics Centre (NIC) of India to furbish the existing cadastral mapping of the land parcels across the states and union territories in the country. Bhunaksha, translating to 'the map of land' in Hindi, aims at digitizing the cadastral maps to depict property boundaries, land parcels and ownership details. The assistance of these maps as the ancillary data for the districts along the River Gomti has enhanced its accuracy and depth, facilitating a comprehensive understanding of the land ownership system and spatial landscape under examination. The cadastral maps utilized in the study engulf the villages of Phulhar, Baigpur and Gopalpur in Pilibhit district, Lakshmipur, Barahara Udaipur and Mahmadpur Sainjania villages of Shahjahanpur district, Lucknow district's Badiya, Duggor, Ardhona Mau and Jiyamau villages.

3.3.2 Census Data

The District Census Handbook (DCHB), published by the state's Census Directorate, is a treasure of information on the demography of Uttar Pradesh, combining its census data with the infrastructure and amenities in the cities and villages. The DCHB contains Part-A focusing on listing the availability of basic facilities, schools, hospitals, water sources, transportation, banks and non-census data in the villages and towns and Part-B diving into the census information providing the breakdown of population characteristics of villages and wards. The data 2011 has been incorporated in the study.

3.3.3 Software

Software is a vital component of modern technology, encompassing a wide array of applications and tools that facilitates various tasks and processes. Software related to GIS (Geographic Information System) encompasses a wide range of tools for mapping, spatial analysis and data visualization.

3.3.3.1 ArcGIS

ArcGIS is a Geographic Information System (GIS) software created by Esri (Environmental System Research Institute) compatible to a suite of GIS environment conducing the client, server and online components. Introduced in 1982 as 'ARC/INFO', a command-line GIS, ArcGIS has evolved over time. The ArcGIS enables geodata sharing and management, fabricates maps and reports with charts, graphs, diagrams and information, collates the data making it logical, comprehensible and correlated, data processing, representation and interpretation and supports tools for spatial modelling, geoprocessing, spatial analysis, proximity analysis and spatial interpolation. It has nurtured the decision-making and evidence-based conclusion in the river encroachment study.

4 Methodology

The methodology adopted in the execution of the aforesaid study has been discussed as follows:

4.1 Data Acquisition

The data used has been collected from distinct, reliable and recognized sources, elaborated as follows:

4.1.1 Secondary Data

The satellite data, obtained from the Landsat 8/SENTINEL-2 satellites, and the toposheets, sourced from the online portal of the Survey of India, National Mapping Agency, make up the secondary data. The toposheets form the basemap for the year 1972 and for the study while the satellite data creates the same for year 2022.

4.1.2 Ancillary Data

The study includes ancillary data in the form of cadastral maps providing the registered plot number of the land marked under the area of river encroachment.

4.2 Data Processing

The collected data has been processed using geospatial techniques to uplift the standardization and applicability. The geospatial techniques and processes implemented are discussed as follows:

The methodology implemented in the study has been represented as a flowchart as follows:

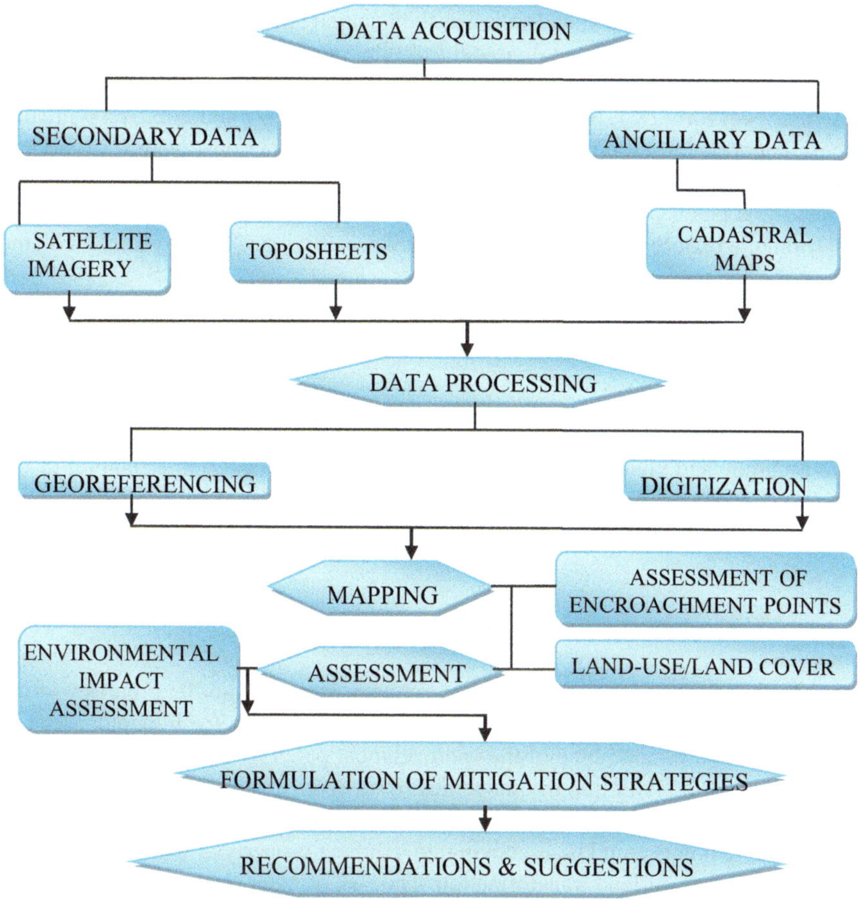

4.2.1 Georeferencing

The process of georeferencing involves allotment of a known coordinate system to the geographic attributes of that particular digital data which allows it to be viewed, queried, analysed and reviewed with other geographic data (*Environmental Systems Research Institute, esri*). The satellite images, toposheets and cadastral maps must be georeferenced before being brought to use. To georeference the secondary and the ancillary data of the study, the ArcMap software of the ArcGIS application package has been used. Each data has been georeferenced separately with utmost accuracy.

4.2.2 Digitization

The features on the satellite images as well as the maps are traced with their x, y coordinates being recorded and stored as their spatial attribute, automatically (*esri*). Numerous features have been digitized from the secondary and ancillary data. The River Gomti has been traced from the mosaic of the toposheets and that of the satellite images to form two separate datasets for the years 1972 and 2022, respectively. These features, representing the river in different time periods, were overlaid and the area left non-overlapping has been marked as an encroachment patch. The land-use of these patches has been incorporated from the satellite data by creating it in the form of a feature. The cadastral maps have been digitized entirely along with the given plot numbers catalogued as an attribute.

4.3 Mapping

The accuracy and vividness of the maps add to the graphical representation of the data and consequently the study. Maps, which are considered an important tool in any research, have been used extensively in the study and described as follows:

4.3.1 Encroachment Point

An encroachment point is recognized as an area, lying within the premises of the River Gomti, demonstrating anthropogenic type of land-use. Each map consists of the detailed address and type of the encroachment patch (*refer to* Table 2) situated on the bank of the river and the shapefiles of the river for the year 1972 as well as 2022 for close comparison, placed on the basemap of the toposheet and the satellite image. These points could be occupying government property as well hence may be termed as pockets of land transformation. The task of mapping has been performed on the ArcMap software. The maps display the pinpointed geographical location,

Table 2 Type of encroachment on the Riverbanks of the Gomti

Type	Description
Built-up	Encroachment dominated by infrastructure covering all types of concrete structures For example—Residences, buildings and roads
Barrier	Encroachment dominated by infrastructure constructed on the river which hinders its free-flow For example—Pontoons, floating and other bridges and ghats
Industrial	Encroachment dominated by infrastructure and set-up for the industrial sector For example—Manufacturing units, factories and workshops
Agricultural	Encroachment dominated by agriculture and allied practices For example—Agricultural land, cropped land and commercial plantations
Vegetation	Encroachment dominated by vegetation and plantations For example—Fallow land, grasslands, pastures and forested land
Sediment	Encroachment dominated by sand and unfertile, salt-affected soil on the banks For example—Unnatural Riverine Islands, coarse riverbanks, broken grounds and wastelands

extent and existence, over the duration of 50 years, of every encroachment patch on the banks of the Gomti.

5 Result and Analysis

Throughout the course of River Gomti, i.e. from the origin (Fulhar Lake) to confluence with Ganga river, i.e. Kaithi village, a total of 441 places are facing the problem of encroachment in different ways. But in this paper, we are addressing only those places where encroachment extent is too large.

The alternations in land utilization along the River Gomti in ***Tanda Madiau*** village of Pilibhit district, Uttar Pradesh which happens to be the origin of the river, named as 'Fulhaar Jheel'. The map portrays three aspects of the data where the first part is the river basin of the Gomti with the village in presentation is highlighted, second displays the data from the 1972 with the help of topographical map and the third part showcases the satellite data of 2022. In the topographic map of 1972, it is evident that the width of the Gomti covered a larger swath of land while in 2022 the area occupied by the river has dwindled as it was transformed into *agricultural land with cropped fields, plantation and roads* (refers Fig. 2). Within the encroached area along the River Gomti, agricultural activities are prevalent with a special reference made to the cadastral map of the village marking *plot no. 223*. The visual representation serves as a comprehensive portrayal of the shift in land-use along the Gomti in the village of Tanda Madiau over the past five decades. The encroachment patch along the river bank stands out as a significant indicator of the impact of human activities on the natural environment. Exhibits the depiction of another encroachment patch situated in ***Tanda Madiau*** village within the Pilibhit district. The point

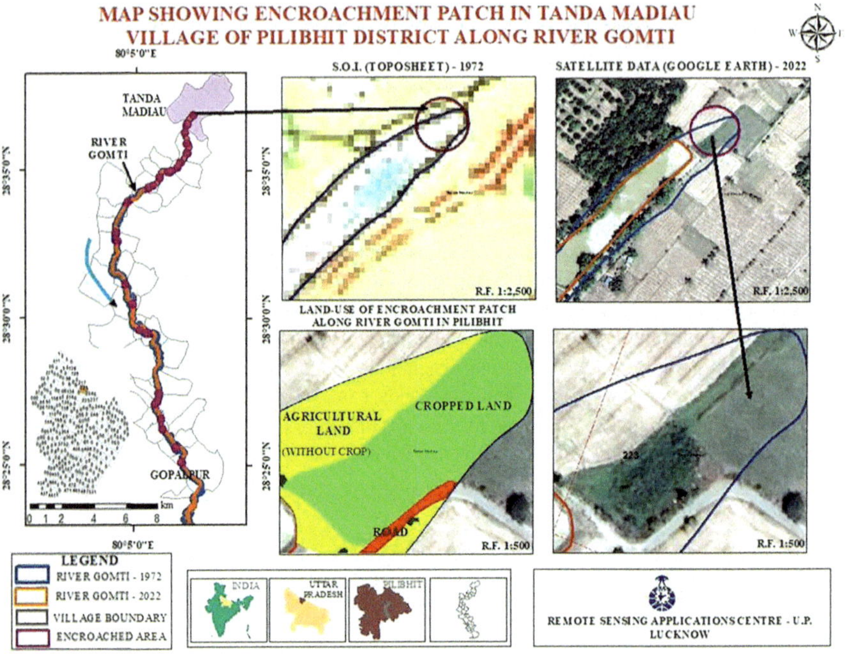

Fig. 2 Map showing the encroachment patch in Tanda Madiau Village of Pilibhit District along the River Gomti

marks the conclusion of the *Fulhaar Jheel*. Encompassing an area adjacent to the Gomti snout, the map delineates the river's course as observed in 1972 and 2022.

Another one illustrates a stark example of encroachment on the River Gomti floodplain in ***Baigpur*** village, Pilibhit district. A comparison between the 1972 topographical map and the 2022 satellite imagery reveals a dramatic shift in the land-use. The area, once part of the Gomti's catchment, is now dominated by human activity. Satellite imagery shows a mosaic of land-uses within the encroached area, including *plantations, built-up areas (structures), moist lands (potentially wetlands) and fallow land*. The cadastral map provides details of village land ownership highlighting *plots no. 100 and 108*. The built-up areas emerge dominantly within the encroached zone. The reasons behind the encroachment are multifaceted with the potential drivers of population growth leading to increased demand for housing and built-up, economic factors influencing land-use decisions increasing the plantations and alterations in the land management practices impacting the natural habitats flourishing in the moist lands and fallow lands (Fig. 3) highlights the complex structure of human needs and environmental consequences on the riverine ecosystems with predominance of moist land and vegetation coupled with cultivated land.

A detailed illustration of an encroached zone within the ***Lakshmipur*** village in Shahjahanpur district. The encroachment area highlights various functional categories of land-use *including plantations, transportation, barren land and built-up*

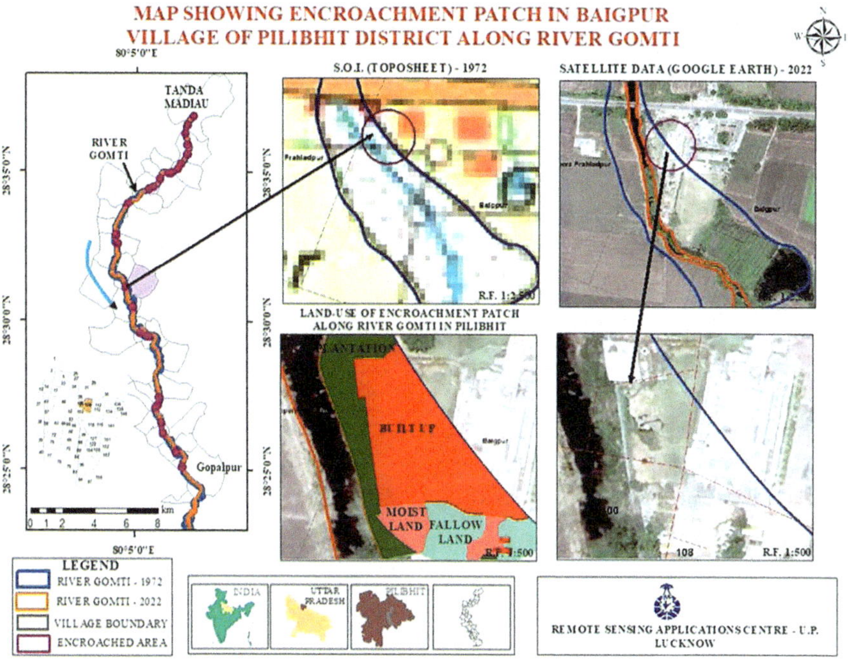

Fig. 3 Map showing the encroachment patch in Baigpur Village of Pilibhit District along the River Gomti

zones. The establishment of human settlements along riverbanks is driven by the *accessibility of freshwater resources for irrigation* of agricultural fields and potable water for *domestic use*. The proximity to fluvial transportation networks facilitates *commercial activities and trade*, making the riverbanks strategically advantageous. The presence of human settlements has a profound negative impact on the surrounding river ecosystem through *wastewater discharge* from domestic and industrial sources, *water pollution* and the *degradation of water quality***.** Deforestation for land clearance and construction activities associated with settlements leads to *soil erosion* which consequently causes *sedimentation* and *siltation*. The map highlights specific plots within the encroachment area, namely *plot No. 241, 234 and 248* playing a significant role in emphasizing the human impact on the river's ecosystem (as shown in Fig. 4).

Encroachment patterns in ***Mahmadpur Saijania***, a village in the Shahjahanpur district serves as a stern evidence of the changing landscape and. The map reveals an array of land-use inclusive of *plantation, road networks and agricultural fields*. This diversity in land utilization reflects the evolving dynamics of human settlement and activity in the region. A meticulous examination is imperative to grasp the undefined catalysts and repercussions of river encroachment. The encroachment patch features agricultural land and functional roads which aggravate habitat fragmentation, disturbed flow patterns and sedimentation. The commercial plantations in the

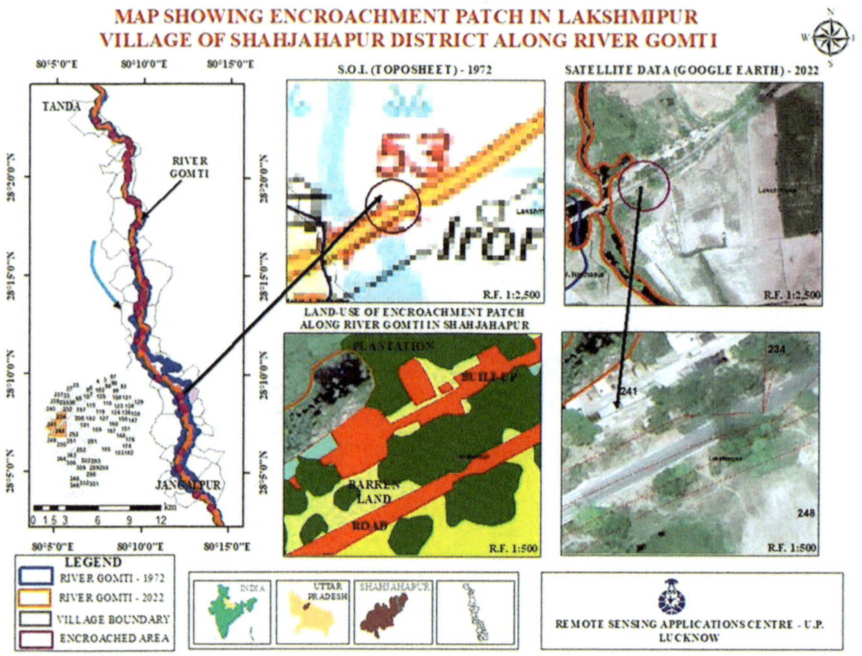

Fig. 4 Map showing the encroachment patch in Lakshmipur Village of Shahjahanpur District along the River Gomti

floodplains, driven by economic incentives, pose significant ecological risks of soil degradation, biodiversity loss and increased vulnerability to flooding. The cadastral map offers valuable information about the growth of the village by demarcating individual plots such as *plot no. 45 and 50* (as shown in Fig. 5).

The encroachment observed in ***Barahara Udaipur*** villages, located in the Shahjahanpur district, is targeted in Fig. 6. The proximity to the River Gomti has attracted the dense settlement due to abundance in natural resources for *irrigation and transportation* leading to gradual encroachment over the years. The demand for residential and commercial spaces driven by population growth and economic activities has resulted in encroachment for *build-up, roads and plantation*.

The encroachment in the ***Badiya*** village of Lucknow district comprehended in Fig. 7 highlights the built-up area of *Nilansh Theme Park* on the riverbed indicating the unlawful and intrusive presence of infrastructure development within the floodplain. Such developments restrict the natural flow of water during floods, exacerbating flood risks and potentially causing damage to the built environment. During normal conditions, floodplains act as natural buffers, *absorbing excess water and allowing it to spread out gradually*, thereby reducing the force and impact of floods downstream. *Recreational infrastructures like Nilansh Water Park* obstruct the natural flow and storage capacity during floods and the water is constricted, leading to

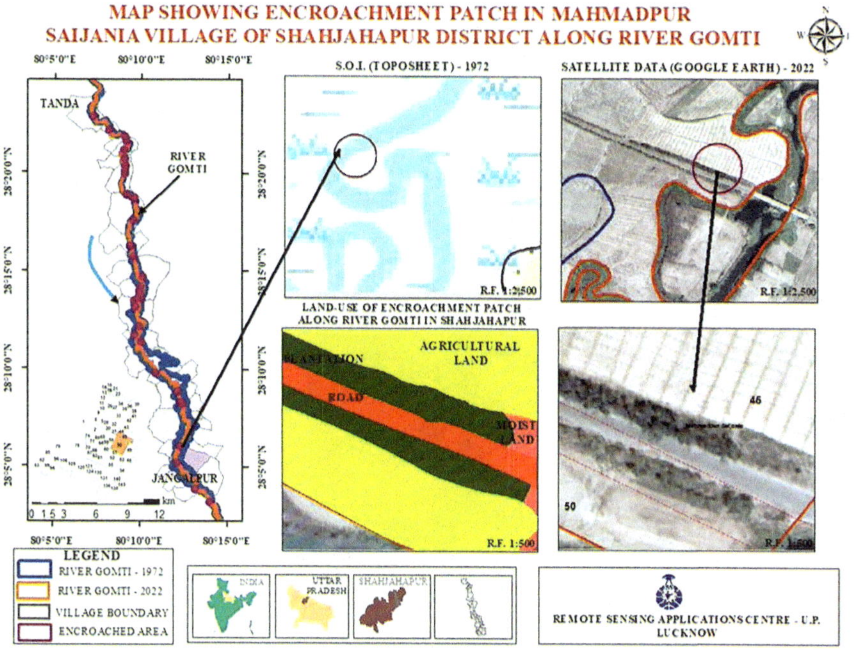

Fig. 5 Map showing the encroachment patch in Mahmadpur SaijaniaVillage of Shahjahanpur District along the River Gomti

increased water velocity, vulnerability to flood and damage posing risk to the safety of the occupants and the structural integrity of the infrastructure.

Showcases the predominant land-uses along the River Gomti in ***Duggor*** village in Lucknow district with an emphasis on the *transportation infrastructure*, visible as a *bridge* over the river. The construction of the bridge can be attributed transportation needs, accessibility and infrastructure development. *Plot no. 127*, as referred to by the cadastral map of the village, provides detailed information about the area (as shown in Fig. 8). The plot lays out *commercial landmixed with built-up and green belt*.

In the areas on the floodplains of the Gomti in ***Manjhi Nikrojpur*** village in Lucknow district shows the drastic changes. The shift highlights the impact of agricultural practices, overuse of water for irrigation, reduced flow of the river and altered land-use patterns. Figure 9 depicts the development in the form of construction and plantation on the riverbanks which has narrowed down the river further as proved by the ground verification. Healthy plant life along the river prevents soil erosion when practised as riparian buffer zones for riverbank stabilization. By deciphering the *plot no. 1691 and 1687*, the map unlocks a treasure trove of details about the village's land ownership and layout.

The encroachment patch in ***Ardhauna Mau*** village, located along the Gomti in the Lucknow district of Uttar Pradesh has a significant impact on the river and its

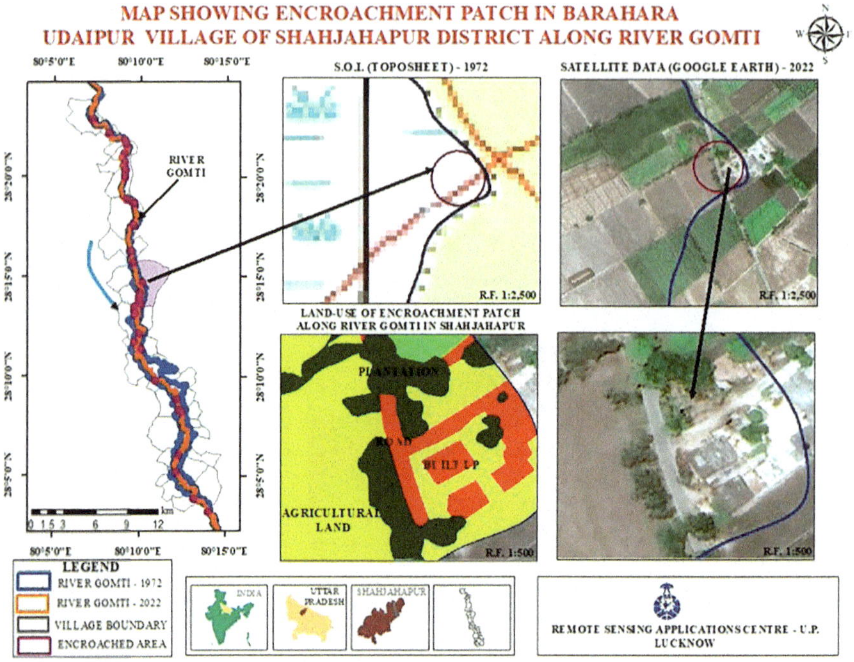

Fig. 6 Map showing the encroachment patch in Barahara Udaipur Village of Shahjahanpur District along the River Gomti

surroundings. The river has been witnessing acute changes in its course observed in the Fig. 10. The encroachment patch shows the area being opted for *residential built-up Plot no. 303, 304, 308, 309* and *310*. The encroachment has several implications on the river and its flow dynamics by increasing the *surface runoff during rainfall*. The built-up *limits the infiltration of rainwater into the ground* which leads to huge runoff into the river, higher water levels, faster flow rates during monsoons, erosion of riverbanks and higher flood risk downstream.

The presence of vegetation in the encroached area of the Gomti in ***Baroliya***, a village falling under the *Lucknow Municipal Corporation*. Increased impervious surface of roads push higher surface runoff, causing soil erosion and river pollution. It is crucial to strike equilibrium between the advantages of road infrastructure and the need to preserve the river's health. *Plot no. 374* in Fig. 11 offers cue for a comprehensive understanding of the land ownership in Baroliya.

The map delineating land-use and encroachment patches along the River Gomti in Jaunpur District provides insights into the changing land-use patterns in the region, respectively. The detailed analysis of the encroachment patch in ***Barahpur*** village observed trends indicate a shift towards *commercial plantation, agriculture and settlement activities* (refer in Fig. 12). It pinpoints settlements established within the critical riparian zone, the ecologically sensitive area bordering the river. The presence of these settlements signifies human encroachment, potentially

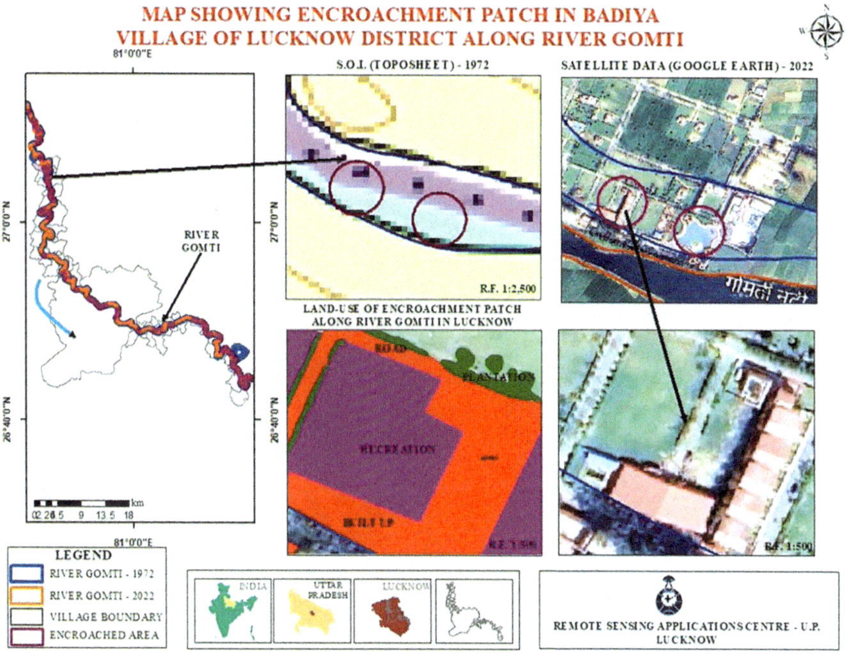

Fig. 7 Map showing the encroachment patch in Badiya Village of Lucknow District along the River Gomti

impacting the river's ecosystem, water quality and flood risk. Riparian zone development can disrupt natural habitats and alter vital hydrological processes.

When old topographical data and new satellite image of ***Bishesarpur*** and ***Hajipur urf Faridabad villages*** in Jaunpur are compared, it is clear that the area along the River Gomti has been changed a lot evident in map. When a water body is transformed into *agricultural land, commercial plantations, built-up settlements and roads*, it indicates the influence of various factors. This transformation is attributed to excessive water usage, leading to a reduction in river flow and altered land-use patterns. The cadastral map plays a crucial role in documenting these changes and providing legal clarity regarding land ownership and boundaries. Specific plot numbers, such as *plot no. 73* in Bishesarpur and *plot no. 41* in Hajipur urf Faridabad, are instrumental in this process (*as shown in* Figs. 13 and 14). These plots likely serve as reference points for identifying the extent of encroachment and determining ownership rights in the affected areas.

Jafarabad (Nagar Palika) situated in Jaunpur district along the River Gomti. Figure 15 visually illustrates this expansion over time, showing changes from historical data to recent satellite imagery. Plot numbers 10 and 18 on cadastral maps likely denote specific land parcels. These plots have experienced encroachment over the years. This encroachment has resulted in the establishment of commercial plantations, vegetation growth and the development of urban infrastructure. The

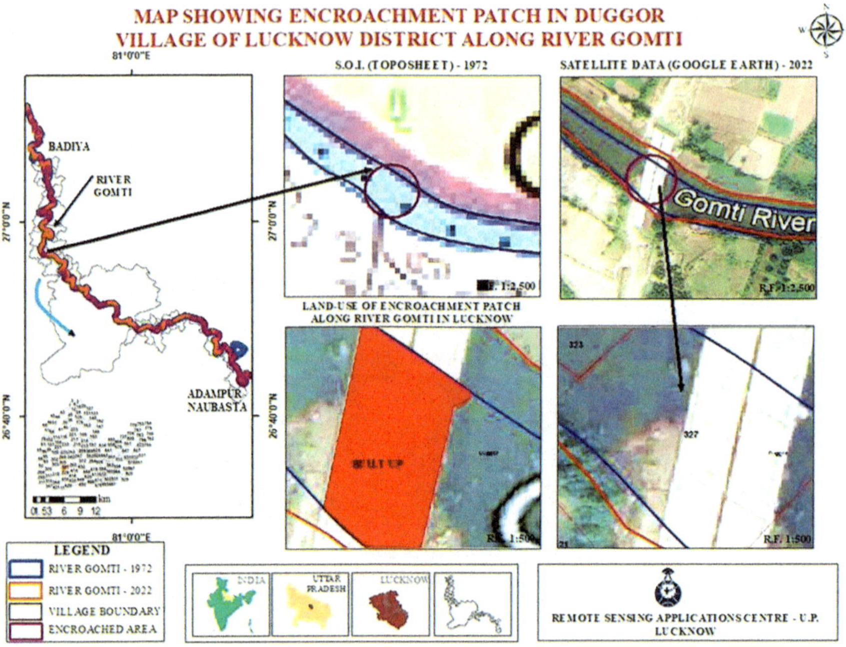

Fig. 8 Map showing the encroachment patch in Duggor Village of Lucknow District along the River Gomti

encroached area has likely grown significantly, indicating the spread of urbanization and human settlement into areas that were previously rural or undeveloped. This shift reflects broader patterns of urban sprawl, changes in land-use and population growth in the region.

The encroachment of the ***Gorkha villages*** in the Ghazipur district, specifically the land-use of *commercial plantations and the expansion of settlements* have several negative impacts on the surrounding environment and the river in various ways: Loss of Natural Habitat: The encroachment leads to the loss of natural habitat for various plant and animal species. Encroachment near the river can disrupt the natural flow and balance of the river ecosystem. This can have consequences for the aquatic species, as changes in water quality, flow and temperature can impact their survival and reproductive patterns. Visually, the map will pinpoint the exact shape and size of *plot no. 322 and 332,* outlining the precise area of ownership in Gorkha village (*as shown in* Fig. 16).

Along the River Gomti in ***Gorry village***, Ghazipur district, suggests in Fig. 17 a notable reliance on *urban development and settlements in the region.* Construction doesn't completely remove habitat; it can fragment it. This means large areas of suitable habitat get broken up into smaller, isolated patches. Fragmentation makes it harder for animals to move between areas, find mates and disperse. Settlement

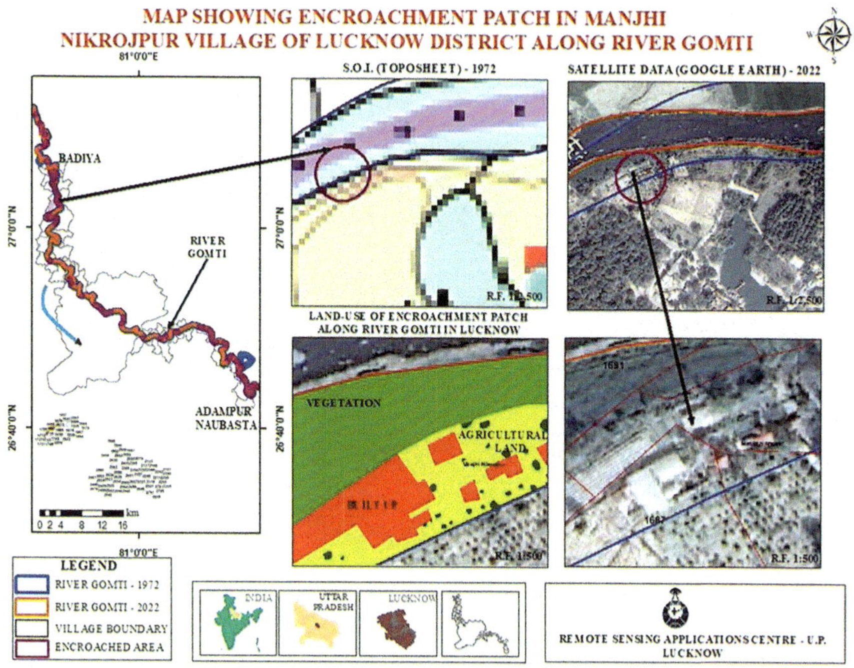

Fig. 9 Map showing the encroachment patch in Manjhi Nikrojpur Village of Lucknow District along the River Gomti

infrastructure can alter the natural flow of the river, affecting how sediment is deposited and how fast the water moves. This can impact the types of plants that can grow and the spawning success of fish. Runoff from roads, buildings and lawns can introduce pollutants like fertilizers, pesticides and oil into the river. This can harm aquatic life and disrupt the food chain. Using *plot no. 1015* as a reference, the map defines the ownership limits by showing the specific dimensions and form of the land.

In Fig. 18, the comparison of historical and recent satellite data of ***Kushi Village*** in Ghazipur district highlights a significant change in the encroachment patch along the River Gomti. Previously, a water body existed in this area, but it has now been transformed into *agricultural land, commercial plantations and built-up settlements*. Agricultural activities in the region likely involve intensive irrigation, leading to the excessive extraction of water from the River Gomti. This excessive usage can result in reduced river flow, causing the water body to shrink or disappear entirely. The presence of built-up settlements suggests urbanization and population growth in the area. As communities expand, they encroach upon nearby water bodies and agricultural land for residential and commercial purposes, further altering the landscape.

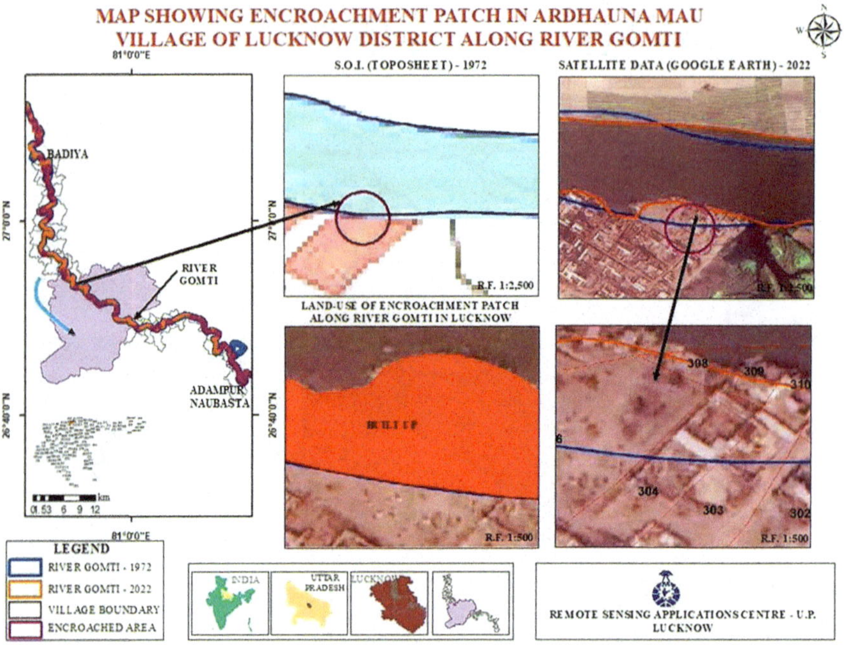

Fig. 10 Map showing the encroachment patch in Ardhauna Mau Village of Lucknow District along the River Gomti

6 Assessment

The assessment of the far-reaching effects of river encroachment has been done based on the data employed, findings and the maps compiled for the study. The ecological consequences, their risks and the havocs caused by river encroachment have been assessed critically. The criteria which lay the cornerstone for assessing the impacts of encroachment have been reviewed as follows:

6.1 Environmental Impact

The assessment of the environmental impact of river encroachment has been directed on the lines of the ecological discomfort caused to the flora and fauna in the catchment of the river. The socio-economic and infrastructural aftermaths of encroachment on the people settled around them have also been considered. The environment has been hugely affected by encroachment as it alters the natural functioning and processes of the river system into artificial setting which disrupts the harmony of the environment, consequently. The ill-effects of river encroachment have been interpreted on the grounds of jeopardy of aquatic ecosystem, biome

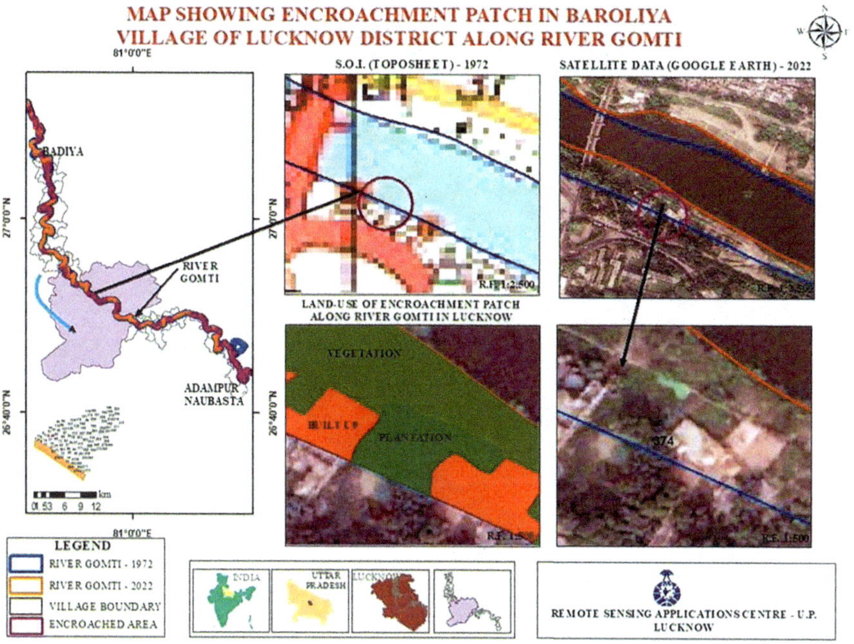

Fig. 11 Map showing the encroachment patch in Baroliya Village of Lucknow District along the River Gomti

destruction, hindrance faced by the natural fluvial processes, increasing toxicity in the water of the river, increased disposal of solid, domestic, industrial and chemical wastes into the river body and other environmental imbalances brought about by ungoverned encroachment on rivers.

6.2 Risks and Hazards

The environmental impacts of river encroachment come along with several dangers and insecurities to the lives dependent on the river for survival. These risks and hazards have been evaluated in association with the man-made disasters like flooding and water-logging, groundwater depletion and contamination, drying of river beds caused by incessant sand mining, damage to the infrastructure built on the floodplain of the rivers because of shifting of silt and other sediments, infrastructural damage by land subsidence in land-fills, loss of habitat, popping of illegal temporary settlements and slums on the edges of the river banks, multiplying numbers of communicable diseases due to vector-breeding in the water-logged areas, evolution of microorganisms spreading epidemics and pandemics and above all threat to life.

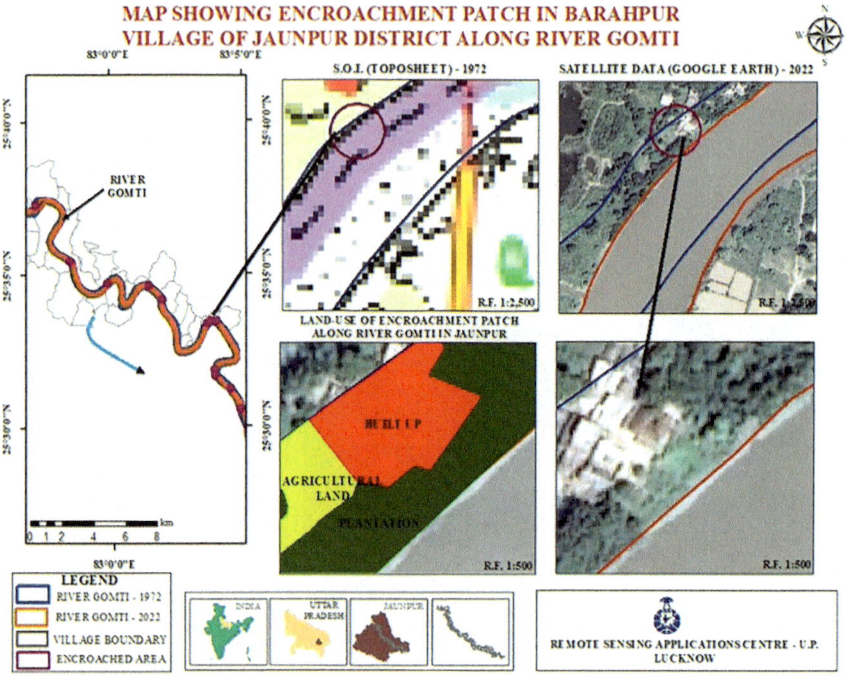

Fig. 12 Map showing the encroachment patch in Barahpur Village of Jaunpur District along the River Gomti

7 Formulation of Mitigation Strategies

The formulation of the mitigation strategies and policies has been devised which cater to the findings of the study and the assessment of hazards evoked by river encroachment. The mitigation policies have been developed to tackle the challenges posed by river encroachment. These strategies have been designed to involve the local communities in gathering the evidences and knowledge about the forthcoming disasters because of the unmanageable rise in the extent of encroachment. The strategies, based on the ground-based data and findings, suggest that the intertwined role of the governmental regulation, local intervention as well as environmentalists' concerns, collaboratively, can attain the desired results and control over the practice of river encroachment.

7.1 Recommendations and Suggestions

The recommendations and suggestions have been derived in accordance with the findings of the study and the mitigation strategies. The recommendations have been suggested inclusive of the policy adjustments and regulatory policies which control

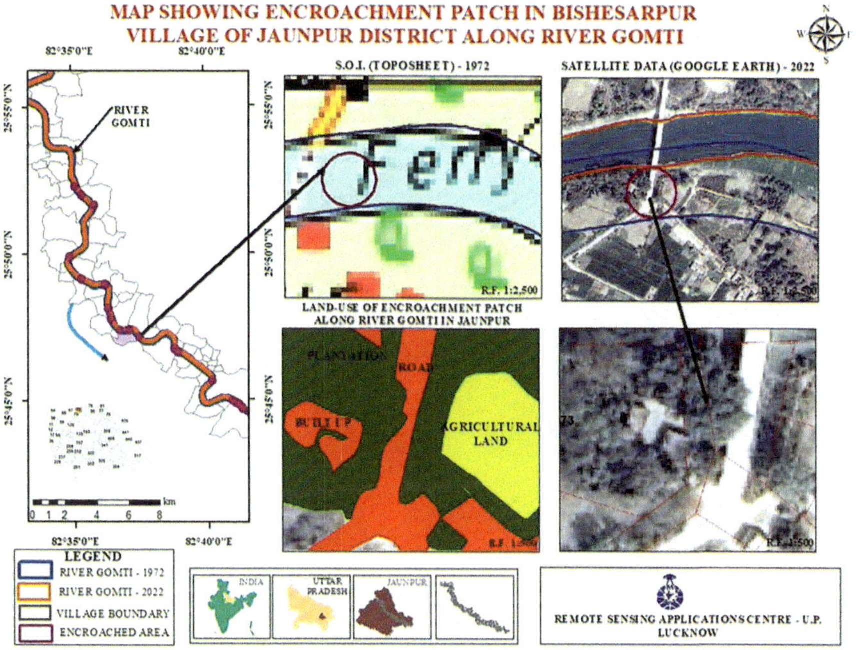

Fig. 13 Map showing the encroachment patch in Bishesarpur Village of Jaunpur District along the River Gomti

present intensity of encroachment and prevent further encroachment. The suggestions have been canvassed in a manner that delivers a permanent, long-lasting and feasible solution to the problem of river encroachment.

8 Results and Discussion

8.1 River Encroachment

River encroachment refers to the intrusion by human-made infrastructure, transportation routes, utilities and other developmental structures into the natural landscapes like floodplains, riverbanks, wetlands and buffer zones of the water bodies. With unnecessary intervention, activities like filling of land, removal of vegetation, changing the natural vegetation and the terrain come into play. Such encroachments have detrimental effects on the ecological functions and significance of these areas, with problems of diminished water quality, habitat loss for aquatic as well as terrestrial species, distribution of natural balance, reduced flood control capacity and disruption of essential ecological processes rising incessantly (*Vermont Department of Environmental Conservation*).

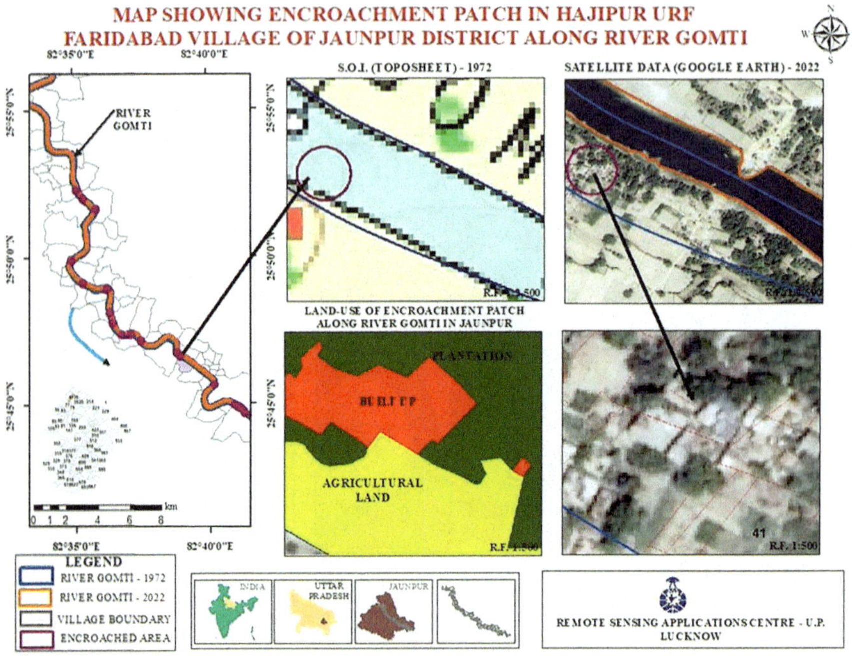

Fig. 14 Map showing the encroachment patch in Hajipur Urf Faridabad Village of Jaunpur District along the River Gomti

The study examines the human encroachment on the River Gomti and its tributary in the northern India, the challenges faced and the consequences of the practices on such significant river. The Gomti, originating from the Gomat Taal in the district of Pilibhit, Uttar Pradesh, crosses several districts before joining the much revered, River Ganga. Rapid urbanization, industrialization and extensive encroachment along the Gomti and its tributaries hamper the fluvial processes of the rivers. Development around the river, comprising of houses, roads and farms, enforced upon the floodplains is incurred by displacing the sand and gravel from the riverbed which forces the river to flow out of proportion. Such instances lead to increased sediment accumulation in the riverbed making the river shallower and thereby a reduced ability in handling floods. The natural habitats along the riverbank have been damaged, the variety of lifeforms reduced in the river and the water turned unhealthy from the untreated sewage and industrial waste. This harms the entire ecosystem and people relying on the river for drinking water, irrigation and livelihood.

The Gomti has been witnessing the brutal repercussions of river encroachment since the expansion of the urban centres flourishing on its banks. The huge urban centres, their out-growths as well as the areas surrounding have the floodplain of the river being absorbed into human activities of farming, manufacturing, settlement, recreation and commercial centres. The River Gomti experiences compounding

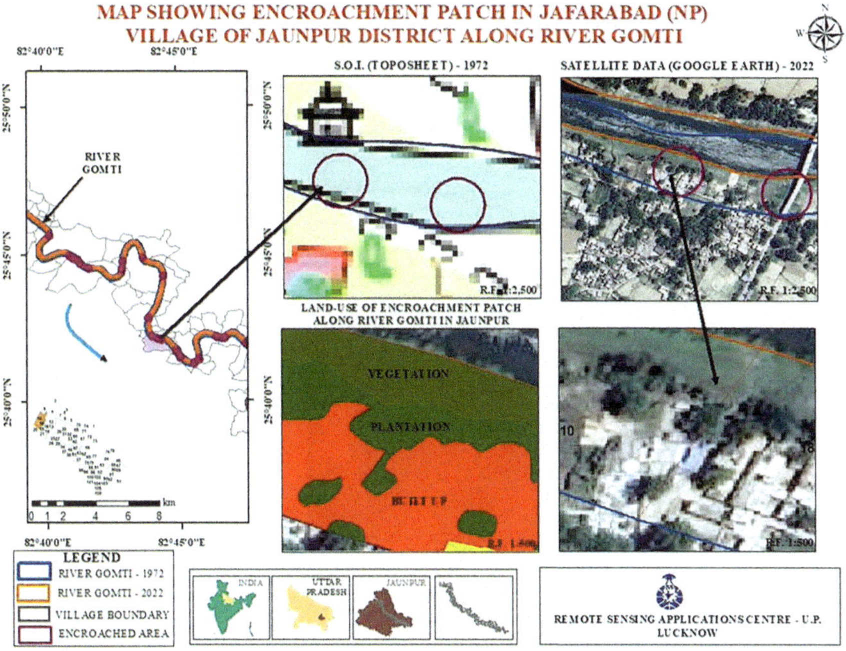

Fig. 15 Map showing the encroachment patch in Jafrabad (NP) Village of Jaunpur District along the River Gomti

frequency of dry patches and seasonality due to the unchecked outflow of its water in addition to the depleting groundwater owing to the private bore and drilling systems in each household. However, the loss of freshwater resources is being caused by deforestation, increased agricultural activity, rural to urban land use changes and rapid population growth [12–15]. The development of riverfronts and other artificial, concrete structures on the edge of the river complimented by the excessive unnatural siltation and sedimentation in the riverbed block the water from percolating into the deeper layers of the soil which worsens the dilemma of the falling underground water levels in the region. The dryness in the river has provoked the development of water hyacinth in the water body which tarnishes the aquatic life as it ceases the oxygen cycle of the river. The endangered life in the Gomti, its stagnant froth-laden waters, foul odour, obnoxious levels of toxicity and chemicals, altered morphology and dying river system are the reactions of the river to the stigma of encroachment.

Monitoring and understanding how land-use changes over time in the course of the river is an integral component of the present study. Remote Sensing and Geographical Information Systems (GIS) are the game changers in this field. These technologies offer powerful tools to analyse the shifting landscapes and land-use patterns. Remote Sensing allow gathering spatial data across vast areas at different points in time. Using satellite images and aerial photographs, aids in aligning a

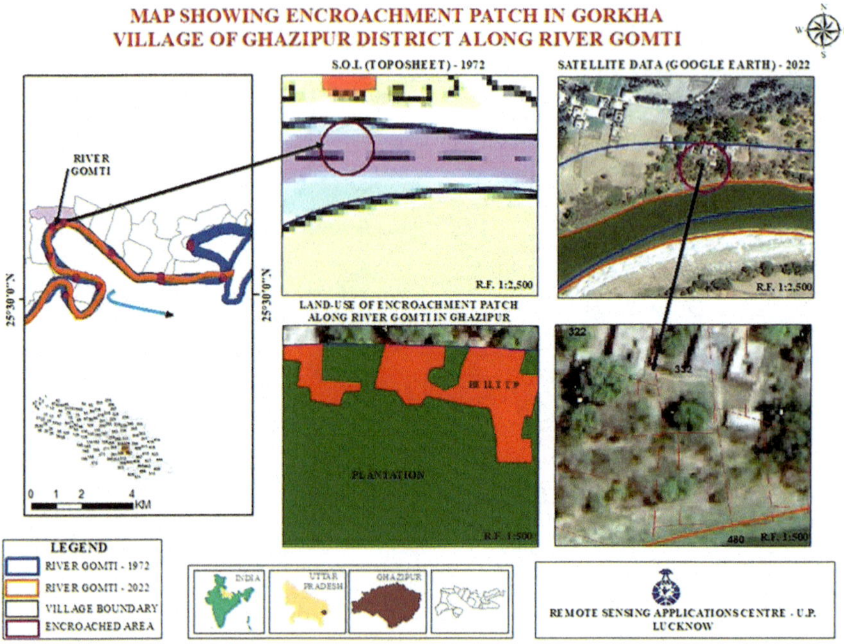

Fig. 16 Map showing the encroachment patch in Gorkha Village of Ghazipur District along the River Gomti

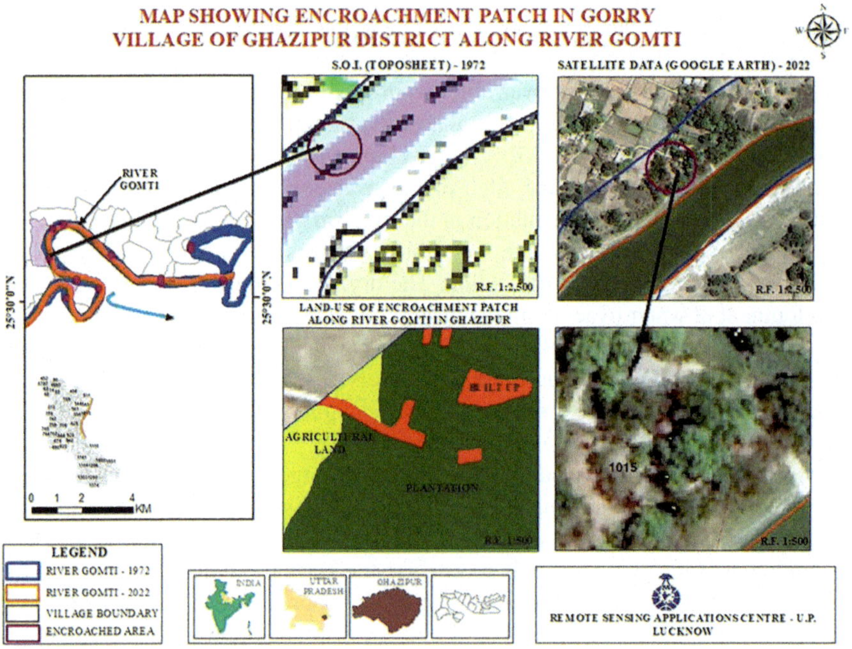

Fig. 17 Map showing the encroachment patch in Gorry Village of Ghazipur District along the River Gomti

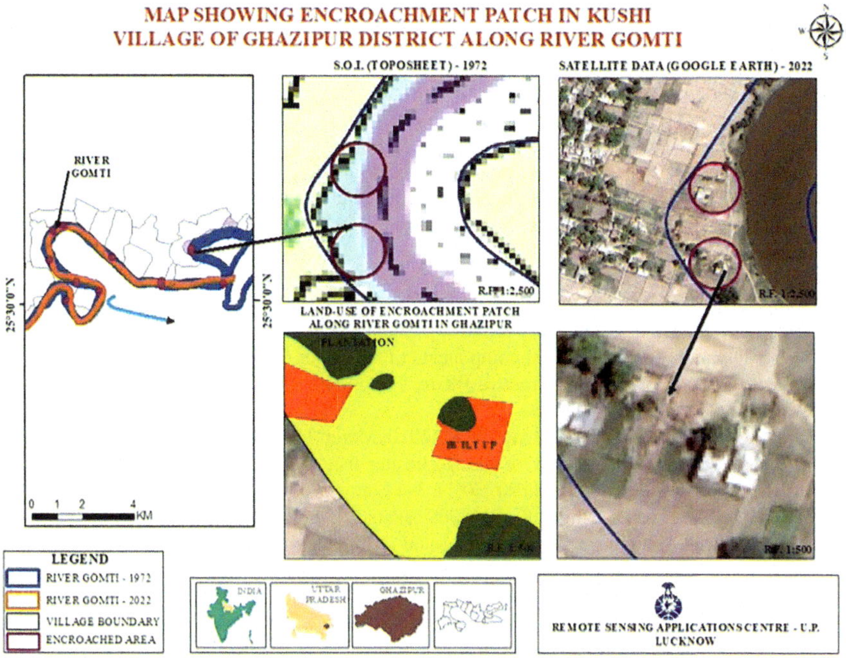

Fig. 18 Map showing the encroachment patch in Kushi Village of Ghazipur District along the River Gomti

detailed insight into the land-cover, vegetational changes, urban sprawl and land-use shifts. This data can be systematically analysed to spot trends, pinpoint areas needing immediate attention and assess the impact of anthropogenic actions on the environment. While, GIS makes it possible to integrate, visualize and analyse spatial data from various sources, layering information like land-cover maps, population density, infrastructure networks and socio-economic data helps researchers and decision-makers identify the spatial relationships and patterns that influence land-use changes. In addition to this, GIS offers tools for modelling and creating scenarios allowing the evaluation of different land-use policies and management strategies.

References

1. Ashraful, M., Ashraful Chowdhury, M., Amir, M., Bhuyain, H., & Kabir, M. M. (2015, January 1). *Assessment of river encroachment and land-use patterns in Dhaka City and its peripheral rivers using GIS techniques* (p. 6)
2. Raihan, F., & Kaiser, N. (2012). *Land use changing scenario at Kerniganj Thana of Dhaka District using remote sensing and GIS.* https://www.semanticscholar.org/paper/Land-Use-

Changing-Scenario-at-Kerniganj-Thana-of-Raihan-Kaiser/274fcc3e5a8124ceed0b069f58a90 60c49d0801b
3. Biswas, B., & Hamada, T. (n.d.). *Relation between Hazaribagh Tannery Industry development and Buriganga River pollution in Bangladesh.*
4. Mahmood, S., Nourin, F. T. J., Siddika, A., & Khan, T. (2017, January 1). Encroachment of the Buriganga River in Bangladesh. *Journal of Minerals and Materials Characterization and Engineering, 05*, 266–273. https://doi.org/10.4236/jmmce.2017.55022
5. Rees, W. E. (2003, July 1). Economic development and environmental protection: An ecological economics perspective. *Environmental Monitoring and Assessment, 86*(1), 29–45. https://doi.org/10.1023/A:1024098417023
6. Tangri, A., Kumar, D., Singh, D., & Dubey, C. A. (2018). *The Gomati river: Lifeline of central Ganga plain* (pp. 135–150). Springer. https://doi.org/10.1007/978-981-10-2984-4_11
7. Shukla, A. K., Ojha, C. S. P., Garg, R. D., Shukla, S., & Pal, L. (2020). Influence of spatial urbanization on hydrological components of the upper Ganga River basin, India. *Journal of Hazardous, Toxic, and Radioactive Waste, 24*(4), 04020028-1-19. https://doi.org/10.1061/(ASCE)HZ.2153-5515.0000508
8. Sarkar, U. K., Gupta, B. K., & Lakra, W. S. (2010, March 1). Biodiversity, ecohydrology, threat status and conservation priority of the freshwater fishes of river gomti, a tributary of river Ganga (India). *The Environmentalist, 30*(1), 3–17. https://doi.org/10.1007/s10669-009-9237-1
9. Malik, A., Ojha, P., & Singh, K. P. (2009, January). Levels and distribution of persistent organochlorine pesticide residues in water and sediments of gomti river (India)—A tributary of the ganges river. *Environmental Monitoring and Assessment, 148*(1–4), 421–435. https://doi.org/10.1007/s10661-008-0172-2
10. Dutta, V. (2015, September 1). Restoration of the river Gomti: How can regulatory interventions from the government contribute towards a cleaner river Ganga? *Global Water Forum, 4*, 6.
11. Singh, K. P., Malik, A., Mohan, D., & Sinha, S. (2004, November). Multivariate statistical techniques for the evaluation of spatial and temporal variations in water quality of gomti river (India)—A case study. *Water Research, 38*(18), 3980–3992. https://doi.org/10.1016/j.watres.2004.06.011
12. Gandhi, V. P., & Bhamoriya, V. (2011). Groundwater irrigation in India: Growth, challenges, and risks. In *India Infrastructure Report, 2011. Water: Policy and performance for sustainable development*. Infrastructure Development Finance Company. Oxford University Press.
13. Saha, D., Marwaha, S., & Mukherjee, A. (2018). Groundwater resources and sustainable management issues in India. In *Clean and sustainable groundwater in India* (pp. 1–11). Springer.
14. Nayak, D., Shukla, A. K., & Devi, N. R. (2024). Decadal changes in land use and land cover: Impacts and their influence on urban ecosystem services. *AQUA - Water Infrastructure, Ecosystems and Society, 73*(1), 57–72. https://doi.org/10.2166/aqua.2024.211
15. Zolekar, R. B., Todmal, R. S., Bhagat, V. S., Bhailume, S. A., Korade, M. S., & Das, S. (2020). Hydrochemical characterization and geospatial analysis of groundwater for drinking and agricultural usage in Nashik district in Maharashtra. *Environment, Development and Sustainability*. https://doi.org/10.1007/s10668-020-00782-2

Advancements in Remote Sensing for Water Quality Assessment: A Comprehensive Exploration

Bishnu Kant Shukla, Lokesh Gupta, Pushpendra Kumar Sharma, Khushi Tyagi, Harshit Yadav, Satvik Singh, and Yogendra Yadav

1 Introduction

The vitality of our ecosystems, cultures, and economies depends critically on the quality of the water that sustains them. Making sure that everyone has access to clean water and maintaining its quality has become crucial due to the growing global population and industrialization. Monitoring water quality has long been acknowledged as being essential to environmental management and sustainable development [1]. It's not just about making sure there is clean water for people to drink; it's also about protecting biodiversity, aquatic habitats, and sustainable industrial and agricultural practices. Due to the degradation of water resources necessary for fisheries, tourism, and agriculture, declining water quality can have a disastrous effect on ecosystems and result in significant economic losses [2]. Numerous toxins are released into our water bodies as a result of modern industrial operations. The variety of contaminants influencing our waters is extensive, ranging from dangerous compounds to heavy metals and organic pollutants [3]. Furthermore, changes in temperature, salinity, and even dissolved oxygen levels brought on by climate change, such as changing rainfall patterns and rising temperatures, are making the

B. K. Shukla (✉)
Department of Civil Engineering, JSS Academy of Technical Education, Noida, India

School of Civil Engineering, Lovely Professional University, Phagwara, India

L. Gupta
College of Pharmacy, JSS Academy of Technical Education, Noida, India
e-mail: lokeshgupta@jssaten.ac.in

P. K. Sharma
School of Civil Engineering, Lovely Professional University, Phagwara, India

K. Tyagi · H. Yadav · S. Singh · Y. Yadav
Department of Civil Engineering, JSS Academy of Technical Education, Noida, India

S. Pathak et al. (eds.), *Geo-Data Revolution*, Advances in Geographic Information Science, https://doi.org/10.1007/978-3-031-94078-1_4

problem of poor water quality worse [4]. Monitoring these variables is essential for future planning and predictive modelling as well as for taking immediate action.

In situ measurements, where samples were physically taken from water bodies and subsequently analysed in labs, were historically the mainstay of water quality monitoring. While precise, these techniques require a lot of work, take a long time, and frequently have a finite range of coverage [5]. Introducing remote sensing, a current world technological marvel, has revolutionized water quality monitoring by enabling large-scale, real-time assessment of surface water bodies with greater spatial and temporal coverage, thereby overcoming many of the limitations associated with traditional in situ methods. Large water bodies may be monitored via remote sensing in a comparatively short amount of time, providing data that is nearly real-time and essential for hasty decision-making [6]. For example, satellite-based sensors can offer detailed information on a variety of water quality indicators over broad regional scales, including chlorophyll content, turbidity, and even salinity [7]. With the introduction of convolutional neural networks, the accuracy with which these characteristics may be approximated has been significantly improved, making remote sensing a potent tool for assessing water quality [8]. Additionally, the spatial resolution of remote sensing data has significantly improved with the advent of platforms like Sentinel and VIIRS, enabling more thorough and localized evaluations of water quality [9]. Remote sensing offers a complete view when combined with ground truth data, fusing the macro with the micro and providing insights that were before thought to be difficult to attain [10]. Tools like remote sensing, which promise precision, efficiency, and thorough coverage, offer a beacon of hope as the problems with water quality worsen. Adoption of these tools isn't an option; it's essential, for building a future that can be sustained.

2 Historical Evolution of Remote Sensing Tools

In the decade's advancements, in remote sensing technology have greatly evolved, expanding the possibilities for scientific studies, such as monitoring water quality. This progress has led to changes in how scientists observe and understand water sources and their components with the advancements, in satellite technologies.

2.1 Satellite Evolution and Its Implication for Water Quality Monitoring

In the days of satellite sensing, the equipment used was not as advanced and had certain limitations. However, with progress satellites have greatly improved in terms of accuracy, coverage, and range. The early satellite sensors were primarily intended for monitoring terrestrial resources, but it soon became clear that they could also be used for aquatic environments [11]. The spectral, geographical, and temporal resolutions have significantly increased with the launching of increasingly advanced satellites. For identifying and tracking the variability of various water constituents, spectral bands in the visible and infrared ranges in particular became crucial [12]. Multispectral and hyperspectral sensors that can provide more specific information regarding water quality, particularly in terms of chlorophyll-a concentrations, have come into play with the advent of satellites like GF2 and Sentinel-2 [13].

Advances in ocean and freshwater studies have been made possible by satellite platforms like the European Space Agency's (ESA) Envisat. Such satellite missions are significant due to their improved observational capabilities as well as their consistent, worldwide coverage. Due to this, hitherto unreachable large and isolated aquatic areas can now be monitored [14]. The measurement of sea surface temperatures has been made possible by the addition of thermal infrared devices to satellites, adding another layer of surveillance. These data, collected from thermal infrared devices on ships, in the air, and in space, provided understanding of coastal dynamics and their impact on water quality [15]. A critical parameter for comprehending different physical and biological processes in the marine environment, salinity measurements have been explored by more modern satellite missions including the ESA's SMOS (Soil Moisture Ocean Salinity) mission [16]. Although measuring salinity from space was a difficult task, successful implementations cleared the way for more thorough evaluations of water quality. In essence, the field of water quality monitoring has undergone a considerable transformation due to the development of satellite remote sensing instruments. These developments provide a unique perspective, assisting scholars and decision-makers in deliberating over and assuring the sustainable management of water resources.

2.2 Transition from Airborne to Spaceborne Sensors

The dominance of airborne sensors in the field of remote sensing has given way to the constantly increasing domain of spaceborne sensors. Both platforms are equipped with distinctive characteristics and meet various demands in the field of environmental monitoring. But comprehending this transition is essential to understanding how water quality assessment instruments have developed and changed over time. Initially, airborne sensors had the benefit of flexibility. Because they were mounted aboard airplanes, these sensors allowed for targeted observations and improved spatial resolution monitoring of particular areas at specific times. They offered abundant details, although for limited areas, and were especially useful for localized research [17]. But the move towards spaceborne sensors, notably satellites, has been prompted by the requirement for worldwide coverage, continuous temporal monitoring, and the capacity to collect wider spatial extents. The benefit of regular and reliable measurements over large areas provided by satellites ensures thorough monitoring. When analysing events with broad effects on marine ecosystems and human activities, such as algal blooms, river plumes, and coastal dynamics, this large-scale monitoring becomes increasingly important [18].

Table 1 gives a brief summary of the satellites that can be used for retrieving water quality data using remote sensing. This table shows a significant growth in the

Table 1 Application of polymeric membranes for purifying water contaminated by PhACs

PhACs Class	Sample	Name	Polymer	Nanomaterial	Additive	Process	Removal %	Reference
Antibiotics	Synthetic	Sulfadiazine	PVDF	TiO_2	PVP	UF	91.4	[20]
		Erythromycin	PA_{TF}	Zeolite		NF	>90	[21]
Antihistamine	Synthetic	Nizatidine	PA_{TF}	Zeolite	PVP	NF	>90	[21]
Hormones and endocrine disruptive compounds	Water treatment plant	Bisphenol A	PES	SiO_2	PVP	UF	87	[22]
Non-steroidal Anti-inflammatory	Synthetic	Diclofenac	PA_{TF}	Zeolite	PVP	RO	95	[23]
Lipid regulator	Synthetic	Gemfibrozil	PA_{TF}	Zeolite	PVP	RO	99.5	[24]
Anti-hypertensives	Synthetic	Propranolol	PVDF	β-CDP	PVP	UF	99.9	[25]

number of satellites having multi-band capability, which is evidence of the improved accuracy and adaptability of contemporary spaceborne sensors. The bands, especially those above five, demonstrate the sophisticated spectral capabilities these satellites bring to the field of determining water quality. Additionally, as the table emphasizes, improvements in spectral, spatial, and temporal resolutions have been made in tandem with the growth of spaceborne sensors, considerably improving the precision and dependability of data on water quality [19]. The development of multi-mission satellites, which contain a variety of sensors, emphasizes the multidisciplinary approach now used in remote sensing.

The switch to spaceborne sensors has ushered in an era of global, comprehensive, and consistent water quality monitoring, while airborne sensors lay the groundwork for focused and high-resolution water quality evaluations. This change underlines the significance of thorough environmental monitoring in light of international issues and shifting aquatic ecosystems.

3 Water Substances and Sensor Interaction

3.1 The Spectrum Transfer Process

The interaction between water substances and sensors is crucial in the field of remote sensing for water quality monitoring. The idea of spectrum transfer is central to this relationship. It entails comprehending how different elements in water, such as chlorophyll, total suspended matter, and dissolved organic matter, interact with and change the electromagnetic radiation detected by the distant sensors [26]. The efficiency of remote sensing systems is fundamentally determined by the spectrum transfer process. Simply explained, electromagnetic radiation from a source (such the sun) that interacts with water bodies experiences absorption, reflection, and scattering by the water and its constituents. These interactions lead to specific spectral signatures that sensors onboard remote sensing platforms detect. Different substances in water yield distinct spectral responses, allowing for their identification and quantification [27].

Figure 1 graphically portrays this spectrum transfer process. As showcased in the figure, the interaction starts with the incident solar radiation. Upon reaching the water surface, a part of it is reflected, while the rest penetrates the water body. The submerged radiation interacts with water constituents and undergoes spectral changes. It's these alterations in the spectrum—resulting from the interaction with water substances—that remote sensors capture. One may also see the layers of interaction within the water body by carefully examining Fig. 1. The type of substance, its concentration, and the wavelength of the incoming radiation are some of the variables that affect how deep and intense the contact is. These observations assist in fine-tuning remote sensing techniques for best results, assuring precise retrieval of water quality metrics [28, 29].

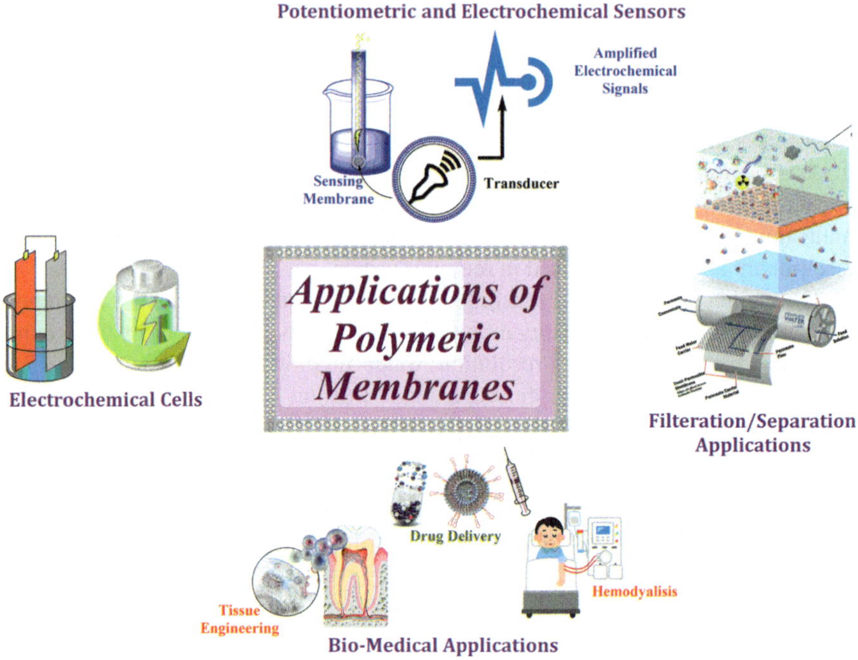

Fig. 1 Utilization of polymeric membrane application

In order to effectively use remote sensing to monitor water quality, it is essential to comprehend the spectrum transfer process. In order to guarantee the accuracy and dependability of remote sensing results, it serves as the foundation on which cutting-edge algorithms and retrieval models are developed.

3.2 Band Positioning and Its Implications

The placement of spectral bands is crucial in the field of remote sensing, notably in the monitoring of water quality. The accuracy and specificity of the data collected can be greatly influenced by where these bands are positioned on the spectrum. The ability of a satellite or sensor to recognize, distinguish, and measure water components such as chlorophyll, dissolved organic matter, and suspended particles rely heavily on the placement of these bands [30]. In the past, bands were placed on sensors based more on progress than a full understanding of waters optical properties. Now band placement is more intentional, tailored for water quality applications, thanks to advancements and a deeper understanding of aquatic optics [31].

For example, instruments like Landsat7/ETM+ have bands strategically located to capture wavelengths where water substances display unique behaviours. This deliberate positioning not only enhances drug discrimination but aids in clearer data interpretation. The arrangement of Landsat7/ETM+ bands on the ASD spectroradiometer spectrum showcases how knowledge, about light water interactions has progressed [32, 33]. Beyond simple substance detection, band location has further ramifications. The impacts of air interference can be reduced, the signal-to-noise ratio can be improved, and even the inherent optical qualities of the water can be compensated for with the use of strategically placed bands [34]. Additionally, the precise choice and placement of bands have opened the door for more comprehensive and in-depth aquatic research with the development of multispectral and hyperspectral photography. While advances in remote sensing technology have increased our capabilities, careful spectral band placement continues to be a key factor in guaranteeing the effectiveness and dependability of water quality monitoring from space.

4 Innovative Models and Algorithms

The scientific community has turned to novel models and algorithms in its never-ending search for remote sensing accuracy. Convolutional Neural Networks (CNNs) are one such ground-breaking technique used to decode complicated water quality data.

4.1 Convolutional Neural Networks for Phytoplankton Pigment Estimation

CNNs, a subset of deep learning algorithms, have lately gained prominence as effective image analysis and recognition tools. CNNs automatically learn hierarchies of features directly from data, in contrast to standard algorithms that require customized feature extraction, making them particularly well-suited for challenging applications like recognizing phytoplankton colours from satellite imagery [35, 36].

A rigorously planned convolutional neural network architecture for phytoplankton pigment estimate is shown in Fig. 2. The network has numerous convolutional layers, followed by pooling layers, as seen in the picture. The input image's attributes are gradually captured and refined by each layer, building up to a dense layer that outputs the estimated pigment concentrations. The brilliance of this architecture rests in its automatic recognition of patterns and characteristics in satellite photos that are indicative of phytoplankton concentrations. Thus, even in complex or murky environments, the CNN offers a robust and trustworthy method for estimating pigment [37].

Previous algorithms frequently struggled when faced with complicated or ambiguous satellite data since they mainly depended on manually created features and linear regressions. On the other hand, the CNN strategy ushers in a new era of

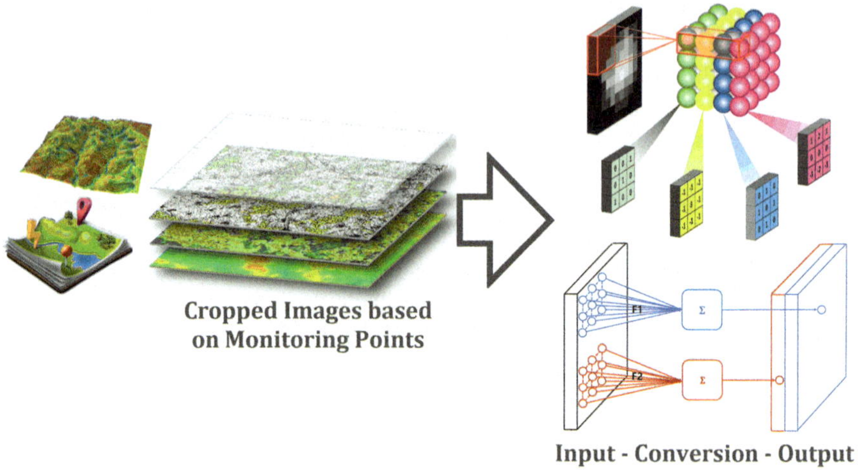

Fig. 2 Phytoplankton pigment estimate using a convolutional neural network

agility and accuracy. The potential of CNNs in remote sensing of water quality increases as datasets expand and algorithms get more precise. The development of CNNs and the effective use of this technology in estimating phytoplankton colours highlight the fascinating developments in remote sensing models and algorithms. It is evidence of how computer science and environmental science work together to provide unmatched insights into our aquatic ecosystems.

4.2 Hybrid Spatio-Temporal-Spectral Fusion Models

Modern issues in remote sensing, particularly in the monitoring of water quality, call for comprehensive modelling approaches that can effectively take advantage of the wealth of satellite data. When dealing with ever changing scenarios the limitations of models that mainly focused on spatial or spectral aspects often became apparent. This has led to the development of models that blend temporal and spectral data to offer more insightful, accurate, and comprehensive assessments. These hybrid fusion models bring forth an approach, to integrating data from different time points diverse spatial resolutions and a wide range of wavelengths. They effectively capture the features of water components and the dynamic behaviours of water systems [38].

The key strength of this fusion lies in its adaptability. For instance, a large body of water hosting blooms and sediment flows could undergo rapid changes within hours due to factors like wind, rainfall, or human intervention. Conventional spatial or spectral models might misinterpret these shifts. In contrast, the hybrid model is capable of identifying, monitoring, and predicting these alterations with precision by incorporating the temporal aspect. Moreover, leveraging variations in reflectance across wavelengths through the spectral component allows these models to detect minor fluctuations in water quality attributes [39].

When combined with information, this allows for the mapping of a range of water features from large-scale characteristics such as river bends and estuary shapes to specifics like floating particles or specific types of phytoplankton. The emergence of hybrid models that fuse spatial, temporal, and spectral data signifies a shift in sensing methods. These models showcase advancements in the field by integrating patterns, changing processes and diverse spectral characteristics to set a higher standard for monitoring water quality from satellite observations.

5 Inversion Models in Remote Sensing

Inversion models play a role, in sensing by converting satellite signals into valuable data, about specific environmental features. These models are used to derive water quality indicators from the reflected or emitted radiation captured by the sensors, especially in aquatic remote sensing. The underlying premise of these models is the complex interaction between the spectral fingerprints of molecules suspended or dissolved in water. Different components of the water absorb, scatter, or reflect sunlight differently across the spectrum as it interacts with water bodies. To decode these spectral interactions and forecast the presence or absence of particular water quality characteristics, inversion models are theoretically developed.

5.1 Comparative Analysis of Different Inversion Models

As shown in Table 2, numerous inversion models have been suggested and used in research conducted all over the world, each one adapted to a particular water quality parameter. For instance, the retrieval models for the concentration of coloured dissolved organic matter (CDOM) show the many techniques used to extract this particular metric. While some models may use simpler algorithms with a single variable, others may use more complicated algorithms with numerous variables [40].

Table 2 Inversion models of CDOM concentration

Models	R^2	Data	Modes	Reference
$C_{CDOM}(400) = 0.2355(R(412)/R(555))^{-1.3423}$	0.85	GOCI	EM	[41]
$C_{CDOM}(400) = 40.75e^{-2.463(R(563)/R(655))}$	0.83	OLI	EM	[42]
$C_{CDOM}(420) = 5.13(R(560)/R(660))^{-2.67}$	0.73	ALI	SEM	[43]
$C_{CDOM}(400) = 10^{(-0.2-0.5\log(R(490)/R(590)) + 0.65(\log (R(490)/R(590)))2}$	0.63	In situ	EM	[44]
$C_{CDOM}(412) = 3.18(R(551)/R(671))^{-1.046}$	0.87	VIIRS	EM	[45]
Neural network (RBF)	0.972	USB2000+	IAM	[46]

Comparing these models demonstrates a wide range of approaches, each having advantages and disadvantages. Even though they are simple and frequently effective, empirical models may not always be relevant to different water bodies or in all situations. The accuracy of complicated models, however, may be higher, but they may also be computationally taxing and necessitate complex calibration. The table's constant expansion and improvement of these models is a startling finding that highlights the scientific community's work to increase the accuracy and utility of remote sensing in water quality monitoring [47]. The variety of models also highlights how crucial it is to choose an appropriate inversion model depending on the particular water quality parameter being studied, the type of water body, and the available remote sensing data.

Inversion models in remote sensing serve as a link between unprocessed satellite data and useful information on the quality of the water. These models will surely continue to develop as technology improves and more accurate data becomes accessible, stretching the limits of what we can learn about the water bodies on our globe from orbit.

6 Diverse Water Quality Parameters and Their Measurement Techniques

6.1 Chlorophyll Analysis

Assessment of water quality has several facets and involves a number of indicators that provide information on the productivity, health, and potential dangers of aquatic ecosystems. The concentration of chlorophyll, specifically chlorophyll-a, which is a crucial indication of phytoplankton biomass in freshwater and marine habitats, is one of these factors that is most important.

The process of photosynthesis, which enables phytoplankton to transform sunlight into energy, depends on the green pigment chlorophyll-a. Increased levels of this pigment may indicate eutrophication, a condition in which an abundance of nutrients, frequently from agricultural runoff or wastewater discharges, promotes excessive phytoplankton development and causes algal blooms. These blooms can disrupt aquatic ecosystems in a number of ways, including by reducing oxygen levels, blocking sunlight, and even by releasing toxins in the event of dangerous algal blooms [48]. Manual sampling is the first step in the traditional chlorophyll analysis process, which is followed by lab-based extraction and quantification. Although these techniques are precise, they require a lot of work, have a restricted range of coverage in both time and space, and might not be timely enough to handle sudden changes in water quality.

We now have remote sensing, which has completely changed our understanding of phytoplankton dynamics by allowing us to estimate chlorophyll concentrations over large spatial scales and on a frequent basis. The distinctive spectral characteristics of chlorophyll-a, which absorbs light in the blue (about 430–450 nm) and red (approximately 640–680 nm) regions while reflecting it in the green (around 550 nm), are used in remote sensing techniques [33]. In large bodies of water, satellites with sensors that can identify these patterns can estimate chlorophyll concentrations. This revolution was methodological as well as technological. To translate remotely observed light into chlorophyll concentrations, a variety of techniques, both empirical and semi-analytical, have been developed, as described in the literature. These methods emphasize the complexity of remote sensing-based chlorophyll analysis by being adapted for certain water types or situations [49].

As a result, while chlorophyll continues to be a crucial parameter in the assessment of water quality, advances in remote sensing technology and methodology have improved the effectiveness, thoroughness, and adaptability of this evaluation to the shifting dynamics of aquatic systems.

6.2 Remote Sensing Frameworks for Predicting Water Quality Variables

In the field of monitoring water quality, remote sensing has shown to be a game-changer, bridging the gap between conventional labour-intensive techniques and the demand for large-scale, real-time data. The technological and spectral capabilities of remote sensors unquestionably contribute to this transition, but equally crucial are the frameworks that support data analysis and interpretation from such sensors [50]. From raw spectrum observations to usable water quality data, these frameworks specify the data processing methods.

Figure 3 explains one such thorough framework in more detail. This picture shows a remote sensing-based framework designed for the forecast and evaluation of different water quality factors. The framework begins with the collecting of remote sensing data and then advances through several processes, including data pre-processing, spectral analysis, algorithm application, algorithm validation, and finally the mapping of water quality characteristics [51]. This framework stands out in particular for its methodical approach, which emphasizes both the gathering and processing of remote data as well as the verification of derived water quality estimations using actual or in situ measurements. This procedure is crucial for assuring the accuracy and dependability of the data obtained from satellite or airborne sensors.

Moreover, Fig. 3 underscores the versatility of remote sensing. While certain steps, like spectral analysis, are consistent across all water quality parameters, the specific algorithms applied can vary based on the parameter in question, be it chlorophyll concentration, turbidity, or dissolved oxygen levels. This modularity allows for tailored approaches depending on the specific needs of a study, ensuring optimal results [52]. Furthermore, the incorporation of ground-truthing and validation within the framework facilitates a feedback loop. Discrepancies between satellite-derived estimates and ground-based measurements can guide adjustments in

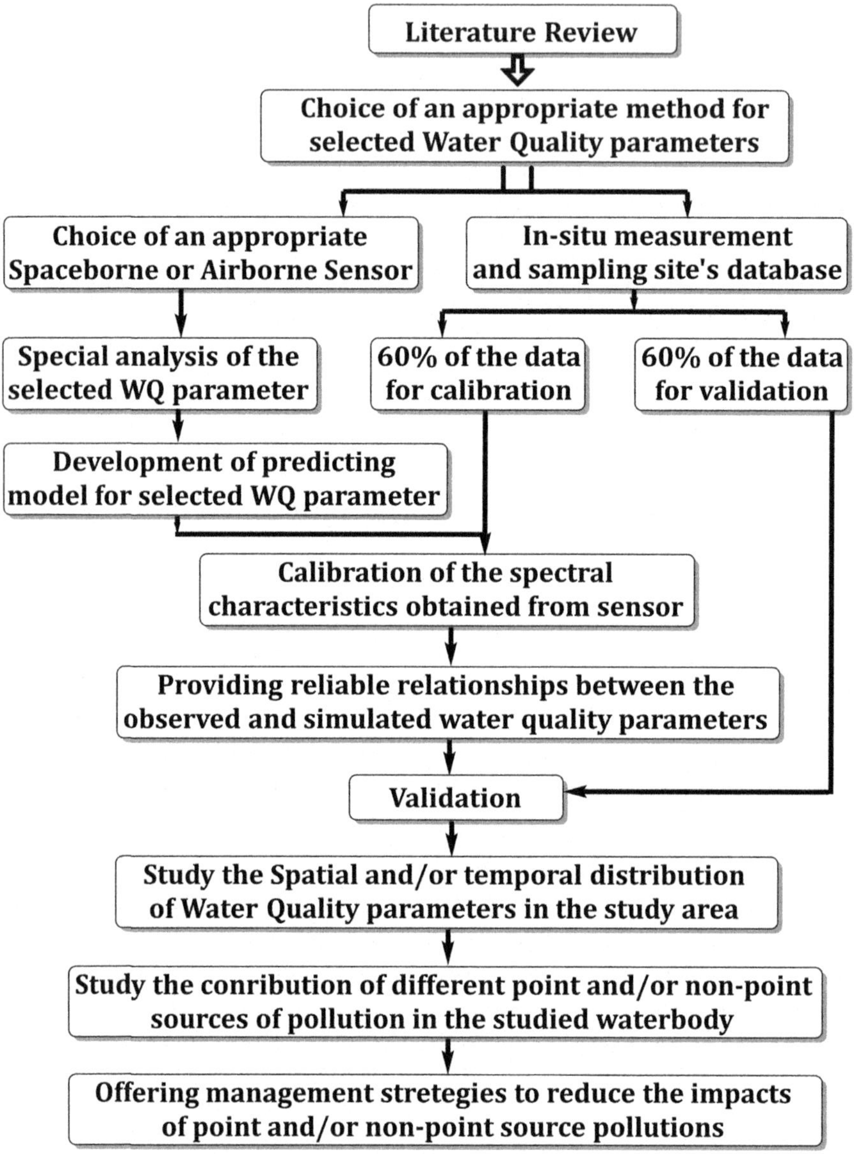

Fig. 3 A proposed system based on remote sensing for predicting and evaluating water quality variables

algorithmic approaches or indicate areas where sensor calibration may be required. While remote sensing offers the technological tools for observing and monitoring extensive aquatic regions, it is the underlying frameworks, like the one shown in Fig. 3, that transform these data into useful, trustworthy insights into water quality.

7 Advanced Applications

7.1 Remote Sensing of Salinity

In order to comprehend oceanic and coastal dynamics, hydrological cycles, and marine ecosystems, it is essential to grasp salinity, a crucial quantity indicating the salt concentration in water bodies. A strategic benefit of being able to feel salinity remotely is that it allows for large-scale, real-time monitoring and extends observations beyond localized measures [53]. As seen in Fig. 4, the trend towards using remote sensing for salinity measurement has seen various novel advancements. The intricate nature of the salinity measuring process is encapsulated in this picture, which emphasizes the sensor's function in determining salinity levels based on the distinctive absorption spectra of salinity-concentrated fluids.

Figure 4 highlights the spectrum diversity of saline solutions as a noteworthy point. Different salt concentrations reflect and absorb light differently, which can be detected and measured by satellite or aerial sensors. To determine exact salinity concentrations, the resulting spectral data is next analysed using specialized algorithms [54].

In regions like estuaries or river mouths, where salinity changes can have a considerable impact on marine ecosystems and local climates, remote sensing's main strength in salinity detection resides in its capacity to offer continuous and large-scale measurements. Additionally, with improvements in sensor technology and the launch of missions specifically focused on measuring salinity, such as SMOS and Aquarius/SAC-D, the accuracy and spatial resolution of distant salinity

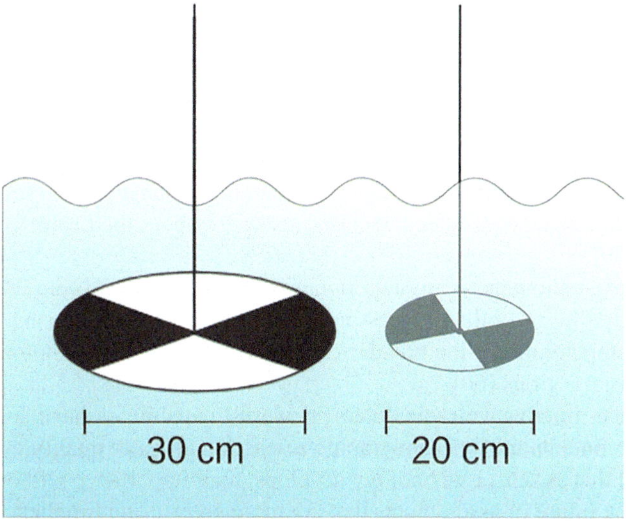

Fig. 4 Two different kinds of Secchi disk

measurements have significantly improved [55]. These technological developments, along with the approaches shown in Fig. 4, could result in more precise, quicker, and larger scale salinity observations, which are essential given our environment's rapid change.

Remote sensing is becoming an increasingly important technology for measuring salinity. The technologies and approaches currently in use not only confirm its promise but also give a signal as to the prospective course it is currently headed in the ensuing years.

7.2 Microwave Radiometer Utility in Oceanography and Water Quality

The development of microwave radiometers has opened up a new area of study for oceanography and the evaluation of water quality. They have a substantial advantage over conventional optical sensors in that they can acquire data in any weather, thanks to their innate ability to pierce clouds, fog, and precipitation [56].

Microwave radiometers primarily work in the microwave region of the electromagnetic spectrum, picking up the planet's surface and atmosphere's natural thermal emissions. These emissions, especially those from the sea surface, are sensitive to salinity and temperature in the context of oceanography. As a result, microwave radiometers can offer vital insights on variations in sea surface salinity (SSS) and other relevant factors [57]. The growing number of airborne and spaceborne sensors created expressly for assessing water quality is a testament to the growing reliance on microwave radiometry, as mentioned in both articles. These sensors can provide a broad range of readings and can pass over large oceanic areas in one pass, allowing for thorough analyses [58].

Additionally, the subtleties of water constituents and how they interact with these sensors are crucial. Indirect cues concerning water quality characteristics are provided by the microwave radiation that is released from the water surface and is absorbed or dispersed by different constituents. For instance, the presence of suspended sediments or phytoplankton in the water may change the microwave signals, which, when properly evaluated, can disclose a lot about the health and quality of the water [59].

Microwave radiometers can support maritime operations and conservation efforts by tracking pollution spills, forecasting algal blooms, and monitoring oceanic currents and temperatures in the broader context of water quality monitoring, particularly in coastal regions [60].

In essence, microwave radiometers' powerful capabilities have solidified their position in contemporary oceanographic research and water quality evaluations. It is projected that as sensor technology develops, their importance will only increase, ushering in a future of assessments that are more precise and timelier.

8 Comparative Results and Discussions

8.1 *Comparing Chl–a Inversion Results Based on Different Images*

Since chlorophyll-a concentration is a crucial factor in determining the health of water bodies, particularly in relation to the presence and extent of phytoplankton, the results of Chl-a (chlorophyll-a) inversion results derived from satellite images can significantly impact the understanding and interpretation of water quality [61]. However, depending on the kind of satellite imaging employed and the methods used, the effectiveness of satellite imaging in determining Chl-a concentrations can vary.

There has been an intriguing comparison between Chl-a inversion results obtained from GF2-like images and those obtained from Sentinel-2 photos in the research mentioned [62]. The accuracy and quality of the results obtained from the derived inversion technique may be influenced by the resolutions and characteristics of both types of images. While images resembling GF2 have been beneficial, in water quality applications they sometimes exhibit discrepancies in Chl-a concentration. This is mainly attributed to their bands and how they interact with water components [63]. In contrast, photos captured by Sentinel 2 offer a number of bands in the near infrared regions, which can play a crucial role in enhancing the precision of estimating chlorophyll [64]. Differences in Chl-a estimation arise from variations in band calibrations and atmospheric corrections applied, as observed through an analysis of inversion models using both types of images. This underscores the importance of refining image-based algorithms to each satellite source, for dependable and accurate results [65].

While both GF2-like and Sentinel-2 photos provide insightful information about Chl-a concentrations, their comparative analysis emphasizes the inherent difficulties and the necessity of developing procedures to produce consistent results. These results highlight the need for ongoing inversion technique improvement to keep them relevant, trustworthy, and compatible across many satellite platforms as remote sensing technology develops and satellite imagery availability rises [66].

9 Challenges and Recommendations

9.1 *Barriers to Satellite Remote Sensing Adoption*

Despite its promising capabilities and game-changing effects on water quality monitoring, satellite remote sensing still confronts obstacles to widespread implementation. To fully realize the potential of remote sensing in water quality assessments, these challenges—often anchored in technical, economic, and perceptual dimensions—require creative solutions and proactive methods [67].

1. **Technical Restrictions:** The quality and granularity of data collected depends on the spectral and spatial resolutions of satellite sensors. The current resolutions frequently fall short when it comes to thorough, localized water quality analyses. For instance, clouds might obstruct the images, making it difficult to continuously monitor water quality [68].
2. **Economic constraints:** Satellite missions are expensive, especially those that use sophisticated sensors. Resources are needed for not just the deployment but also the data collecting, processing, and interpretation. This is a challenge for areas or nations with little resources, resulting in differences in the uptake of satellite remote sensing technology [69].
3. **Perceptual Barriers:** Local government officials and administrators of water quality may not always be familiar with the capabilities and nuances of satellite remote sensing. This lack of knowledge or experience may be the cause of scepticism or resistance to incorporating satellite data into conventional monitoring systems [70].
4. **Data overload:** Due to the numerous satellite missions and the enormous amount of data they produce, handling, processing, and interpretation of the data may present difficulties. This can be overwhelming if you don't have the necessary resources or knowledge [71].
5. **Concerns with calibration and validation**: It is critical to guarantee the accuracy of the satellite-derived estimates. This calls for ground-truthing, which entails gathering in-person measurements and contrasting them with figures obtained from satellites. Although important, these exercises need a lot of resources [72].
6. **Recommendations:** The following actions are advised in order to get beyond these obstacles and encourage a wider usage of satellite remote sensing for water quality monitoring:

    - **Capacity Building:** The process of capacity building involves organizing workshops and training sessions to equip water quality managers with the skills and expertise to interpret and utilize satellite data effectively [73].
    - **Collaborative Ventures:** Promoting collaboration by facilitating the sharing of resources, knowledge, and satellite data. This can be particularly beneficial, for regions facing challenges [74].
    - **Technical Upgrades:** Enhancing the spectral and spatial resolution of satellite sensors is a technical upgrade. Additionally, the creation of cloud-penetrating sensors has the potential to revolutionize the game [75].
    - **Public Awareness:** Educate the public about the possibilities of satellite remote sensing for water quality monitoring by launching campaigns. Demand from the general public may hasten the deployment of such technology [76].

The acceptance and usefulness of satellite remote sensing for water quality monitoring can be significantly increased by confronting these issues head-on and taking the suggested actions.

9.2 Solutions and Future Outlook

New methods, tools, and paradigms are consistently emerging in the quickly developing field of satellite remote sensing, providing creative responses to current problems and broadening the scope of what is feasible for water quality monitoring.

1. **Integrative Data Approaches:** A multi-modal, integrative data strategy may hold the key to helping remote sensing overcome many of its difficulties. Data gaps can be filled, temporal, and spatial resolution can be improved, and a more complete picture of water quality characteristics can be provided by combining data from several satellite missions, sensors, and platforms. A deeper, more comprehensive understanding of water systems can be attained by combining optical and microwave data or by integrating airborne and spaceborne datasets [77].
2. **Advanced Machine Learning and AI:** The use of deep learning and other advanced machine learning algorithms can considerably improve the ability of water quality assessments to analyse data, spot anomalies, and develop predictive models. There is a tonne of untapped potential with these computational tools, as shown by the effective application of convolutional neural networks for phytoplankton pigment estimation [78].
3. **Collaborative International Initiatives:** Water issues frequently have a transnational component. International partnerships can more efficiently solve these issues by combining their resources, knowledge, and databases. Global water quality monitoring initiatives can become more regular, trustworthy, and comprehensive with the use of shared satellite missions or data-sharing agreements [79].
4. **User-Friendly Data Platforms:** The democratization of remote sensing can be achieved by developing platforms that make it simple to access satellite data, processed insights, and interpretation tools. Local governments, water managers, and other stakeholders can benefit greatly from these platforms since they can use satellite remote sensing without needing to have highly developed technical skills [80].
5. **Future Prognosis:** Looking ahead, it seems clear that satellite remote sensing will be used more and more frequently to monitor water quality. Granularity and accuracy of data will significantly increase with the anticipated launch of more advanced satellites with stronger sensors and coverage over a wider range.

Real-time water quality monitoring can be transformed by combining remote sensing data with other cutting-edge technologies like edge computing and the Internet of Things (IoT). When there are changes or anomalies in the water quality, such integrated systems can deliver immediate notifications and enable prompt mitigating steps.

Additionally, governmental and private sector investments in satellite remote sensing for environmental monitoring, particularly water quality, are anticipated to soar as the world community becomes more aware of environmental concerns, spurring additional innovations and successes in the field [81].

10 Conclusion

In the field of monitoring water quality, satellite remote sensing, a multidimensional and constantly developing technology, has become an essential tool. The developments, difficulties, and solutions discussed in this paper give a comprehensive picture of its potential and potential.

1. **Multispectral and Hyperspectral Imaging**: Multispectral and hyperspectral imaging techniques have made it possible to distinguish and analyse water quality characteristics at scales and resolutions never before possible. The level of detail in the data obtained by such tools emphasizes their importance, from the calculation of suspended sediments to the measurement of chlorophyll concentrations.
2. **Innovative Models and Algorithms:** Deep learning and machine learning algorithms, particularly convolutional neural networks, have shown remarkable potential in improving predictions and data interpretations. These models herald a new age in satellite remote sensing, one that promises improved data analysis efficiency and accuracy.
3. **Salinity Sensing:** Monitoring the salinity levels of oceans and large water bodies from a distance marks a progress in this area. Detecting salinity on a scale, which is crucial, for assessing water quality and climate variations used to be a challenging task until recently. Thanks to the advancements in sensor technology and dedicated satellite missions' comprehensive maps detailing salinity levels are now achievable.
4. **Challenges and Barriers:** Despite the benefits of satellite sensing, there are still obstacles that need to be addressed. These challenges, spanning from limitations in accessing data to hurdles underscore the importance of research, investments, and partnerships, in this field.
5. **Future Prospects:** There are several opportunities for using satellite remote sensing to monitor water quality. Satellite remote sensing is positioned to be a key factor in directing efforts to monitor and control the quality of the world's water as technical developments and environmental concerns grow.

The integration of science, technology, and environmental studies through satellite remote sensing signals a fundamental transformation in how we understand, evaluate, and respond to problems with water quality. The findings of this study highlight the value of ongoing investigation, cooperation, and innovation in this field in order to achieve a sustainable and closely regulated water future.

Monitoring water quality continues to be a top priority for international environmental and public health projects. The breadth and depth of monitoring requirements are frequently not addressed by traditional techniques of water quality evaluation due to growing anthropogenic demands and changing climate circumstances. Remote sensing provides a creative, scalable, and effective solution in this situation.

First off, ground-based or manual approaches cannot match the spatial coverage provided by remote sensing sensors. A hitherto unequalled view of water quality at global scales is now possible, thanks to satellite instruments' ability to cover large areas in a single acquisition, from small lakes to vast ocean basins, as noted in earlier studies. This macroscopic approach is essential for large-scale aquatic phenomena and transboundary water challenges, as well as regional water management techniques.

Furthermore, continuous and real-time monitoring is made possible by satellite systems' consistent data gathering and temporal resolution. This makes it possible to identify anomalies quickly and take preventative action against potential environmental problems. For instance, it becomes possible to identify algal blooms before they damage aquatic ecosystems and human health.

Furthermore, the advancement of satellite sensors in terms of spectral resolution and sensitivity offers the possibility of future retrievals of enhanced and nuanced water quality parameters. This analytical power is further enhanced by the incorporation of machine learning and deep learning algorithms, opening the door to analyses of water quality that are more precise, effective, and in-depth.

But when combined with other data sources, remote sensing has the most potential for future water quality monitoring. A thorough, multifaceted approach to water quality monitoring can be produced by fusing satellite data with on-site measurements, hydrological models, and citizen science efforts.

In conclusion, the importance of remote sensing is growing as the world struggles with previously unheard-of water issues. It directs our efforts to protect the sacredness of our water bodies and ensure a sustainable water future by acting as a beacon of technological innovation.

References

1. Shenoy, A., Shukla, B. K., & Bansal, V. (2022). Sustainable design of textile industry effluent treatment plant with constructed wetland. *Materials Today: Proceedings, 61*, 537–542.
2. Liu, Y., Stanturf, J., & Goodrick, S. (2017). Remote sensing-driven climatic/environmental variables for modeling malaria transmission in sub-Saharan Africa. *International Journal of Health Geographics, 16*(1), 29.
3. Shukla, B. K., Khan, A., Saikiran, G., & Sriram, M. (2019). A comparative study on effect of variation in opening shape on oxygenation performance of surface jet aerators used in water and wastewater treatment. *Journal of Green Engineering, 9*(3), 427–440.
4. Xiao, M., Zhou, S., & Cheng, J. (2025). Comparison of three temperature and emissivity separation algorithms for graybodies with low spectral contrast: A case study on water bodies. *Remote Sensing, 17*(3), 455.
5. Kang, K.-M., Park, S., & Kim, D. (2014). Comparison of coastal sea surface temperature derived from ship-, air-, and space-borne thermal infrared systems. In *Proceedings of the 2014 IEEE International Geoscience and Remote Sensing Symposium (IGARSS)*.
6. Tripathi, A., Langhi, M., Shukla, B. K., Kumar, A., & Bharti, G. (2023, September). Assessment of sedimentation rate in Sipu reservoir using remote sensing & GIS technique. In *AIP Conference Proceedings* (Vol. 2800, No. 1). AIP Publishing.

7. Morozov, E., Pozdnyakov, D., Smyrnova, I., Grassl, H., & Pettersson, L. (2015). A space-borne assessment of cyclone impacts on Barents sea surface temperature and chlorophyll. *International Journal of Remote Sensing, 36*, 1921–1941.
8. Shukla, B. K., Sharma, P. K., & Khan, M. A. (2020, May). Physico-chemical study of some surface water bodies of Punjab. In *Journal of Physics: Conference Series* (Vol. 1531, No. 1, p. 012121). IOP Publishing.
9. Vytla, V., Baduru, B., Kolukula, S. S., Ragav, N. N., & Kumar, J. P. (2025). Forecasting of sea surface temperature using machine learning and its applications. *Journal of Earth System Science, 134*(1), 25.
10. Baek, S., & Kim, W. (2025). Review on hyperspectral remote sensing of tidal zones. *Ocean Science Journal, 60*(1), 1–21.
11. Xiao, H., Shao, H., Long, J., Zhang, S., He, S., & Wang, D. (2023). Spatial-temporal pattern evolution and geological influence factors analysis of ecological vulnerability in Western Sichuan mountain region. *Ecological Indicators, 155*, 110980.
12. Rao, D., Sun, D., & Chen, J. (2016). Estimating spatial distribution of soil moisture based on Bayesian maximum entropy method with auxiliary data from remote sensing. *International Journal of Applied Earth Observation and Geoinformation, 46*, 167–175.
13. Salim, A., Rajeev, A., Perumkuni, S. P., Raj, R., Sankaran, A., Pillai, I. M. S., & Pullukkuchalil, F. (2025). Developing the recommendations for restoration of Ashtamudi Lake, Kerala, India, by data analysis based on a novel water body index using Google Earth Engine. *Environmental Science and Pollution Research, 32*, 1–19.
14. Brando, V., Dekker, A., Park, Y., & Schroeder, T. (2015). High-resolution satellite turbidity and sea surface temperature observations of river plume interactions during a significant flood event. *Ocean Science, 11*, 909–920.
15. Shukla, B. K., Teeli, M. A., Shukla, S. K., Chandra, R., Bharti, N., & Singh, U. (2024). A comprehensive overview of vital water quality parameters. *Lecture Notes in Civil Engineering, 439*, 1–20.
16. De Andrés, E., Umbert, M., Sánchez-Urrea, M., González-Gambau, V., Olmedo, E., Gabarró, C., & Elosegui, P. (2025). Sea ice meltwater in the Beaufort Gyre: A comprehensive analysis using sea surface salinity data from SMOS. *Journal of Geophysical Research: Oceans, 130*(3), e2023JC020733.
17. Mohan, S., Kumar, B., & Nejadhashemi, A. P. (2025). Integration of machine learning and remote sensing for water quality monitoring and prediction: A review. *Sustainability, 17*(3), 998.
18. Shukla, B. K., Maurya, N., & Sharma, M. (2023). Advancements in sensor-based technologies for precision agriculture: An exploration of interoperability, analytics and deployment strategies. *Engineering Proceedings, 58*(1), 22.
19. Jarugula, S., Fournier, S., Reager, J. T., & Pascolini-Campbell, M. (2025). Intercomparison of in situ and satellite sea surface salinity products for global coastal ocean studies. *Journal of Atmospheric and Oceanic Technology, 42*(1), 3–16.
20. Gani, A., Singh, M., Pathak, S., & Hussain, A. (2024). Groundwater quality index development using ANN model of Delhi Metropolitan City, India. *Environmental Science and Pollution Research* (pp. 1–16).
21. Dong, L. X., Huang, X. C., Wang, Z., Yang, Z., Wang, X. M., & Tang, C. Y. (2016). A thin-film nanocomposite nanofiltration membrane prepared on a support with in situ embedded zeolite nanoparticles. *Separation and Purification Technology, 166*, 230–239.
22. Muhamada, M. S., Salima, M. R., Lauc, W. J., Yuzira, M. A., Yunusa, S., & Bahru, U. J. (2015). Fabrication of mixed matric membrane incorporated with modified silica nanoparticles for bisphenol a removal. *Journal Teknologi, 74*, 23–31.
23. Kimura, K., Amy, G., Drewes, J. E., Heberer, T., Kim, T. U., & Watanabe, Y. (2003). Rejection of organic micropollutants (disinfection by-products, endocrine disrupting compounds, and pharmaceutically active compounds) by NF/RO membranes. *Journal of Membrane Science, 227*(1–2), 113–121.

24. Urtiaga, A. M., Pérez, G., Ibáñez, R., & Ortiz, I. (2013). Removal of pharmaceuticals from a WWTP secondary effluent by ultrafiltration/reverse osmosis followed by electrochemical oxidation of the RO concentrate. *Desalination, 331*, 26–34.
25. Wang, Z., Zhang, B., Fang, C., Liu, Z., Fang, J., & Zhu, L. (2019). Macroporous membranes doped with micro-mesoporous β-cyclodextrin polymers for ultrafast removal of organic micropollutants from water. *Carbohydrate Polymers, 222*, 114970.
26. Gitelson, A., Dall'Olmo, G., Moses, W., Rundquist, D., Barrow, T., Fisher, T., Gurlin, D., & Holz, J. (2008). A simple semi-analytical model for remote estimation of chlorophyll-a in turbid waters: Validation. *Remote Sensing of Environment, 112*(9), 3582–3593.
27. Smith, R. C., & Baker, K. S. (1981). Optical properties of the clearest natural waters (200–800 nm). *Applied Optics, 20*(2), 177–184.
28. Shukla, B. K., Bashir, M., Sharma, P. K., & Abhimanyu. (2021). An analytical investigation of surface water quality and pollution status in Srinagar, Jammu and Kashmir, India. *Journal of Green Engineering, 11*, 952–962.
29. Cevallos-Mendoza, J., Amorim, C. G., Rodríguez-Díaz, J. M., & Montenegro, M. D. C. B. (2022). Removal of contaminants from water by membrane filtration: A review. *Membranes, 12*(6), 570.
30. Yang, F., Feng, Q., Zhou, Y., Li, W., Zhang, X., & He, B. (2024). Monitoring total phosphorus concentration in the middle reaches of the Yangtze River using Sentinel-2 satellites. *Remote Sensing, 16*(9), 1491.
31. Matthews, M. W. (2011). A current review of empirical procedures of remote sensing in inland and near-coastal transitional waters. *International Journal of Remote Sensing, 32*(21), 6855–6899.
32. Yang, Z., Gong, C., Lu, Z., Wu, E., Huai, H., Hu, Y., Li, L., & Dong, L. (2023). Combined retrievals of turbidity from Sentinel-2A/B and Landsat-8/9 in the Taihu Lake through machine learning. *Remote Sensing, 15*(17), 4333.
33. Hussain, A., Deshwal, A., Priyadarshi, M., Pathak, S., Sambandam, G., Chand, S., & Shukla, A. K. (2024). Landfill leachate analysis from selected landfill sites and its impact on groundwater quality, New Delhi, India. *Environment, Development and Sustainabilty* (pp. 1–26). https://doi.org/10.1007/s10668-023-04403-6
34. Pitarch, J., & Brando, V. E. (2025). A hyperspectral and multi-angular synthetic dataset for algorithm development in waters of varying trophic levels and optical complexity. *Earth System Science Data, 17*(2), 435–460.
35. Hu, C., Lee, Z., & Franz, B. (2012). Chlorophyll a algorithms for oligotrophic oceans: A novel approach based on three-band reflectance difference. *Journal of Geophysical Research: Oceans, 117*(C1). https://doi.org/10.1029/2011JC007395
36. Yuan, D., Tang, B. H., Wang, D., Chen, J., Zhang, Z., & Fu, Z. (2025). An enhanced three-band algorithm for retrieving phycocyanin concentration from satellite data in plateau inland lake. *IEEE Journal of Selected Topics in Applied Earth Observations and Remote Sensing., 18*, 3654.
37. Giardino, C., Bresciani, M., Cazzaniga, I., Schenk, K., Rieger, P., Braga, F., et al. (2019). Airborne hyperspectral data to assess suspended particulate matter and aquatic vegetation in a shallow and Turbid lake. *Remote Sensing of Environment, 227*, 1–12.
38. Ye, H., Tang, S., Bai, Y., & Kalhoro, M. A. (2024). Estimation of sea surface salinity from MODIS-Aqua data in the northern South China Sea. *Regional Studies in Marine Science, 77*, 103684.
39. Cracknell, A. P. (2007). *Introduction to remote sensing*. CRC Press.
40. Zhao, X., Hu, S., Wang, Y., Zhang, J., Hu, Y., Liu, R., .Haibin Ling, Yin Li, Renshu Li, Kun Liu & Li, J. (2024). Biodrone: A bionic drone-based single object tracking benchmark for robust vision. International Journal of Computer Vision, 132(5), 1659–1684.
41. Kim, E. H., & Park, S. J. (2023). Emerging tick-borne Dabie bandavirus: Virology, epidemiology, and prevention. *Microorganisms, 11*(9), 2309.

42. Chen, J., Zhu, W.-N., Tian, Y. Q., & Yu, Q. (2017). Estimation of colored dissolved organic matter from Landsat-8 imagery for complex inland water: Case study of Lake Huron. *IEEE Transactions on Geoscience and Remote Sensing, 55*, 2201–2212.
43. Kutser, T., Pierson, D. C., Kallio, K. Y., Reinart, A., & Sobek, S. (2005). Mapping Lake CDOM by satellite remote sensing. *Remote Sensing of Environment, 94*, 535–540.
44. Pathak, S., Gupta, S., & Ojha, C. S. P. (2021). Assessment of groundwater vulnerability to contamination with modified DRASTIC index: A case study in Haridwar, Uttarakhand, India. *ASCE Journal of Hazardous, Toxic, and Radioactive Waste, 25*(2), 04020081.
45. Joshi, I., D'Sa, E. J., Osburn, C. L., Bianchi, T., Ko, D. S., Oviedo-Vargas, D., Arellano, A. R., & Ward, N. D. (2017). Assessing chromophoric dissolved organic matter (CDOM) distribution, stocks, and fluxes in Apalachicola Bay using combined field, VIIRS ocean color, and model observations. *Remote Sensing of Environment, 191*, 359–372.
46. Sun, H., Yu, R., Liu, X., Cao, Z., Li, X., Zhang, Z., Wang, J., Zhuang, S., Ge, Z., Zhang, L., & Sun, L. (2022). Drivers of spatial and seasonal variations of CO_2 and CH_4 fluxes at the sediment water interface in a shallow eutrophic lake. *Water Research, 222*, 118916.
47. Gani, A., Hussain, A., Pathak, S., & Banerjee, A. (2024). An empirical investigation on the elimination of heavy metals using bioremediation method for selected plant species. *Physics and Chemistry of the Earth, 134*, 103568.
48. Qu, J. J., Wang, W., Dasgupta, S., & Hao, X. (2008). Active fire monitoring and fire danger potential detection from space: A review. *Frontiers of Earth Science in China, 2*, 479–486.
49. Zhang, X. X., Sharratt, B., Chen, X., Wang, Z. F., Liu, L. Y., Guo, Y. H., Li, J., Chen, H. S., & Yang, W. Y. (2017). Dust deposition and ambient PM 10 concentration in Northwest China: Spatial and temporal variability. *Atmospheric Chemistry and Physics, 17*(3), 1699–1711.
50. Rawat, S., Shukla, B. K., Sihag, P., Srivastav, Y., & Singh, H. K. (2024, February). Environmental effects and sustainable hydropower development: A review of deployment research, environmental valuation and IoT sensors integration. In *AIP Conference Proceedings* (Vol. 3050, No. 1). AIP Publishing.
51. Shukla, B. K., Sharma, P. K., & Goel, A. (2020, May). Study on oxygenation performance of solid jet aerator having circular opening corresponding to variable jet length and flow area. In *Journal of Physics: Conference Series* (Vol. 1531, No. 1, p. 012117). IOP Publishing.
52. Wong, M. S., Kwan, S., Young, J., Nichol, J., Zhangging, L., & Emerson, N. (2007). Modeling of suspended solids and sea surface salinity in Hong Kong using Aqua/MODIS satellite images. *Korean Journal of Remote Sensing, 23*, 161–169.
53. Gholizadeh, M. H., Melesse, A. M., & Reddi, L. (2016). A comprehensive review on water quality parameters estimation using remote sensing techniques. *Sensors, 16*(8), 1298.
54. Font, J., Lagerloef, G. S., Le Vine, D. M., Camps, A., & Zanife, O.-Z. (2004). The determination of surface salinity with the European SMOS space mission. *IEEE Transactions on Geoscience and Remote Sensing, 42*, 2196–2205.
55. Lagerloef, G., & Font, J. (2010). SMOS and Aquarius/SAC-D missions: The era of spaceborne salinity measurements is about to begin. In *Oceanography from space* (pp. 35–58). Springer.
56. Yang, Q., Wang, Z., Liu, S., & Li, Z. (2024, July). Research on improved u-net based remote sensing image segmentation algorithm. In *2024 6th International Conference on Internet of Things, Automation and Artificial Intelligence (IoTAAI)* (pp. 686–689). IEEE.
57. Zhang, Z., Li, C., Yang, P., Xu, Z., Yao, L., Wang, Q., Chen, G., & Tan, Q. (2024). Enhancing remote sensing water quality inversion through integration of multisource spatial covariates: A case study of Hong Kong's coastal nutrient concentrations. *Remote Sensing, 16*(17), 3337.
58. Shukla, A. K., Ojha, C. S. P., Shukla, S., & Garg, R. D. (2021). Water quality challenges in Ganga River basin, India. In M. S. Chauhan & C. S. P. Ojha (Eds.), *The Ganga River basin: A hydrometeorological approach* (Society of earth scientists series). Springer. https://doi.org/10.1007/978-3-030-60869-9_1
59. Singh, S., Singh, V. P., Garg, P., & Shukla, B. K. (2023, February). Physico-chemical analysis of groundwater in Noida-Ghaziabad region. In *IOP Conference Series: Earth and Environmental Science* (Vol. 1110, No. 1, p. 012028). IOP Publishing.

60. Heron, M., Prytz, A., Stieglitz, T., & Burrage, D. (2004). Remote sensing of sea surface salinity: A case study in the Burdekin River, North-Eastern Australia. *Gayana (Concepción), 68*, 278–283.
61. Shukla, A. K., Ojha, C. S. P., & Garg, R. D. (2017). Application of overall index of pollution (OIP) for the assessment of the surface water quality in the upper Ganga river basin, India. In *Development of water resources in India* (pp. 135–149). Springer International Publishing.
62. Singh, J., Ahirwal, S. K., Ramteke, K., Kantharajan, G., & Sarma, K. (2025). Remote sensing techniques for monitoring aquatic ecosystems. In *Information technology in fisheries and aquaculture* (pp. 71–107). Springer.
63. Dionisi, D., Bucci, S., Cesarini, C., Colella, S., D'Alimonte, D., Di Ciolo, L., Di Girolamo, P., Di Paolantonio, M., Franco, N., Gostinicchi, G., & Liberti, G. L. (2024). Exploring the potential of Aeolus lidar mission for ocean color applications. *Remote Sensing of Environment, 313*, 114341.
64. Shukla, A. K., Ojha, C. S. P., & Garg, R. D. (2017). Application of overall index of pollution (OIP) for the assessment of the surface water quality in the upper Ganga River basin, India. In V. Garg, V. Singh, & V. Raj (Eds.), *Development of water resources in India. Water Science and Technology Library* (Vol. 75). Springer. https://doi.org/10.1007/978-3-319-55125-8_12
65. Munoz-Martin, J. F., Rodriguez-Alvarez, N., Bosch-Lluis, X., & Oudrhiri, K. (2025). Integrated retrieval of sea-ice salinity, density, and thickness using polarimetric GNSS-R. *Remote Sensing of Environment, 318*, 114617.
66. Keller, P. A. (2001). *Imaging spectroscopy of lake water quality parameters*. Remote Sensing Laboratories, Dept. of Geography, University of Zürich.
67. Dong, S., Wang, L., Du, B., & Meng, X. (2024). ChangeCLIP: Remote sensing change detection with multimodal vision-language representation learning. *ISPRS Journal of Photogrammetry and Remote Sensing, 208*, 53–69.
68. Font, J., Lagerloef, G. S., & Le Vine, D. M. (2010). SMOS: The challenging sea surface salinity measurement from space. *Proceedings of the IEEE, 98*, 649–665.
69. Robinson, I. S. (2010). *Discovering the ocean from space: The unique applications of satellite oceanography*. Springer.
70. Schaeffer, B. A., Schaeffer, K. G., Keith, D., Lunetta, R. S., Conmy, R., & Gould, R. W. (2013). Barriers to adopting satellite remote sensing for water quality management. *International Journal of Remote Sensing, 34*, 7534–7544.
71. Shukla, B. K., Gupta, L., Parashar, B., Sharma, P. K., Sihag, P., & Shukla, A. K. (2025). Integrative assessment of surface water contamination using GIS, WQI, and machine learning in urban–industrial confluence zones surrounding the national capital territory of the Republic of India. *Water, 17*(7), 1076. https://doi.org/10.3390/w17071076
72. Specter, C., & Gayle, D. (1990). Managing technology transfer for coastal zone development: Caribbean experts identify major issues. *Remote Sensing, 11*, 1729–1740.
73. Bygate, M., & Ahmed, M. (2024). Monitoring water quality indicators over Matagorda Bay, Texas, using landsat-8. *Remote Sensing, 16*(7), 1120.
74. Vossepoel, F. C., Reynolds, R. W., & Miller, L. (1999). Use of sea level observations to estimate salinity variability in the tropical Pacific. *Journal of Atmospheric and Oceanic Technology, 16*, 1401–1415.
75. White, S., Lopez, E. M., Silva, T., Spyrakos, E., Martin, A., & Amoudry, L. (2025). Exploring the link between spectra, inherent optical properties in the water column, and sea surface temperature and salinity. *Remote Sensing Applications: Society and Environment, 37*, 101454.
76. Imtiyaz, I., Krishnakant, Shukla, B. K., Varadharajan, S., & Bharti, G. (2023). A comprehensive review of mass transfer phenomenon in gas-liquid phase flow during aeration in wastewater treatment. *Lecture Notes in Civil Engineering, 281*, 127–135.
77. Wang, Y., Heron, M. L., Prytz, A., Ridd, P. V., Steinberg, C. R., & Hacker, J. M. (2007). Evaluation of a new airborne microwave remote sensing radiometer by measuring the salinity gradients across the shelf of the great barrier reef lagoon. *IEEE Transactions on Geoscience and Remote Sensing, 45*, 3701–3709.

78. Shukla, A. K., Ojha, C. S. P., Mijic, A., Buytaert, W., Pathak, S., Garg, R. D., & Shukla, S. (2018). Population growth, land use and land cover transformations, and water quality nexus in the upper Ganga River basin. *Hydrology and Earth System Science, 22*(9), 4745–4770. https://doi.org/10.5194/hess-22-4745-2018
79. Dutta, A., Lall, B., & Sharma, S. (2025). Potential of satellite hyperspectral imaging technology in soil health analysis: A step towards environmental sustainability. *Environmental Monitoring and Assessment, 197*(3), 314.
80. Markogianni, V., Dimitriou, E., & Tzortziou, M. (2013). Monitoring of chlorophyll-a and turbidity in Evros River (Greece) using Landsat imagery. In *Proceeding of the 1st International Conference on Remote Sensing and Geoinformation of Environment*, Paphos, Cyprus, 8–10 April 2013.
81. Chen, L., Liu, L., Liu, S., Shi, Z., & Shi, C. (2025). The application of remote sensing technology in Inland water quality monitoring and water environment science: Recent progress and perspectives. *Remote Sensing, 17*(4), 667.

Evolution of Morphometric Features and River Course of Ganges Basin: A Case Study in the Upper Ganga Basin

Vandana Narwal and Shray Pathak

1 Introduction

The River Ganga holds immense cultural, economic, and ecological significance in India. *Upper Ganga*, which originates from the Gangotri Glacier (Gangotri), Bandarpunch Glacier (Yamunotri), Chorabari Glacier (Kedarnath), and Satopanth Glacier (Badrinath) in the Himalayas. The Yamunotri and Gangotri lies in the Uttarkashi District, Kedarnath in Rudraprayag district and Badrinath in Chamoli district of Uttarakhand. Its journey is marked by a series of significant confluences known as the Panch Prayag, where the river merges with other tributaries, enriching its waters and cultural significance. The Panch Prayag consists of five sacred confluences: Vishnuprayag (Dhauliganga), Nandaprayag (Nandakini river), Karnaprayag (Pindar river), Rudraprayag (Mandakini river), Devprayag (where Bhagirathi and Alaknanda river unites). The Ganga unites with the Yamuna at the Triveni Sangam is a sacred confluence. It traverses through the northern plains, supporting a diverse range of ecosystems and human activities before emptying into the Bay of Bengal. The geomorphology of the Ganga has been a subject of extensive study due to its dynamic nature and the significant impact it has on the surrounding landscapes and communities.

The field of geomorphology has ancient roots, but it wasn't until the period between 1884 and 1899 that the first formal geomorphological model was introduced. This significant development was credited to an American geographer [1].

Geomorphology extends beyond merely describing land features like mountains, rivers, glaciers, and dunes. It delves into understanding the processes that shape

V. Narwal (✉)
Department of Geography, Panjab University, Chandigarh, India

S. Pathak
Department of Civil Engineering, Indian Institute of Technology Ropar, Rupnagar, Punjab, India

S. Pathak et al. (eds.), *Geo-Data Revolution*, Advances in Geographic Information Science, https://doi.org/10.1007/978-3-031-94078-1_5

these features, the history of their formation, and the dynamics of their ongoing evolution [2].

Like Geology and Biology, Geomorphology is regularly depicted as an authentic science, as it manages the investigation of changes after some time. In authentic sciences, for example, Geomorphology, an event may have altogether different results depending upon the states of that specific time and area.

This dual nature of geomorphology makes it both captivating and challenging. It is often difficult to fully explain and manage the complex processes that shape the Earth's surface [3].

Between 2012 and 2024, the geomorphological changes of the Ganga have been influenced by various factors including climatic variations, anthropogenic activities, and natural processes. This period has witnessed considerable shifts in the river's course, sediment load, and bank erosion patterns. These changes have had profound implications for floodplain management, agriculture, and urban planning in the densely populated Ganga Basin.

Understanding the recent geomorphological changes of the Ganga is crucial for sustainable river management practices. This thesis not only documents the physical transformations of the river but also discusses their socio-economic and environmental impacts. The findings aim to contribute to the broader discourse on riverine health and management, offering insights that could guide policy makers, conservationists, and local communities in fostering a balanced coexistence with this vital waterway.

2 Morphology of River Basin

A river basin is a region of the Earth that is drained by a river and its tributaries. It features elements such as tributaries, watersheds, junctures, sources, and mouths. The terms river basin, watershed, drainage basin, and catchment area are often used interchangeably. Morphometry is the discipline that focuses on the quantitative analysis of the form of natural features such as rivers, forests, and mountains. This field involves the numerical study of the Earth's surface configuration, shapes, and dimensions of its landforms.

The significance of geomorphometric studies in providing a quantitative description of drainage basins, which is crucial for understanding their characteristics [4].

Bank erosion is a geomorphic process that aids in the growth of riparian vegetation and forms dynamic habitats crucial for both aquatic and riparian species. To implement effective engineering modifications, detailed information on the properties of the banks and the hydraulic conditions is essential [5]. It plays a crucial role in the functioning of river ecosystems and can occur even in densely forested regions. Researchers have identified various changes, including deposition, erosion, erosion-deposition, alterations in landforms, and areas with no change.

Thematic maps created using Landsat Thematic Mapper (TM) and Shuttle Radar Topography Mission (SRTM)-Digital Elevation Model (DEM) help categorize geomorphological and hydrological features to assess inundation and erosion intensity. This analysis identified zones prone to erosion and flooding at the Ken-Betwa River Linking site in India [6].

3 Statement of Problem

The Ganga River Basin faces numerous challenges, including severe pollution from industrial waste, sewage, and agricultural runoff, which degrade water quality and harm aquatic life. Over-extraction of water for various uses reduces river flow, impacting its ecological balance [7]. Rapid urbanization and industrialization lead to habitat destruction and increased flood risks. Additionally, sedimentation from deforestation and construction alters the river's course and depth. Climate change affects the basin's hydrology, exacerbating issues like flooding and water scarcity.

4 Objectives

The objectives of the study area

1. To delineate the geomorphological changes of the River Ganga from 2012 to 2024.
2. To analyze the driving factors behind these changes, focusing on both natural and anthropogenic influences.
3. To prepare a drainage map and thus compute all geomorphological parameters of Upper Ganga basin.
4. Estimation of left and right bank shifting amount w.r.t. base year and appropriate graphical plotting of these shifting.
5. Estimation and comparison of each bank erosion for different reaches in term of erosion length and erosion area of the river w.r.t. base year using appropriate GIS tool.
6. Identification of critical and vulnerable reaches—locations where shifting is maximum.

5 Geological Profile of the Area

The Upper Ganga Basin, located in northern India, primarily lies within the state of Uttarakhand or Uttarkashi. This region forms the upper catchment area of the Ganga River, originating in the Himalayas and extending through a variety of terrains including glacial areas, rugged mountains, and forested valleys [8].

5.1 Location

Geographical Coordinates: The basin spans latitudes approximately from 29° N to 31° N and longitudes from 78° E to 81° E.

Boundaries:

1. North: The basin is bordered by the Tibetan Plateau.
2. West: It is flanked by the Yamuna Basin.
3. East: The Ghaghara River Basin marks the eastern boundary.
4. South: The Siwalik Hills and the northern parts of the Indo-Gangetic plains delineate the southern boundary.
5. Key Areas: The basin includes notable areas such as the Garhwal and Kumaon regions. Major towns and cities within the basin include Rishikesh, Haridwar, Uttarkashi, and Tehri.

5.2 Extraction and Utilization

Water Resources The Upper Ganga Basin is vital for its water resources, which are used for various purposes:

1. Domestic Use: Water is extracted from the Ganga and its tributaries for drinking and household uses by local communities.
2. Agriculture: Irrigation is a major use, with numerous canals and traditional irrigation systems supporting farming activities in the region and downstream.
3. Hydropower: The steep gradients and perennial flow of the rivers make the basin ideal for hydropower projects. Several dams and hydroelectric stations harness the river's energy, such as the Tehri Dam on the Bhagirathi River.
4. Industrial Use: Limited industrial activity in the region also relies on water from the basin though this is more prevalent in the downstream areas [9].

5.3 Methods of Extraction

Traditional Methods:

1. Gravity Flow Systems: Utilizing natural slopes to channel water for irrigation and domestic use.
2. Step Wells and Springs: Historically, local communities have used step wells and natural springs for water extraction.

Modern Techniques Dams and Reservoirs: Large-scale dams like Tehri store and manage water for multiple uses.

6 Geomorphology of River

Geomorphology of rivers involves the study of the physical features and processes that shape river channels and floodplains. This field focuses on understanding how rivers form, evolve, and interact with their environment. River geomorphology examines aspects such as the river's flow patterns, sediment transport, channel morphology, and the impact of geological and climatic factors. Rivers typically develop through erosion, transportation, and deposition processes [10]. Erosion occurs when flowing water removes material from the riverbed and banks. Transport refers to the movement of this sediment downstream, while deposition happens when the river's energy decreases, causing it to drop its sediment load. These processes contribute to the formation of various river features, such as meanders, oxbow lakes, deltas, and floodplains. Understanding river geomorphology is essential for effective river management and conservation, as it helps predict how rivers will respond to natural changes and human interventions [11]. This knowledge is crucial for mitigating flood risks, maintaining water quality, and preserving aquatic habitats.

6.1 *Linear Aspects of Morphometric Analysis*

Morphometric analysis of a region involves examining several linear aspects, including stream order, stream length, mean stream length, stream length ratio, and bifurcation ratio as shown in Table 1.

6.1.1 Basin Area (Ab)

The area of a basin is a fundamental parameter in watershed studies. Schumm (1956) emphasized the correlation between the basin area and the total stream length.

Table 1 Segment in a drainage pattern

Stream order	Number of segments
1	2458
2	530
3	117
4	24
5	9
6	2
7	1
Total	3141

6.1.2 Stream Order (Nu)

Stream order is a method of classifying the hierarchical ranking of streams within a drainage basin. The primary stream in a basin, known as the trunk stream, is the one with the highest order. Strahler (1968) identified a fifth-order stream as the trunk stream in his study of the Warangal district.

6.1.3 Stream Length (Lu)

Stream length is the total length of streams of a specific order. It provides insight into the characteristics of a region's drainage network, measured from the stream's mouth to its drainage divide. ArcMap 10.2 software was utilized to establish the relationship between stream length and stream order.

6.1.4 Stream Numbers (N)

The stream number denotes the count of streams within a particular order in a basin, and it decreases as stream order increases.

6.1.5 Bifurcation Ratio (Rb)

The bifurcation ratio is the ratio of the number of streams of one order to the number of streams in the next lower order.

6.1.6 Areal Aspects

Areal aspects of morphometric analysis include metrics such as drainage density, stream frequency, texture ratio, form factor, circulatory ratio, length of overland flow, and the constant of channel maintenance. These metrics are essential for understanding the relationship between stream discharge and basin area.

6.1.7 Drainage Density

Drainage density is the total length of streams divided by the total drainage area. It is an important indicator of landscape dissection and peak runoff potential, classified into coarse, medium, or fine patterns. The drainage density for the study area was calculated using ArcGIS 10.2 software.

6.1.8 Stream Frequency (NF)

Stream frequency, also known as drainage frequency, is calculated by dividing the total number of streams by the total area (Horton, 1932). It mainly depends on rainfall and the region's physiography.

6.1.9 Compactness Coefficient or Compactness Constant (CC)

The compactness constant is the ratio of the catchment perimeter to the perimeter of a circle with the same area as the basin, influenced primarily by slope and independent of watershed size.

7 Study Area

The Ganga River flows through a significant stretch from Rudraprayag to Rishikesh in the Indian state of Uttarakhand. Starting at Rudraprayag, where the Alaknanda and Mandakini rivers converge, the river continues its journey through the Himalayan foothills.

7.1 Geographic Location

Rudraprayag is positioned at an elevation of approximately 895 m (2936 ft) above sea level. From here, the Ganga moves southward, winding through a series of steep valleys and mountainous terrain [12]. This region is characterized by its rugged landscape, with the river cutting through deep gorges and flowing past rocky cliffs.

7.1.1 Path and Surrounding Areas

As the river progresses downstream, it passes through several smaller towns and villages that rely on its waters for agriculture and daily use. The surrounding landscape is lush with vegetation, ranging from dense forests to terraced fields [13].

Upon reaching Devprayag, another significant confluence occurs where the Alaknanda meets the Bhagirathi River, officially forming the Ganga. Devprayag sits at an altitude of about 830 m (2723 ft). The river continues its southwesterly course from Devprayag towards Rishikesh.

7.1.2 Rishikesh: Gateway to the Garhwal Himalayas

Rishikesh, located at an elevation of approximately 372 m (1220 ft), is a prominent spiritual and cultural hub. Known as the "Gateway to the Garhwal Himalayas," Rishikesh is situated on the banks of the Ganga, where the river exits the mountainous region and begins its journey across the plains of northern India.

7.2 Significance and Accessibility

This stretch of the Ganga River, from Rudraprayag to Rishikesh, is not only geographically significant but also holds immense cultural, spiritual, and ecological importance [14]. The river supports numerous communities along its banks, provides a habitat for diverse wildlife, and serves as a focal point for religious and recreational activities. The region is accessible by road, with well-connected highways and scenic routes that offer breathtaking views of the river and surrounding mountains. This accessibility makes it a popular destination for pilgrims, tourists, and adventure enthusiasts alike.

8 Methodology

To accomplish the objectives of the present study, the methodology is broadly divided into three steps as given below:

(a) Visual interpretation of satellite data
(b) Stream profile analysis
(c) Digitization to delineate the river

8.1 Stream Profile Analysis

The process begins with acquiring a Digital Elevation Model (DEM), which is a 3D representation of a terrain's surface. DEM data can be collected using various methods, including satellite imagery, LiDAR, or other remote sensing technologies [15]. After obtaining the DEM data, it is necessary to combine multiple DEM tiles into a single, continuous mosaic. This step ensures a seamless representation of the terrain over the entire area of interest (Fig. 1).

The DEM mosaic is transformed from the Geographic Coordinate System (GCS) to a Projected Coordinate System (PCS). This reprojection is essential to improve the accuracy of spatial analyses and make the data suitable for further geospatial processing [16].

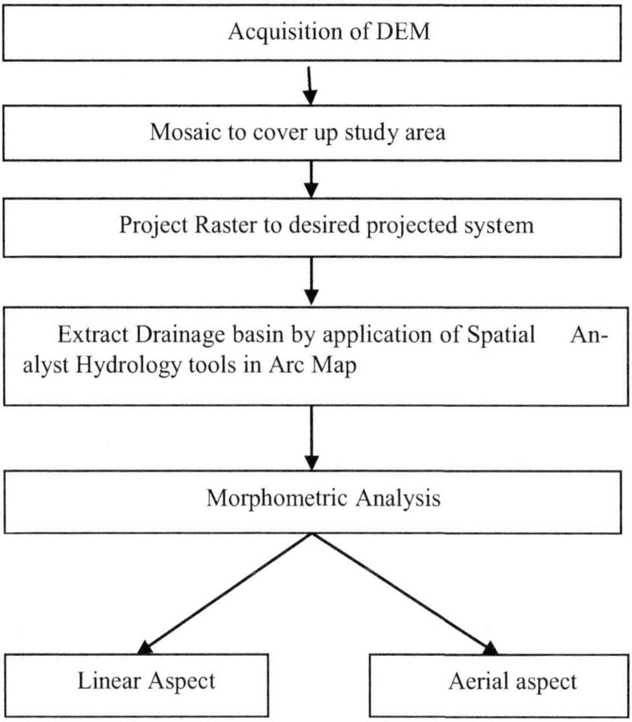

Fig. 1 Stream profile analysis

In ArcMap, the Spatial Analyst Hydrology tools are then used to delineate the drainage basin. This process involves several key steps:

1. **Fill Sinks**: This step modifies the DEM to remove any depressions or pits that could disrupt the flow direction calculations.
2. **Flow Direction**: Determines the direction of water flow from each cell to its steepest downslope neighbor.
3. **Flow Accumulation**: Calculates the accumulated flow as it traverses the landscape, identifying the drainage pattern.
4. **Watershed**: Defines the boundaries of the drainage basin based on the flow accumulation and direction data.

Once the drainage basin is delineated, morphometric analysis is conducted. This analysis involves calculating various quantitative measures of the basin's characteristics, such as its area, perimeter, drainage density, and stream frequency [17]. These metrics provide valuable insights into the geomorphological and hydrological properties of the drainage basin, aiding in understanding its behavior and supporting effective resource management and planning.

8.2 Delineate the River

Delineation of a river involves accurately defining its course, boundaries, and associated features. This process encompasses the identification and mapping of the river's main channel, tributaries, floodplains, and surrounding landscapes [18]. Techniques often involve field surveys, remote sensing, and the use of geographic information systems (GIS) to capture spatial data. The goal is to create a comprehensive and precise representation of the river system, which is essential for effective water resource management, environmental protection, and urban planning. Delineation helps in understanding the river's hydrology, potential areas of flooding, and the ecological significance of its riparian zones [19].

9 Result and Discussion

Figure 2 displays the DEM fill. After applying the fill function, the elevation values were adjusted to eliminate the pits, effectively resolving the issue.

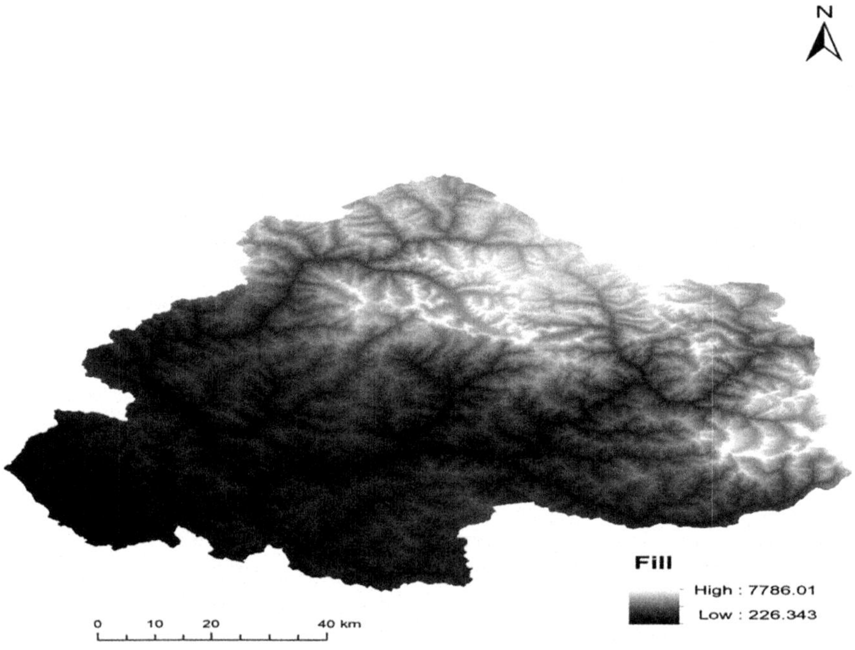

Fig. 2 DEM fill

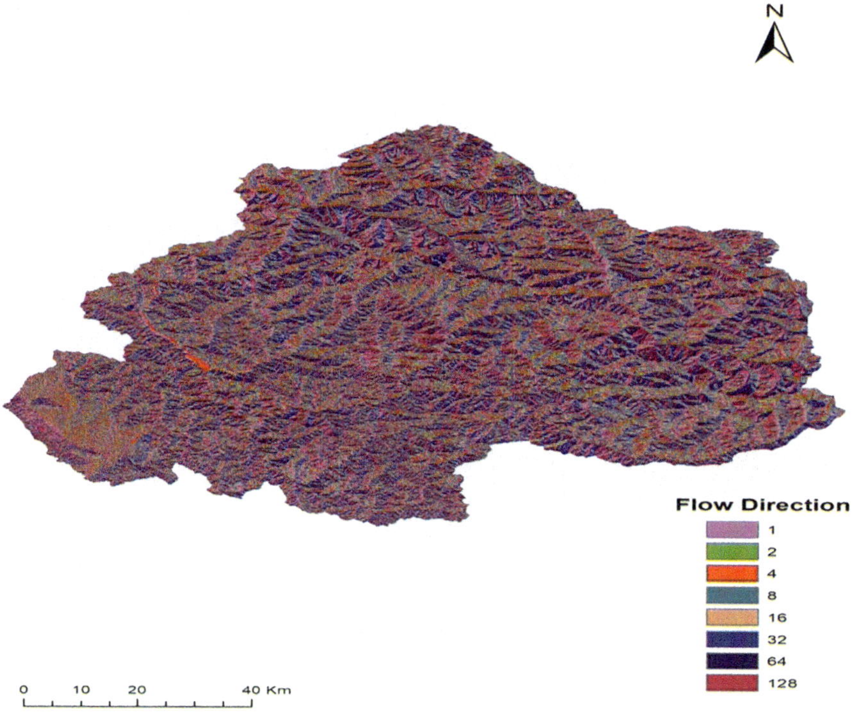

Fig. 3 Flow direction

Figure 3 illustrates that flow direction is influenced by filled sinks, which level the surrounding terrain. Values ranging from 1 to 128 indicate the direction of flow.

As water flows downstream, a stream begins to form. Flow accumulation calculates this by counting the number of upstream cells that flow into a given cell. Figure 4 illustrates flow accumulation.

The extracted stream network of Upper Ganga Basin is shown in Fig. 5.

From Strahler stream ordering scheme, the total number of segments of each order in Upper Ganga drainage basin have been computed and is given in Table 1.

To delineate the left and right banks of river from the Rudraprayag to Rishikesh for the years 2012, 2014, and 2024. The 2012 riverbanks serve as the baseline for comparison. By comparing the 2012 and 2014 riverbanks, we identified changes resulting from the 2013 disaster in Mandakini River. This comparison involves overlaying the digitized banks from both years and analyzing the differences in their positions. Any significant shifts or alterations in the river's course can be attributed to the impact of the disaster.

The riverbanks from 2012 and 2024 are compared to evaluate changes due to human activities and natural climatic variations. By overlaying and analyzing the banks from these 2 years, it is possible to identify changes in the river's course and

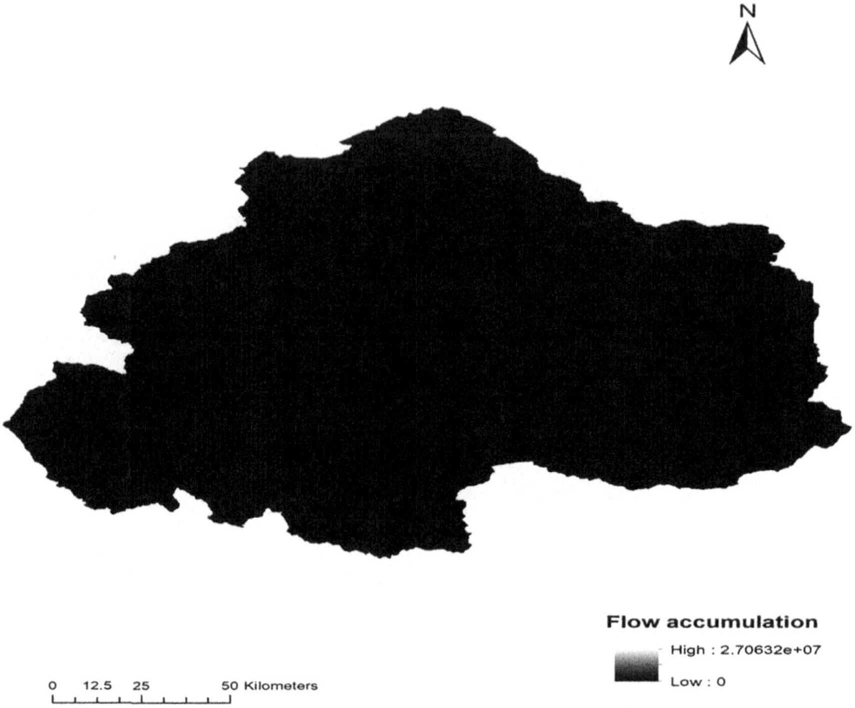

Fig. 4 Flow accumulation

morphology that may have resulted from construction, land use changes, erosion, and other environmental factors [20].

This comparative analysis provides insights into the extent and nature of changes in the riverbanks over time, highlighting the impacts of both natural disasters and human activities [21].

In Fig. 6 from the year 2012, a baseline is established at Rudraprayag. This point is significant as it marks where the Mandakini River, which originates from Kedarnath, converges with the Alaknanda River. As these rivers flow downstream, they reach Devprayag, where the Alaknanda merges with the Bhagirathi River, forming the Ganga. This entire geographic trajectory, from Rudraprayag to Rishikesh.

Figures 7 and 8 illustrate the comparison of riverbank positions between the base year 2012 and the year 2014 to identify lateral shifts.

Figures 9 and 10 illustrate the comparison of riverbank positions between the base year 2012 and the year 2024 to identify lateral shifts.

As the river is shifted from its baseline, we have calculated the quantitative amount of lateral shift with respect to left as well as right bank with base year 2012 of the years 2014 and 2024. The distance is measured with the help of the ruler in the Google Earth Pro. Tables 2 and 3 show the quantitative lateral shift with left and

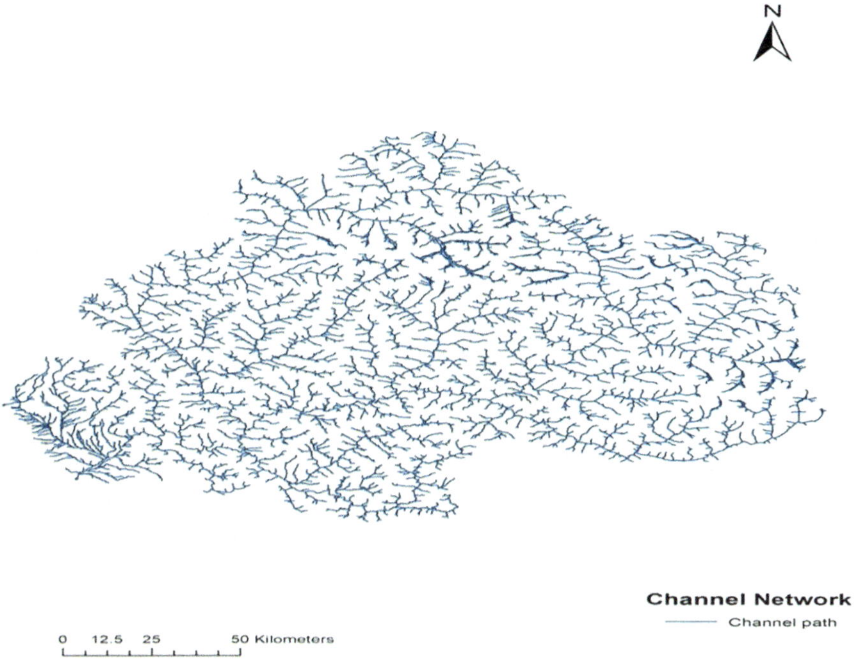

Fig. 5 Stream network

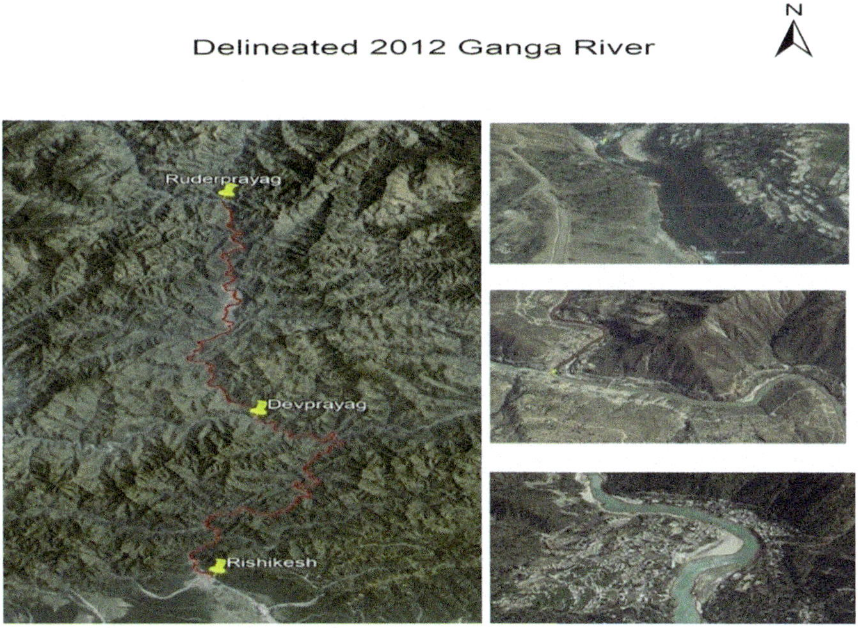

Fig. 6 Ganga River in 2012

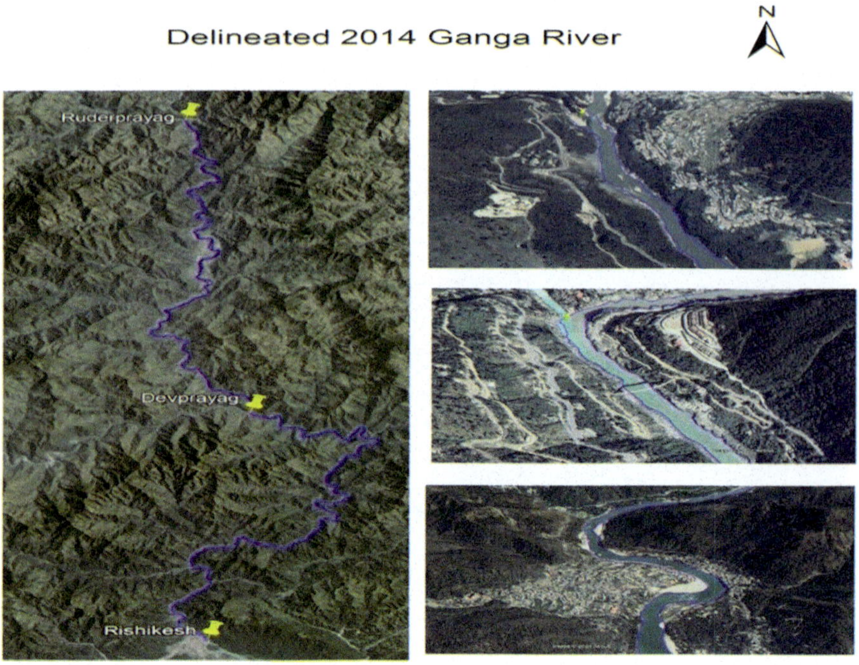

Fig. 7 Ganga River in 2014

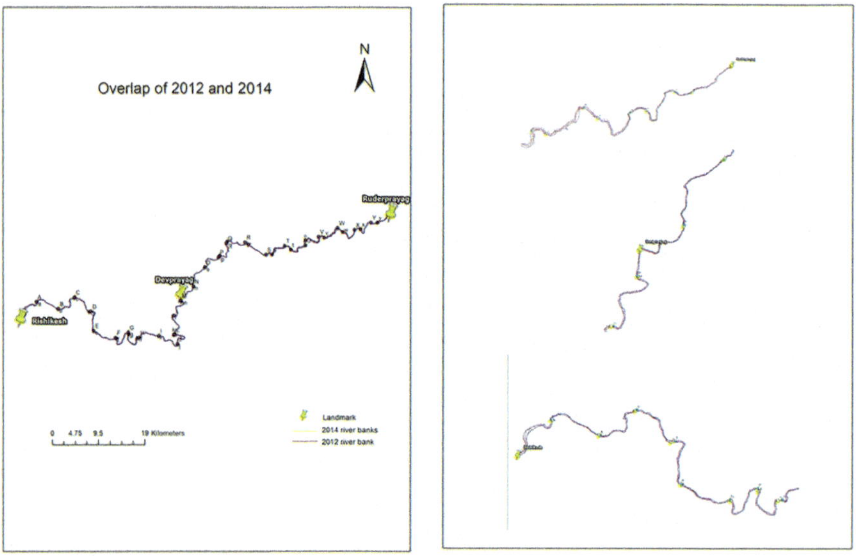

Fig. 8 Lateral shift of 2012 and 2014

Fig. 9 Ganga River in 2024

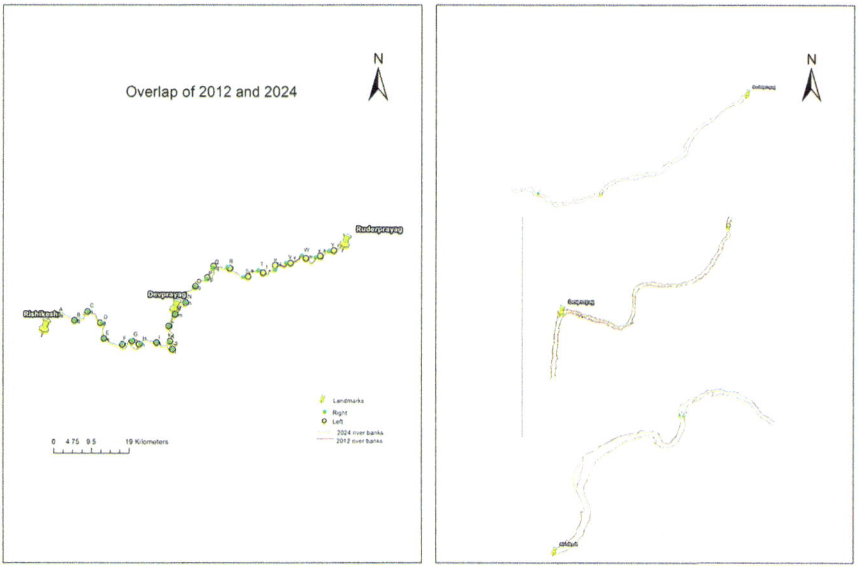

Fig. 10 Lateral shift of 2012 and 2024

Table 2 Lateral shifting of right bank

Points	Distance (km)	Right bank shifting (m) of year 2014 and 2024 (with reference to base year 2012)	
		2014 (m)	2024 (m)
A	5	+3.95	−0.32
B	5	−9.02	+6.94
C	5	+1.43	+21.74
D	5	+18.32	+30.75
E	5	+4.65	+3.59
F	5	+7.22	+13.61
G	5	+5.16	+10.98
H	5	+3.15	−4.42
I	5	+8.25	+11.89
J	5	+1.75	−2.94
K	5	−3.33	−9.82
L	5	+13.16	+11.63
M	5	+5.13	−2.28
N	5	−10.53	−2.88
O	5	+2.60	−0.46
P	5	−3.43	−2.83
Q	5	+2.63	+4.41
R	5	−24.54	−9.28
S	5	+2.91	+5.58
T	5	+14.78	−1.96
U	5	+15.73	−5.59
V	5	+38.56	+72.55
W	5	+58.43	+100.75
X	5	+9.70	+28.88
Y	5	+26.18	+23.86

right banks, respectively [22]. A "+" symbol indicates that the bank has moved outward, while a "-" symbol signifies an inward shift, as detailed in Tables 2 and 4. To determine these shifts, 25 points were marked along both the left and right banks of the river.

10 Conclusions

The study has provided comprehensive insights into the dynamic geomorphological processes shaping one of the world's most significant river systems. By analyzing morphometric features and river course changes, the study highlights the intricate interplay between natural forces and anthropogenic influences in the Upper Ganga Basin. Key findings include significant alterations in river morphology due to sediment deposition, erosion, and human interventions, which have profound

Table 3 Lateral shift of left bank

Points	Distance (km)	Left bank shifting (m) of year 2014 and 2024 (with reference to base year 2012)	
		2014 (m)	2024 (m)
a	5	+4.66	+4.22
b	5	+13.41	+23.38
c	5	+04.49	−2.64
d	5	+0.61	−3.26
e	5	+13.87	+6.52
f	5	+22.52	+6.65
g	5	−15.37	—2.94
h	5	−0.87	−1.68
i	5	−37.72	−12.35
j	5	+8.95	+16.63
k	5	−9.28	−4.91
l	5	−19.46	−2.82
m	5	−6.77	−1.15
n	5	−21.18	−26.76
o	5	+14.85	+14.40
p	5	+16.36	−6.89
q	5	+1.31	−8.63
r	5	+2.07	−3.80
s	5	+2.39	+2.96
t	5	+29.44	+13.96
u	5	−2.97	−25.77
v	5	+67.32	+91.18
w	5	+53.72	+78.89
x	5	+0.65	+28.84
y	5s	−1.83	+18.43

Table 4 Geomorphological parameters

Parameters	Value
Stream order	7
Basin area (A)	22,292 km^2
Basin perimeter (P)	838.289 km
Bifurcation ratio (Rb)	3.31561672

implications for flood management, water resource planning, and ecological sustainability. This research underscores the necessity for continuous monitoring and adaptive management strategies to preserve the ecological balance and ensure the sustainable development of the Ganga Basin. Future studies should focus on integrating advanced remote sensing technologies and hydrological models to enhance the understanding and predictive capabilities concerning river dynamics and morphometric evolution.

References

1. Avtar, R., Singh, C. K., Shashtri, S., & Mukherjee, S. (2011). Identification of erosional and inundation hazard zones in Ken–betwa river linking area, India, using remote sensing and GIS. *Environmental Monitoring and Assessment, 182*(1–4), 341–360. https://doi.org/10.1007/s10661-011-1880-6
2. Kale, V. S. (2012). Tropical geomorphology by Avijit Gupta. *Journal of the Geological Society of India, 79*(1), 94–95. https://doi.org/10.1007/s12594-012-0010-6
3. Goudie, A., & Viles, H. (2010). 9. Unseen landscapes. In *Landscapes and geomorphology* (pp. 111–125). https://doi.org/10.1093/actrade/9780199565573.003.0009
4. Strahler, A. N. (n.d.). Slope analysis. In *Encyclopedia of earth science* (pp. 998–1002). https://doi.org/10.1007/3-540-31060-6_335
5. Florsheim, J. L., Mount, J. F., & Chin, A. (2008). Bank erosion as a desirable attribute of rivers. *Bioscience, 58*(6), 519–529. https://doi.org/10.1641/b580608
6. Rozo, M. G., Nogueira, A. C., & Castro, C. S. (2014). Remote sensing-based analysis of the planform changes in the upper Amazon river over the period 1986–2006. *Journal of South American Earth Sciences, 51*, 28–44. https://doi.org/10.1016/j.jsames.2013.12.004
7. Nandini, V. (2015). *Geomorphological study of river Mandakini at some critical locations using geospatial tools*.
8. Basha, U., Pandey, M., Nayak, D., Shukla, S., & Shukla, A. K. (2024). Spatial–temporal assessment of annual water yield and impact of land use changes on Upper Ganga Basin, India, using InVEST model. *Journal of Hazardous, Toxic, and Radioactive Waste, 28*(2), 04024003. https://doi.org/10.1061/JHTRBP.HZENG-1245
9. Bali, R., Agarwal, K. K., Nawaz Ali, S., Rastogi, S. K., & Krishna, K. (2011). Drainage morphometry of himalayan glacio-fluvial basin, India: Hydrologic and neotectonic implications. *Environmental Earth Sciences, 66*(4), 1163–1174. https://doi.org/10.1007/s12665-011-1324-1
10. Chopra, R., Verma, V. K., & Sharma, P. K. (2001). Mapping, monitoring and conservation of Harike wetland ecosystem, Punjab, India, through remote sensing. *International Journal of Remote Sensing, 22*(1), 89–98. https://doi.org/10.1080/014311601750038866
11. Gao, B. (1996). NDWI—A normalized difference water index for remote sensing of vegetation liquid water from space. *Remote Sensing of Environment, 58*(3), 257–266. https://doi.org/10.1016/s0034-4257(96)00067-3
12. Pareta, K., Reddy, T., & Singh, H. (2022). Geomorphological analysis of Ayad river basin, Udaipur (India) using remote sensing and GIS techniques. *Advances in Image and Video Processing, 10*(5). https://doi.org/10.14738/aivp.105.12962
13. Li, L., Lu, X., & Chen, Z. (2007). River channel change during the last 50 years in the middle Yangtze River, the Jianli reach. *Geomorphology, 85*(3–4), 185–196. https://doi.org/10.1016/j.geomorph.2006.03.035
14. Sharma, C., Shukla, A. K., & Zhang, Y. (2021). Climate change detection and attribution in the Ganga-Brahmaputra-Meghna river basins. *Geoscience Frontiers, 12*(5), 101186. https://doi.org/10.1016/j.gsf.2021.101186
15. Robertson, N., & Bakimchandra, O. (2024). Comparative assessment of land use land cover at different temporal scales in Imphal-iril river catchment, Manipur, India. *Remote Sensing Letters, 15*(6), 655–666. https://doi.org/10.1080/2150704x.2024.2365810
16. RE, & Rank, H. D. (2013). Assessment of morphometric characteristics of Shetrunji river basin using remote sensing and geographical information system (GIS). *African Journal of Agricultural Research, 8*(18), 2003–2015. https://doi.org/10.5897/ajar2013.7031
17. Saeedrashed, Y., & Guven, A. (2012). Estimation of geomorphological parameters of lower Zab river-basin by using GIS-based remotely sensed image. *Water Resources Management, 27*(1), 209–219. https://doi.org/10.1007/s11269-012-0179-x
18. Shukla, A. K., Ojha, C. S. P., Mijic, A., Buytaert, W., Pathak, S., Garg, R. D., & Shukla, S. (2018). Population growth, land use and land cover transformations, and water quality nexus

in the Upper Ganga River basin. *Hydrology and Earth System Sciences, 22*(9), 4745–4770. https://doi.org/10.5194/hess-22-4745-2018

19. Sarma, J., & Phukan, M. (2004). Origin and some geomorphological changes of Majuli Island of the Brahmaputra river in Assam, India. *Geomorphology, 60*(1–2), 1–19. https://doi.org/10.1016/j.geomorph.2003.07.013
20. Palchowdhuri, Y., & Roy, P. S. (2018). Driver based statistical model for simulating land use land cover change in Indus River Basin India. *Remote Sensing of Land, 2*(1), 15–30. https://doi.org/10.21523/gcj1.18020102
21. Kaushal, N. (2013). Environmental flows for river Ganga. In *Our National River Ganga* (pp. 223–237). https://doi.org/10.1007/978-3-319-00530-0_9
22. Richardson, W., & Thorne, C. R. (2001). Multiple thread flow and channel bifurcation in a braided river: Brahmaputra–jamuna river, Bangladesh. *Geomorphology, 38*(3–4), 185–196. https://doi.org/10.1016/s0169-555x(00)00080-5

Integrated Urban Water Management: Optimizing Rainwater Harvesting and Stormwater Infrastructure for Sustainable Urban Development

Bishnu Kant Shukla, Pushpendra Kumar Sharma, Chitrangi Singh, Nitu Kumari, Ayush Chaudhary, and Aditya Garg

1 Introduction

Urban environments are becoming more complex, marked by expanding infrastructure and growing populations [1]. This has drawn attention to the difficulties in managing urban water resources. The natural hydrological processes in such places, which are interspersed with impermeable surfaces like concrete roadways and high-rise structures, are interfered with, frequently increasing surface runoff while also reducing groundwater recharge [2]. In the past, the supply of water and the removal of trash were the main considerations when designing urban water systems [3, 4]. Given the evolving challenges, at hand there is a pressing need for water management practices that consider environmental and socioeconomic impacts alongside immediate water-related concerns. Insufficient management of stormwater in cities can lead to repercussions varying from localized problems such as flooding and soil erosion to widespread issues, like water pollution and the loss of aquatic ecosystems [2, 5]. Moreover, with the expansion of areas the volume of stormwater generated also increases, highlighting the importance of efficient management practices [6]. Stormwater is no longer regarded as a waste product in current urban planning paradigms; rather, it is valued as a crucial resource with possibilities for harvesting, storing, and reusing. Such actions might reduce the need for conventional water sources [7, 8]. Effective stormwater management can also relieve the strain on

B. K. Shukla (✉)
Department of Civil Engineering, JSS Academy of Technical Education, Noida, India

School of Civil Engineering, Lovely Professional University, Phagwara, India

P. K. Sharma
School of Civil Engineering, Lovely Professional University, Phagwara, India

C. Singh · N. Kumari · A. Chaudhary · A. Garg
Department of Civil Engineering, JSS Academy of Technical Education, Noida, India

S. Pathak et al. (eds.), *Geo-Data Revolution*, Advances in Geographic Information Science, https://doi.org/10.1007/978-3-031-94078-1_6

urban sewage systems, reducing the likelihood of overflows and the ensuing ecological contamination [4]. Given a number of benefits which range from urban attractiveness to community well-being [9, 10], green infrastructure choices like bioretention systems, permeable pavements, and rain gardens are quickly gaining popularity.

2 Causes and Impacts of Urban Flooding

Urban flooding, a major problem in modern cities, involves more than just water overflow. Its origins can be found in the intricate interaction of environmental, infrastructural, and societal elements. Such flooding catastrophes can have long-lasting socioeconomic effects on the impacted areas in addition to upsetting daily life. While a sizeable portion of the urban population is becoming more susceptible to flood threats, it is crucial to identify the root causes in order to develop effective and long-lasting solutions.

2.1 Overview of the Main Factors Leading to Urban Flooding

Urban environments are associated with quick development and expansion. Although economically advantageous, this increase has unintentionally resulted in changes to the natural landscape. The natural ability of the ground to absorb rainwater is substantially diminished, and surface runoff is increased, as vast swaths of green are replaced with impermeable surfaces like asphalt and concrete [11]. Another significant factor is the outdated or inadequate drainage system in many cities, which is frequently unprepared to handle the volume of water during heavy downpours [12]. Furthermore, a lot of metropolitan areas have expanded along or near rivers or other naturally flood-prone places, which has made them more vulnerable [13]. The problem has been made worse by climate change, which has unpredictable weather patterns and more frequent extreme events [14]. Flooding can also be significantly influenced by ineffective waste disposal practices where trash and debris plug drains [15]. Figure 1 provides a visual summary of the several factors that have been identified as the primary causes of urban flooding. Urban planners and politicians can prioritize areas of intervention and take a more pro-active approach to flood management by being aware of these aspects [16].

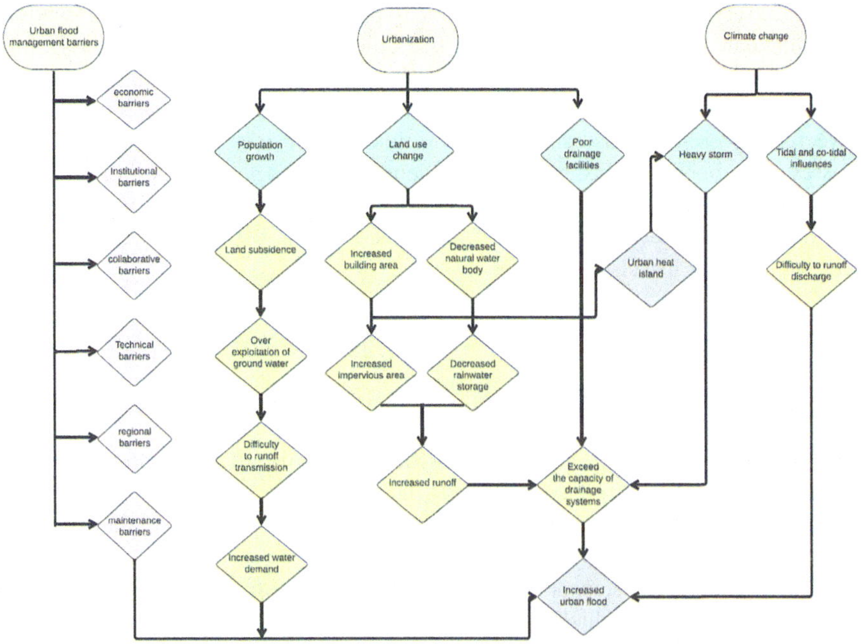

Fig. 1 Primary factors causing urban flooding

2.2 *Effects of Urban Floods: Infrastructure, Public Safety, and Environment*

Cities all around the world are becoming more and more concerned about urban flooding, which has serious consequences for the environment, infrastructure, and public safety. Rapid urbanization, climatic change, and inadequate stormwater management systems are some of the causes that amplify these negative effects.

Infrastructure Impact Floods have a significant influence on the city's physical infrastructure, including its roads, bridges, and buildings. Roads under water can get degraded, develop potholes, and in extreme cases, completely wash out, posing serious risks to transportation networks [17]. Bridges, which are necessary for urban connectivity, are vulnerable to scouring, in which a swift water flow erodes their structural support, threatening their soundness [18]. Buildings may also experience fundamental problems, the growth of mold, and considerable material degradation, especially those built without taking flood concerns into account. The ensuing costs for repairs and maintenance due to damage from flooding can significantly strain municipal finances and impede urban growth.

Public Safety Concerns The direct threat to public safety posed by urban floods is its immediate effect. Floodwaters that rise quickly have the potential to turn into dangerous streams and pose a drowning risk, especially in places without effective drainage systems [19]. Accident risk is increased by the ability of rushing water to sweep away cars. Floods can also pose serious health risks due to waterborne illnesses and other toxins because they frequently mix with sewage [20]. Floodwaters that come into contact with electricity lines provide a risk of electrocution, and stagnant waters that follow a flood serve as a haven for mosquitoes and other disease-carrying insects [12].

3 Environmental Ramifications

Urban floods have a variety of diverse environmental impacts. As an illustration, the rushing runoff may introduce pollutants like pesticides, heavy metals, and lubricants into natural reservoirs, harming aquatic ecosystems [21]. Flooding episodes can cause soil erosion, which alters the landscape and makes the ground less able to absorb subsequent rains, increasing the severity of following floods [13]. Floods can alter urban hydrology, which can have a negative impact on the biodiversity of the area. Additionally, solid wastes are dispersed during floods by waste management system overflows, further polluting the ecosystem and endangering both land and aquatic organisms [22].

In conclusion, the numerous effects of urban flooding highlight the need for thorough stormwater management methods, public education campaigns, and forward-thinking urban planning that takes into account the increased flood dangers brought on by anthropogenic climate change.

3.1 Urban Flood Modeling Techniques

Urban flood modeling has changed significantly over the last few decades as a result of technological breakthroughs and a growing understanding of urban hydrological dynamics. The development of successful flood mitigation strategies has benefited greatly from the improved prediction and management of urban floods made possible by modeling techniques.

Traditional Flood Modeling Methods One-dimensional (1D) hydrodynamic models have traditionally been used extensively in flood simulation in metropolitan areas. These models did not take into account the difficulties of surface runoff in urban contexts and instead concentrated primarily on the flow of water in river channels [23]. These models mainly made use of the Saint-Venant equations, which describe how water moves in open channels [24]. Such models' main benefit is their computational effectiveness, which makes them appropriate for situations where accurate representation of spatial variability was not crucial. These approaches were less suited for urban environments where drainage infrastructure and terrain variability significantly affect flood dynamics due to their generalized nature.

Contemporary Flood Modeling Methods Modern urban flood modeling tools, as opposed to conventional approaches, are capable of capturing the complexity of urban terrains. Due to its capacity to offer a spatially precise depiction of floodwater movements over the urban landscape, including flow paths, depths, and velocities, two-dimensional (2D) and three-dimensional (3D) hydrodynamic models have grown in popularity [25].

In exercises involving flood modeling, metropolitan topography, land use, and infrastructure elements can be more accurately represented, thanks to the combination of Geographic Information Systems (GIS) and hydrodynamic models [26]. Real-time weather data is now incorporated into modern models, enabling early warning and real-time flood forecasts [16]. A comprehensive knowledge of flood dynamics in urban contexts is also made possible by the development of integrated urban flood modeling methodologies, which combine simulations of drainage networks in one dimension with those of overland flows in another dimension [27].

The intricate interactions between different urban water components, such as stormwater systems, sewers, and natural water bodies, have begun to be taken into account by advanced modeling techniques, such as the integrated urban flood forecasting and warning modeling frameworks, to provide a comprehensive view of urban flood scenarios [28]. Modern modeling methods have advanced significantly compared to flood modeling techniques providing more detailed insight into urban flood behaviors. This knowledge plays a role, in devising flood prevention plans amidst shifting climate trends and rapid urban growth.

4 Mitigation Strategies: Gray Vs Green Infrastructure

Urban flooding is worsening as a result of climate change and growing urbanization prompting a reevaluation of flood control tactics. The ongoing argument between gray and green infrastructure techniques lies at the core of this progression. Both approaches have particular benefits and address certain facets of flood management.

4.1 Gray Infrastructure

Gray infrastructure is a term used to describe traditional, constructed stormwater management and conveyance systems that largely use concrete, metal, and other man-made elements. For many years, this model has dominated urban water management.

Primary Measures In order to prevent flooding in metropolitan areas, storm drains and sewers are often subterranean pipes that collect and transport stormwater away from such places [29]. Large concrete or earthen structures called retention and detention basins are used to temporarily store stormwater and release it at controlled rates in order to avoid upstream areas from becoming overburdened [30].

Channelization This technique involves redesigning and lining existing streams with concrete to increase water flow and ensure that it quickly exits populated areas [31]. Figure 2 shows these and other gray infrastructure practices that are frequently used in urban contexts. Even though they are frequently effective, these systems can be expensive to build and operate. Additionally, they frequently interfere with normal hydrological processes, which has an adverse effect on the ecosystem [27].

4.2 Green Infrastructure

The management of stormwater through green infrastructure, on the other hand, emphasizes solutions that resemble natural processes and takes a more comprehensive approach. Due to its dual advantages of flood mitigation and environmental conservation, this technology has gained popularity recently.

Modern Urban Planning's Function The importance of green infrastructure in sustainable urban development is becoming more widely acknowledged. While addressing stormwater management, it increases urban biodiversity, promotes the quality of the air and water, and offers recreational spaces for locals [16].

Benefits Infiltration and Groundwater Recharge: By allowing water to permeate the soil, features like rain gardens and permeable pavements can replenish groundwater while lowering surface runoff [32].

Natural Filtration Stormwater contaminants are filtered out by wetlands, vegetated swales, and bioretention zones, which enhances the quality of the water [33].

Fig. 2 Gray infrastructure measures for stormwater control examples

Habitat Creation Numerous elements of green infrastructure, such as green roofs and urban wetlands, offer habitats for urban wildlife, hence promoting biodiversity [34].

The functionality of various green infrastructure solutions is shown in Fig. 3.

Despite their effectiveness, these systems require a specialized approach to maintenance and might not be appropriate for all metropolitan environments due to space and soil limitations. But because of their numerous advantages, they are an important tool for urban planners [35]. Overall, green infrastructure is becoming an appealing choice even though gray infrastructure provides reliable, tried-and-true solutions. The best course of action for long-term urban flood management may be a hybrid strategy that draws on the advantages of both approaches.

5 Techniques in Focus: Rainwater Harvesting and Infiltration Systems

In urban settings, the sustainable management of stormwater is crucial, with a focus on the preservation and effective use of water resources. This section describes two distinct methods: rainwater harvesting and infiltration systems.

Fig. 3 Examples of green infrastructure strategies for stormwater management

5.1 Rainwater Harvesting

Although rainwater harvesting has its origins in ancient techniques, its importance in modern urban design has increased significantly. This is a sustainable strategy for effective urban water management, not merely water collecting.

Meaning and Approaches Rainwater harvesting can be used as a stormwater management strategy as well as a solution to the shortage of drinkable water caused by rising urban flooding incidents [36]. Rainwater harvesting systems typically involve collecting rainwater from roofs and storing it in tanks (like the underground tank shown in Fig. 4 before using it for non-potable uses like flushing, gardening, and occasionally, after treatment, even drinking [37].

5.2 Infiltration Systems

While rainwater harvesting focuses on storage, infiltration systems prioritize allowing water to permeate into the ground, mimicking natural hydrological cycles and reducing surface runoff.

Role in Flood Control and Stormwater Management Infiltration systems serve dual purposes—firstly, they assist in reducing the instantaneous peak runoff during rain events, thereby mitigating urban flood risks. Secondly, they aid in recharging groundwater tables, which is crucial in urban areas witnessing a decline in their

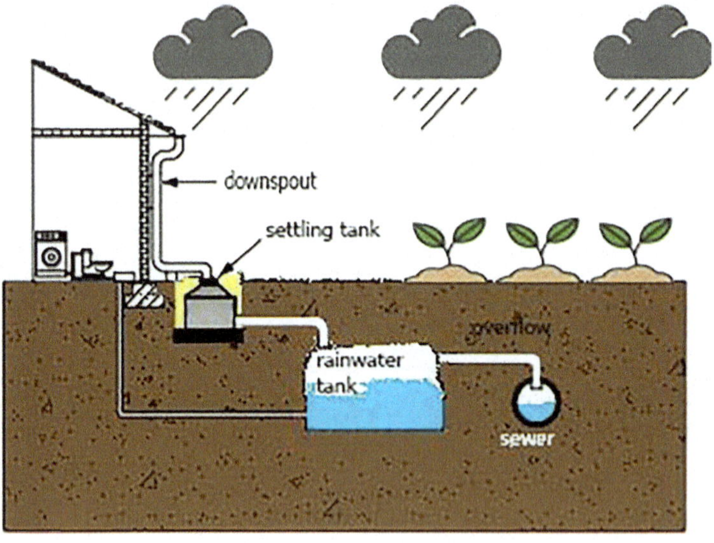

Fig. 4 Rainwater from the above-sealed surfaces is collected in an underground tank for later non-potable purposes. The tank's overflow is directed to the nearby sewers

groundwater levels. The second benefit is that they assist in replenishing groundwater tables, which is important in metropolitan areas where groundwater levels are declining [38]. Permeable pavements are one of the common infiltration methods, as seen in Fig. 5. In order to reduce surface runoff and encourage groundwater recharge, these pavements are made to let rainwater travel through them.

Additionally, Table 1, which provides a summary of several optimization model characteristics in this domain, can be used to understand the conceptual and practical aspects of stormwater management. Through the fusion of strategies such as rainwater harvesting and infiltration systems, these models offer frameworks that cities can adopt for effective stormwater treatment [39].

6 Economic and Environmental Considerations

The choice to choose a particular flood mitigation strategy depends not only on its technical effectiveness but also heavily on its economic viability and wider environmental effects. Striking a balance between sustainability and economic viability has long been a key problem in urban landscapes.

Cost Comparisons Between Gray and Green Infrastructures Gray infrastructure is conventional and frequently successful, but it has high initial capital expenditures and may need a lot of upkeep over time [53]. On the other side, green infrastructure may have lower upfront costs, especially when incorporated into ongoing urban planning initiatives, but the effectiveness over the long term and maintenance costs would vary depending on the particular technologies and their use [54]. In Table 2, some of the economic factors that are taken into account when using these systems are explained in depth with regard to the implementation of green and gray infrastructure solutions.

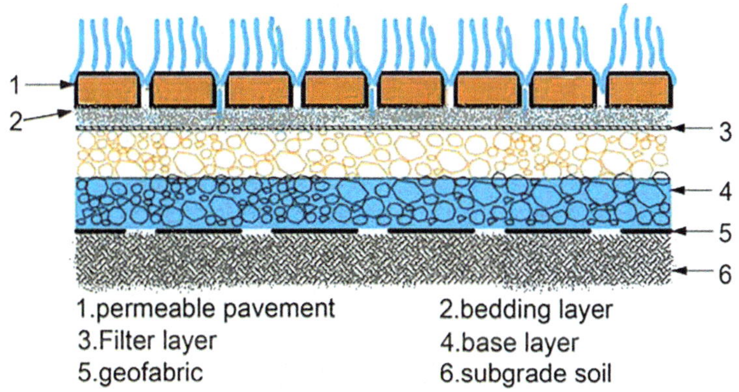

Fig. 5 The base of a permeable pavement is where precipitation infiltrates after falling, acting as a reservoir to catch and infiltrate rainwater

Table 1 Summary of Stormwater Management Optimization Models and Their Key Characteristics

Flood mitigation measures	Model	Optimization objects				Optimization algorithm	References
		Type	Location	Size	Combination way		
Green roofs and permeable pavement nondominated sorting	SWMM		✓			GA	[40]
Green roof and bioretention cell	SWMM		✓			SA	[41]
Pipe system, permeable pavement, green roof, etc.	SWMM	✓	✓		✓	GA	[42]
Grassed swale, green roof, bioretention, etc.	L-THIA-LID	✓	✓			Multi-algorithm genetically adaptive multiobjective	[43]
Retention pond, rain barrel, porous pavement, etc.	L-THIA-LID	✓	✓		✓	Multi-algorithm genetically adaptive multiobjective	[44]
Rain barrels, infiltration trenches and sewer system	SWMM	✓	✓	✓	✓	Hanging garden algorithm	[45]
Bioretention cells and porous pavement	SWMM		✓	✓		GA	[46]
Bioretention cell, permeable pavement, green roof	SWMM	✓			✓	GA	[47]
Bioretention, porous pavement, and green roof epsilon-non-dominated sorting	SWMM	✓	✓	✓		GA	[48]
Permeable pavement, concave-down greenbelt	SWMM		✓			GA	[49]
Stormwater detention	SWMM		✓			GA	[50]
Bioretention cell, permeable pavement	SWMM		✓			GA	[51]
Rainwater harvesting and permeable pavement	SWMM HEC-HMS		✓		✓	GA	[22]
Porous pavement, bioretention, rain barrel/ cistern	L-THIA-LID	✓	✓	✓		GA	[52]

Table 2 Summary of RTC applications based on urban food model data from the research

References	RTC operation objectives	Model tool	Control pattern	Control measure
[56]	Mitigating urban flood	EPA-SWMM	Optimization-based	Urban drainage system
[57]	Reducing flood peak flows	EPA-SWMM	Optimization-based	Detention pond
[58]	Improving runoff quantity and TSS concentration	SWIFT	Optimization-based	Urban drainage system
[59]	Mitigating the urban runoff and improving water quality	EPA-SWMM	Rule-based	Detention pond
[60]	Mitigating urban flood	EPA-SWMM	Optimization-based	Pumping station
[61]	Reducing peak flows and improving the water quality	PCSWMM	Optimization-based	Detention pond
[62]	Minimizing the pollution impacts of CSO	UNSTDY	Optimization-based	Urban drainage system
[63]	Increasing the retention tank capacity	EPA-SWMM	Optimization-based	Urban drainage system
[56]	Flood mitigation efficiency	EPA-SWMM	Optimization-based	Urban drainage system
[64]	Delaying basin peak flows and improving water quality	PCSWMM	Rule-based	Detention pond
[65]	Improving basin water level, detention time, and water quality	EPA-SWMM	Rule-based	Detention pond

Real-Time Control (RTC) Based on Urban Flood Models A lens for evaluating the economic viability of different solutions is provided by an understanding of the expenses related to the installation and upkeep of these RTC systems [55].

The Overall Effect on the Environment and Sustainability Beyond the financial sphere, it is critical to take the environmental impact of these initiatives into consideration. In addition to assisting in flood control, green infrastructure also encourages biodiversity, lowers urban heat islands, and provides recreational places, all of which improve the quality of urban life [66]. Traditional infrastructure, although often successful, may not always align with the values of creating cities, especially when considering aspects, such as carbon emissions and impacts on ecosystems [67]. While both traditional and nature-based infrastructures offer benefits a comprehensive approach, to addressing flooding should integrate both types taking into consideration constraints and environmental priorities. The goal is to strike a balance that addresses requirements without sacrificing long-term sustainability.

7 Conclusion and Future Directions

Undoubtedly urbanization and its myriad challenges underscore the importance of water management. The need for a coordinated strategy that combines the durability of modern, environmentally conscious procedures with the dependability of conventional ways is highlighted by the duality of gray and green infrastructures, each with its own strengths and limits.

Floods in urban environments not only offer a challenge to infrastructure but also pose serious risks to human safety and environmental health. The discussions that are highlighted in this paper stress the value of strategic planning and effective stormwater management, utilizing strategies like rainwater collecting and infiltration systems. But creating long-lasting solutions requires more than just technological fixes; it also requires a broad grasp of economic viability and environmental stewardship.

Our methods for addressing the numerous problems caused by urban floods must change as cities develop. In order to ensure that policies are not only theoretically sound but also realistically implementable, future research directions must incorporate empirical investigations with simulation models. To ensure resilient, sustainable, and welcoming urban landscapes, collaborations between researchers, urban planners, and policymakers are crucial. These partnerships bring together knowledge from several fields.

Integrated urban water management will definitely continue to be of great interest given the intricate interplay of hydrological, infrastructure, and social elements. Our common duty is to make sure that our cities are capable of handling both today's issues and those that will arise in the future.

References

1. Almeida, J., Proaño, O., Avila, C., Villacres, D., & Rivera-Tapia, E. D. (2025). Island ecosystems unleashed: Rain gardens transforming flood management in Galápagos. *Urban Water Journal, 22*(2), 186–199.
2. Shukla, B. K., Bhowmik, A. R., Raj, R. B., & Sharma, P. K. (2019). Physico-chemical parameters and status of ground water pollution in Jalandhar-Phagwara region. *Journal of Green Engineering, 9*(2), 212–223.
3. Shukla, B. K., Khan, A., Saikiran, G., & Sriram, M. (2019). A comparative study on effect of variation in opening shape on oxygenation performance of surface jet aerators used in water and wastewater treatment. *Journal of Green Engineering, 9*(3), 427–440.
4. Muleta, T. N., & Knolmar, M. (2025). Ecological impacts of combined sewer overflows on receiving waters. *Discover Water, 5*(1), 24.
5. Shukla, B. K., Sharma, P. K., & Khan, M. A. (2020). Physico-chemical study of some surface water bodies of Punjab. In *Journal of Physics: Conference Series* (Vol. 1531, No. 1, p. 012121). IOP Publishing.
6. Ou, Y., Deng, Y., Yan, L., Cui, Q., Yan, B., Shang, M., Liu, H., & Wang, J. (2025). Assessment of the major factors influencing the efficiency of vegetated filter strips under concentrated flow conditions—A case study in Northeast China. *Journal of Soils and Sediments, 25*, 1–16.

7. Otto, S., Vianello, M., Infantino, A., Zanin, G., & Di Guardo, A. (2008). Effect of a full-grown vegetative filter strip on herbicide runoff: Maintaining of filter capacity over time. *Chemosphere, 71*, 74–82.
8. Saleh, I., Kavian, A., Habibnezhad Roushan, M., & Jafarian, Z. (2018). The efficiency of vegetative buffer strips in runoff quality and quantity control. *International journal of Environmental Science and Technology, 15*, 811–820.
9. Shukla, B. K., Bhowmik, A. R., & Kumar, P. (2020). Use of waste synthetic fiber reinforcement in environmental friendly and economic pavement construction. *Journal of Green Engineering, 10*(1), 62–65.
10. Zhou, Y. N., He, J., Feng, L., Wang, B., Chen, Y., & Miao, L. (2025). Multiscale impacts of landscape metrics on water quality based on fine-grained land use maps. *Frontiers in Environmental Science, 13*, 1544078.
11. Luell, S. K., Winston, R. J., & Hunt, W. F. (2021). Monitoring the water quality benefits of a triangular swale treating a highway runoff. *Journal of Sustainable Water in the Built Environment, 7*(1), 05020004.
12. Winston, R. J., Hunt, W. F., Jadlocki, S. J., & Hathaway, J. M. (2009). Runoff volume reduction by a level spreader—Vegetated filter strip system in suburban Charlotte, NC. *Journal of Hydrologic Engineering, 15*, 499–503.
13. Line, D. E., & Hunt, W. F. (2009). Performance of a bioretention area and a level spreader-grass filter strip at two highway sites in North Carolina. *Journal of Irrigation and Drainage Engineering, 135*, 217–224.
14. Khadka, A., Kokkonen, T., Niemi, T. J., Lähde, E., Sillanpää, N., & Koivusalo, H. (2020). Towards natural water cycle in urban areas: Modelling stormwater management designs. *Urban Water Journal, 17*, 587–597.
15. Winston, R. J., Dorsey, J. D., & Hunt, W. F. (2016). Quantifying volume reduction and peak flow mitigation for three bioretention cells in clay soils in Northeast Ohio. *Science of the Total Environment, 553*, 83–95.
16. Roy-Poirier, A., Champagne, P., & Filion, Y. (2010). Review of bioretention system research and design: Past, present, and future. *Journal of Environmental Engineering, 136*(9), 878–889.
17. Singh, S., Shukla, B. K., & Sharma, P. K. (2020). Use of system dynamics forecasting model in transportation system to reduce vehicular emissions-a mathematical approach. In *Journal of Physics: Conference Series* (Vol. 1531, No. 1, p. 012118). IOP Publishing.
18. Hathaway, J. M., & Hunt, W. F. (2008). Field evaluation of level spreaders in the Piedmont of North Carolina. *Journal of Irrigation and Drainage Engineering, 134*, 538–542.
19. Muthanna, T. M., Viklander, M., & Thorolfsson, S. T. (2008). Seasonal climatic effects on the hydrology of a rain garden. *Hydrological Processes, 22*, 1640–1649.
20. Autixier, L., Mailhot, A., Bolduc, S., Madoux-Humery, A.-S., Galarneau, M., Prévost, M., & Dorner, S. (2014). Evaluating rain gardens as a method to reduce the impact of sewer overflows in sources of drinking water. *Science of the Total Environment, 499*, 238–247.
21. Shukla, B. K., Bashir, M., Sharma, P. K., & Abhimanyu. (2021). An analytical investigation of surface water quality and pollution status in Srinagar, Jammu and Kashmir, India. *Journal of Green Engineering, 11*(1), 952–962.
22. Pathak, S., Gupta, S., & Ojha, C. S. P. (2021). Assessment of groundwater vulnerability to contamination with modified DRASTIC index: A case study in Haridwar, Uttarakhand, India. *ASCE Journal of Hazardous, Toxic, and Radioactive Waste, 25*(2), 04020081.
23. Jato-Espino, D., & Pathak, S. (2021). Geographic location system for identifying urban road sections sensitive to runoff accumulation. *Hydrology, 8*(2), 72.
24. Saraçoğlu, K. E., & Kazezyılmaz-Alhan, C. M. (2025). Development of a hydrological model for vegetated swales under different rainfall and swale properties. *Water Environment Research, 97*(2), e70014.
25. Zhang, K., Chui, T. F. M., & Huang, P. (2025). Impact of climatic factors and groundwater level on the hydrologic performance of pervious concrete pavements. In *Pervious concrete pavements* (pp. 251–269). Woodhead Publishing.

26. Jato-Espino, D., Sillanpää, N., & Pathak, S. (2019). Flood modelling in sewer networks using dependence measures and learning classifier systems. *Journal of Hydrology, 578*, 124013.
27. Bouwer, H. (2002). Artificial recharge of groundwater: Hydrogeology and engineering. *Hydrogeology Journal, 10*(1), 121–142.
28. Roldin, M., Fryd, O., Jeppesen, J., Mark, O., Binning, P. J., Mikkelsen, P. S., & Jensen, M. B. (2012). Modelling the impact of soakaway retrofits on combined sewage overflows in a 3 km2 urban catchment in Copenhagen, Denmark. *Journal of Hydrology, 4522–453*, 64–75.
29. Pathak, S., Ojha, C. S. P., Zevenbergen, C., & Garg, R. D. (2017). Assessing stormwater harvesting potential in Dehradun city using geospatial technology. In *Development of water resources in India* (pp. 47–60). Springer.
30. Yang, H., Florence, D. C., McCoy, E. L., Dick, W. A., & Grewal, P. S. (2009). Design and hydraulic characteristics of a field-scale bi-phasic bioretention rain garden system for storm water management. *Water Science and Technology, 59*, 1863–1872.
31. Winston, R. J., Hunt, W. F., Osmond, D. L., Lord, W. G., & Woodward, M. D. (2011). Field evaluation of four level spreader-vegetative filter strips to improve urban storm-water quality. *Journal of Irrigation and Drainage Engineering, 137*, 170–182.
32. Gani, A., Pathak, S., Hussain, A., Shukla, A. K., & Chand, S. (2024). Emerging pollutant in surface water bodies: A review on monitoring, analysis, mitigation measures and removal technologies of micro-plastics. *Environmental Geochemistry and Health, 46*(7), 214. https://doi.org/10.1007/s10653-024-01992-7
33. Shenoy, A., Shukla, B. K., & Bansal, V. (2022). Sustainable design of textile industry effluent treatment plant with constructed wetland. *Materials Today: Proceedings, 61*, 537–542.
34. Zhang, L., Lu, Q., Ding, Y., Peng, P., & Yao, Y. (2018). Design and performance simulation of road bioretention media for sponge cities. *Journal of Performance of Constructed Facilities, 32*, 04018061.
35. Gowda, S., Kunjar, V., Gupta, A., Kavitha, G., Shukla, B. K., & Sihag, P. (2024). Prediction of the subgrade soil California bearing ratio using machine learning and neuro-fuzzy inference system techniques: A sustainable approach in urban infrastructure development. *Urban Science, 8*(1), 4.
36. Imteaz, M. A., Shanableh, A., Rahman, A., & Ahsan, A. (2011). Optimisation of rainwater tank design from large roofs: A case study in Melbourne, Australia. *Resources, Conservation & Recycling, 55*, 1022–1029.
37. Domènech, L., & Saurí, D. (2011). A comparative appraisal of the use of rainwater harvesting in single and multi-family buildings of the metropolitan area of Barcelona (Spain): Social experience, drinking water savings, and economic costs. *Journal of Cleaner Production, 19*, 598–608.
38. Hansda, A., Chand, S. K., Pradhan, B., Chand, S., Shukla, A. K., & Rout, P. R. (2025). Toxicological impacts and microbial-mediated degradation processes of microplastics. *Journal of Hazardous, Toxic, and Radioactive Waste, 29*(2), 04024044. https://doi.org/10.1061/JHTRBP.HZENG-1400
39. Pathak, S., Shukla, A. K., Ojha, C. S. P., & Garg, R. D. (2016, July). Identifying hot spots for stormwater harvesting and reuse in urban areas applying remote sensing, GIS and system optimization techniques. In *2016 IEEE International Geoscience and Remote Sensing Symposium (IGARSS)* (pp. 967–970). IEEE.
40. Giacomoni, M. H., & Joseph, J. (2017). Multi-objective evolutionary optimization and Monte Carlo simulation for placement of low impact development in the catchment scale. *Journal of Water Resources Planning and Management, 143*(9), 04017053.
41. Huang, C. L., Hsu, N. S., Liu, H. J., & Huang, Y. H. (2018). Optimization of low impact development layout designs for megacity flood mitigation. *Journal of Hydrology, 564*, 542–558.
42. Zhou, Q., Lai, Z., & Blohm, A. (2019). Optimising the combination strategies for pipe and infiltration-based low impact development measures using a multiobjective evolution approach. *Journal of Flood Risk Management, 12*(2), e12457.

43. Yadav, N., Jianping Wu, J., Banerjee, A., Pathak, S., Garg, R. D., & Yao, S. (2024). Climate uncertainty and vulnerability of urban flooding associated with regional risk using multi-criteria analysis in western coast of India. *Environmental Research, 244*, 117962.
44. Liu, Y., Theller, L. O., Pijanowski, B. C., & Engel, B. A. (2016). Optimal selection and placement of green infrastructure to reduce impacts of land use change and climate change on hydrology and water quality: An application to the Trail Creek Watershed, Indiana. *Science of the Total Environment, 553*, 149–163.
45. Bakhshipour, A. E., Dittmer, U., Haghighi, A., & Nowak, W. (2019). Hybrid green-blue-gray decentralized urban drainage systems design, a simulation-optimization framework. *Journal of Environmental Management, 249*, 109364.
46. Hussain, A., Deshwal, A., Priyadarshi, M., Pathak, S., Sambandam, G., Chand, S., & Shukla, A. K. (2024). Landfill leachate analysis from selected landfill sites and its impact on groundwater quality, New Delhi, India. *Environment, Development and Sustainability*, 1–26. https://doi.org/10.1007/s10668-023-04403-6
47. Zeng, J., Huang, G., Mai, Y., & Chen, W. (2020). Optimizing the cost-effectiveness of low impact development (LID) practices using an analytical probabilistic approach. *Urban Water Journal, 17*(2), 136–143.
48. Guoshun, Z., Patrick, R., & Yong, T. (2013). Multi-objective optimization of low impact development designs in an urbanizing watershed. *Open Journal of Optimization, 2013*. https://doi.org/10.4236/ojop.2013.24013
49. Xu, T., Jia, H., Wang, Z., Mao, X., & Xu, C. (2017). SWMM-based methodology for block-scale LID-BMPs planning based on site-scale multi-objective optimization: A case study in Tianjin. *Frontiers of Environmental Science & Engineering, 11*, 1–12.
50. Tao, T., Wang, J., Xin, K., & Li, S. (2014). Multi-objective optimal layout of distributed storm-water detention. *International journal of Environmental Science and Technology, 11*, 1473–1480.
51. Liang, C. Y., You, G. J. Y., & Lee, H. Y. (2019). Investigating the effectiveness and optimal spatial arrangement of low-impact development facilities. *Journal of Hydrology, 577*, 124008.
52. Xu, H., Xu, K., Lian, J., & Ma, C. (2019). Compound effects of rainfall and storm tides on coastal flooding risk. *Stochastic Environmental Research and Risk Assessment, 33*, 1249–1261.
53. Pathak, S., Sharma, S., Banerjee, A., & Kumar, S. (2024). A methodology to assess and evaluate sites with high potential for stormwater harvesting in Dehradun, India. *Big Data Research, 35*, 100415.
54. Roebuck, R. M., Oltean-Dumbrava, C., & Tait, S. (2011). Whole life cost performance of domestic rainwater harvesting systems in the United Kingdom. *Water Environment Journal, 25*, 355–365.
55. Morales-Pinzón, T., Rieradevall, J., Gasol, C. M., & Gabarrell, X. (2015). Modelling for economic cost and environmental analysis of rainwater harvesting systems. *Journal of Cleaner Production, 87*, 613–626.
56. Jafari, F., Mousavi, S. J., Yazdi, J., & Kim, J. H. (2018). Long-term versus real-time optimal operation for gate regulation during flood in urban drainage systems. *Urban Water Journal, 15*(8), 750–759.
57. Di Matteo, M., Liang, R., Maier, H. R., Thyer, M. A., Simpson, A. R., Dandy, G. C., & Ernst, B. (2019). Controlling rainwater storage as a system: An opportunity to reduce urban flood peaks for rare, long duration storms. *Environmental Modelling & Software, 111*, 34–41.
58. Shukla, A. K., Pathak, S., Pal, L., Ojha, C. S. P., Mijic, A., & Garg, R. D. (2018). Spatio-temporal assessment of annual water balance models for upper Ganga Basin. *Hydrology and Earth System Sciences, 22*(10), 5357–5371. https://doi.org/10.5194/hess-22-5357-2018
59. Gaborit, E., Muschalla, D., Vallet, B., Vanrolleghem, P. A., & Anctil, F. (2013). Improving the performance of stormwater detention basins by real-time control using rainfall forecasts. *Urban Water Journal, 10*(4), 230–246.

60. Jafari, F., Jamshid Mousavi, S., Yazdi, J., & Kim, J. H. (2018). Real-time operation of pumping systems for urban flood mitigation: Single-period vs. multi-period optimization. *Water Resources Management, 32*, 4643–4660.
61. Pathak, S., Ojha, C. S. P., Garg, R. D., Liu, M., Jato-Espino, D., & Singh, R. P. (2020). Spatiotemporal analysis of water resources in the Haridwar region of Uttarakhand, India. *Sustainability, 12*(20), 8449.
62. Fang, J., Liu, M., Liu, W., Pathak, S., Li, S., Tang, X., Zhou, L., & Sun, F. (2020). Piloting a capital-based approach for characterizing and evaluating drivers of Island sustainability-an application in Chongming Island. *Journal of Cleaner Production, 261*, 121123. https://doi.org/10.1016/j.jclepro.2020.121123
63. Abou Rjeily, Y., Abbas, O., Sadek, M., Shahrour, I., & Chehade, F. H. (2018). Model predictive control for optimising the operation of urban drainage systems. *Journal of Hydrology, 566*, 558–565.
64. Mishra, P., Jena, D., Thakur, R. R., Chand, S., Javed, B., & Shukla, A. K. (2024). Peri-urban floodscapes: Identifying and analyzing flood risk areas in North Bhubaneswar in Eastern India. *Water, 16*(21), 3019. https://doi.org/10.3390/w16213019
65. Sharma, S., Pathak, S., & Kumar, S. (2024). A planning-support tool for spatial suitability assessment of harvesting sites for stormwater infrastructure. *Journal of Hazardous, Toxic, and Radioactive Waste ASCE, 28*(1), 04023044.
66. Opher, T., & Friedler, E. (2024). Sustainability assessment of harvesting rainwater and air-conditioning condensate water in multi-family residential buildings under various conditions in Israel—A simulation study. *Sustainability, 16*(19), 8369.
67. Schubert, J. E., Burns, M. J., Fletcher, T. D., & Sanders, B. F. (2017). A framework for the case-specific assessment of green infrastructure in mitigating urban flood hazards. *Advances in Water Resources, 108*, 55–68.

Enhancing Environmental Impact Assessment Practices: A Comparative Analysis of International and National Approaches Focusing on India

Sandeep Budde, Sandeep Agrawal, and P. S. Chani

1 Introduction

The Industrial Revolution, which led to substantial social stratification, paved the way for more facilities and the upliftment of the lower classes. Industries are the primary contributors to the Gross Domestic Product (GDP) of the nation. The industrial accidents that have happened recently all over the world suggest that since the 1800s, when the industrial era first began, the developing world has experienced more misfortunes [1]. However, the actions taken to address this situation appear to be a new paradigm in every discussion, which has caused the sector to stagnate. Because land, labour, and capital are more readily available in urban areas than rural ones, industrialization has resulted in making. Most metropolitan cities and other urban regions exhibit socioeconomic stratification [2]. Growing industries have both positive and harmful effects on the environment and society. These implications include increased employment possibilities, population growth, an increase in the pollution index, greenhouse gas emissions that pose health risks, and an impact

S. Budde (✉)
Urban Environment Observatory, School of Urban and Regional Planning Program, University of Alberta, Edmonton, AB, Canada

Bharti Institute of Public Policy, Indian School of Business, Hyderabad, India
e-mail: sbudde@ualberta.ca

S. Agrawal
Urban Environment Observatory, School of Urban and Regional Planning Program, University of Alberta, Edmonton, AB, Canada
e-mail: sagrawal@ualberta.ca

P. S. Chani
Built Environment Lab, Department of Architecture and Planning, Indian Institute of Technology, Roorkee, India
e-mail: p.chani@ar.iitr.ac.in

S. Pathak et al. (eds.), *Geo-Data Revolution*, Advances in Geographic Information Science, https://doi.org/10.1007/978-3-031-94078-1_7

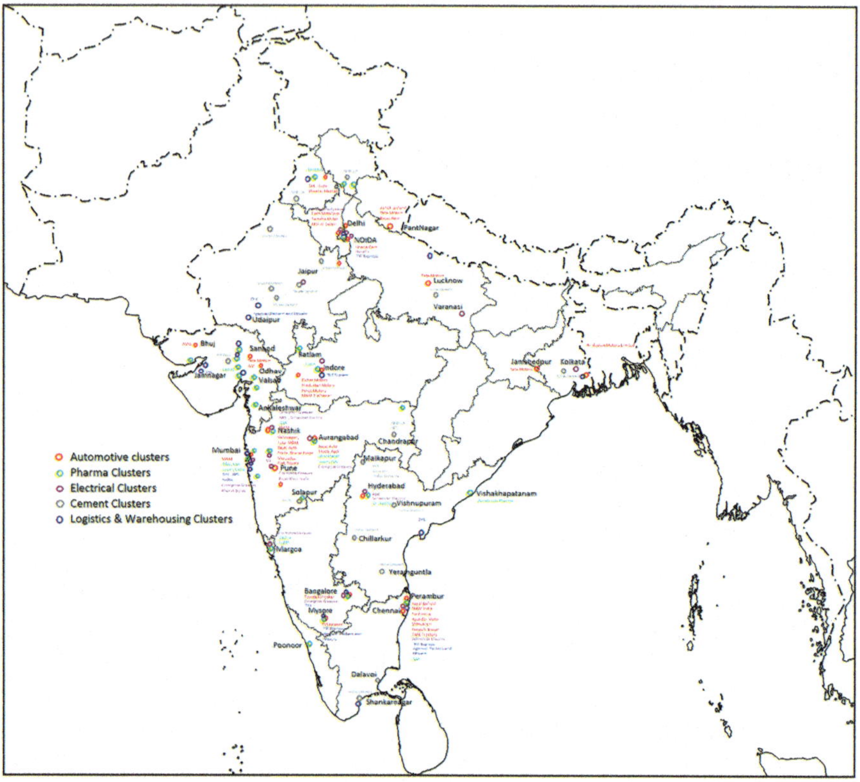

Fig. 1 Map depicting the location and type of cluster. (Source: Ref. 3)

on communities and way of life. A process-oriented framework is required to quantify the adverse effects of industrial environments on their immediate surroundings, as many industrialized, emerging, and impoverished nations lack one.

India has over 388 industrial development areas as shown in Fig. 1, which has automobile, Pharma, Electrical, Cement, Logistics, giving urban local governments' full authority. Research is needed to establish a connection between industry and pollution within buffers of India's industrial clusters.

2 Background

Indoor pollution is blamed for about seven million fatalities globally, with China and India having the highest death tolls. According to estimates from the World Health Organization (WHO), over 4.7 lakh deaths in India alone are attributed to indoor pollution [4]. India is quickly becoming an urban nation. Hence, more research in this field is required. Concerns concerning industrial areas and establishments are raised because the two countries with the most significant death rates are those whose economies are focused on manufacturing and services. The ESPOO Convention was established in Finland in 1991 and broadly addresses the impact of industrialization on health (ESPOO: UNECE Convention on Environmental Impact Assessment in a Transboundary Context) neither of India nor China are members [5].

Air quality measurements for fine particulate matter (PM2.5) in different nations and areas increased from 2018 to 2022. As seen in Fig. 2, the countries and regions are ranked according to their PM2.5 levels in 2022. India had the highest PM2.5 level (89.7 μg/m^3), followed by Bangladesh (65.8 μg/m^3) and Pakistan (70.9 μg/m^3). The World Health Organization's (WHO) recommended limit for PM2.5 levels (5.1–10 μg/m^3) was surpassed by these three countries by one to two times. Other nations that went above the limit by one or two times include Tajikistan, Kuwait, Iraq, Burkina Faso, Chad, and Kuwait. PM2.5 levels in the USA, Canada, and Australia were below the WHO threshold (<5.1 μg/m^3). Additionally, the data demonstrates how PM2.5 levels have changed over time in certain countries. For example, the PM2.5 level in Pakistan grew from 46.8 μg/m^3 in 2019 to 70.9 μg/m^3 in 2022, while the PM2.5 level in India declined from 83.3 μg/m^3 in 2019 to 53.3 μg/m^3 in 2022. Similarly, Bangladesh had a decrease in PM2.5 levels from 58.1 μg/m3 in 2019 to 65.8 μg/m^3 in 2022. There are a number of factors contributing to these countries' high PM2.5 levels, including industry, urbanization, and heavy traffic. High PM2.5 exposure can cause a few health concerns, such as cardiovascular diseases, respiratory disorders, and early mortality. Thus, it is imperative that these

0-5 Meets WHO guideline | 5.1-10 Exceeds by 1 to 2 times | 10.1-15 Exceeds by 2 to 3 times | 15.1-25 Exceeds by 3 to 5 times | 25.1-35 Exceeds by 5 to 7 times | 35.1-50 Exceeds by 7 to 10 times | >50.1 Exceeds by over 10 times

Rank	Country/Region	2022	2021	2020	2019	2018	Population
1	Chad	89.7	75.9	--	--	--	17,179,740
2	Iraq	80.1	49.7	--	39.6	--	43,533,592
3	Pakistan	70.9	66.8	59	65.8	74.3	231,402,117
4	Bahrain	66.6	49.8	39.7	46.8	59.8	1,463,265
5	Bangladesh	65.8	76.9	77.1	83.3	97.1	169,356,251
6	Burkina Faso	63	--	--	--	--	22,100,683
7	Kuwait	55.8	29.7	34	38.3	56	4,250,114
8	India	53.3	58.1	51.9	58.1	72.5	1,407,563,842
9	Egypt	46.5	29.1	--	18	--	109,262,178
10	Tajikistan	46	59.4	30.9	--	--	9,750,064

Fig. 2 Pollution based on PM2.5 concentration (μg/m^3) (Polluted countries, 2022)

nations act to lower PM2.5 emissions and enhance air quality. It can be guaranteeing a more sustainable and healthy future for everybody if people take this action.

When compared to other countries, India has the most polluted cities worldwide. While the third interim WHO limit is only 15, there are claims that the PM2.5 concentration in some cities can approach 135 εg/m^3. Also, according to the WHO, 15 of India's top 20 most polluted cities are isolated. The reasons for the increased particulate matter are growing industry and vehicle development [6]. The industrial GDP in future economic plans is increasing in the coming years, as per Niti Ayog [7]; it caused the nation with the youngest population to have different economic and environmental policies where the northern strip of the Indian subcontinent has the majority of the highly polluted cities. The most current draft EIA, which considered air-generated toxins in its risk assessment reports, demonstrated the necessity of environmental activism [8]. Interactive global map of 2022 PM2.5 concentrations by city is shown in Fig. 3.

The objectives of this paper are multifaceted, seeking to comprehensively investigate and compare EIA practices at both international and national levels, with a particular emphasis on India. Firstly, the study aims to provide an in-depth analysis of the historical development and evolution of international EIA standards and frameworks. It will delve into the examination of key principles and methodologies outlined in international EIA guidelines, shedding light on the global best practices that form the foundation of environmental impact assessments. Additionally, the research will investigate the legal frameworks and institutional structures

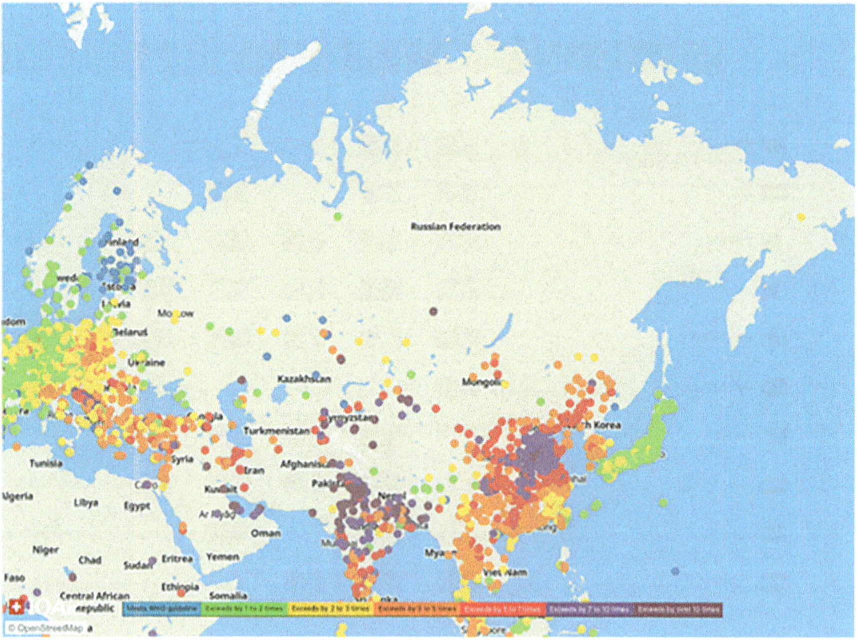

Fig. 3 Interactive global map of 2022 PM2.5 concentrations by city (PM2.5 concentrations, 2022)

supporting EIA at the international level to understand the overarching governance mechanisms. Furthermore, the study will assess the application and implementation of these principles within the Indian regulatory context, identifying challenges and successes. Through a comparative lens, the research will scrutinize the extent to which international EIA standards have been integrated into India's national regulatory framework. Ultimately, the paper seeks to offer recommendations for enhancing EIA practices in India by leveraging insights from international best practices, contributing to the broader discourse on sustainable development and environmental governance.

2.1 *Acts and Policies About Environment and Industries*

Acts discussed in Fig. 4 delineate a comprehensive timeline spanning from 1853 to 2020, encapsulating India's trajectory of environmental legislation and initiatives. Figure 4a, b represents the timeline of Indian laws and development goals, respectively. The timeline unfolds with milestones such as the "Shore Nuisance (Bombay

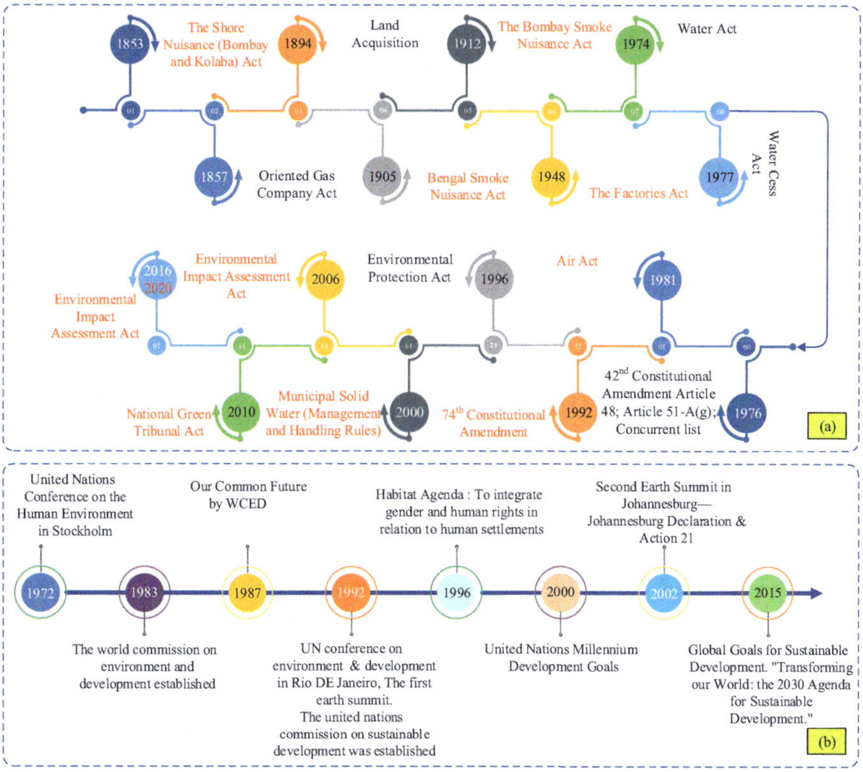

Fig. 4 Timeline of (**a**) Indian laws and (**b**) development goals

and Kolaba) Act" in 1853, embedding environmental considerations into the constitutional framework.

Noteworthy legislation, including the "Air Act" (1987) and the "Environmental Protection Act" (1986), underscores India's commitment to addressing diverse environmental challenges. Concurrently, global engagements are evident in commencing with the United Nations Conference on the Human Environment in 1972 and culminating in the formulation of the Sustainable Development Goals in 2015. The timeline concludes with a focus on twenty-first-century legislation, featuring acts such as the "Municipal Solid Waste (Management and Handling) Rules" (1992), "Environmental Impact Assessment Amendment" (2006), "National Green Tribunal Act" (2010), and the "Environmental Impact Assessment Act" (2016/2020). Temporal progression provides a visual narrative of India's legislative journey, dynamically responding to evolving environmental imperatives.

The EIA has become an essential process worldwide for evaluating the potential environmental effects of proposed development projects. The genesis of EIA as a formal procedure dates back to the 1960s, with the USA leading the way by requiring EIAs for all major federal projects through the National Environmental Policy Act of 1969. Over the following decades, several countries adopted EIA legislation, with Australia doing so in 1974, France in 1976, Brazil in 1977, and many others following. The World Bank's adoption of EIA in its policies in 1984 was a significant endorsement, demonstrating the international community's growing recognition of environmental considerations in development projects. In 1992, the United Nations emphasized the importance of EIA by declaring it a national instrument, further solidifying its role in environmental management the whole timeline is explained in Fig. 5a. In Fig. 5b, the Evolution of EIA: India's response to environmental change is shown.

In India, the trajectory of EIA has been shaped by its response to environmental challenges, notably after the Bhopal gas tragedy. The Department of Environment was established in the 1980s and became a full-fledged ministry in 1985. The Environmental Protection Act passed in 1986 laid the groundwork for environmental governance. The 1990s saw further developments with the Ministry of Environment and Forests issuing guidelines for EIAs, particularly for river valley projects. The legislation continued to evolve with the National Environment Tribunal Act in 1995, and the process was strengthened by making public hearings mandatory for all projects to which EIA notification applied in 1997. India continued to refine its EIA framework with subsequent notifications to update and improve the EIA process in 2006, 2016, and 2020, demonstrating an ongoing commitment to environmental management.

In regularizing the pharmaceutical industry, the following regulations and no objection certificate should be obtained from different national level and state-level agencies; the existing industrial clusters in the developing nations are still following the age-old traditional industrial policy where the process of environmental activism is still paper-oriented which made the laws and rules to go in a slow pace, the establishment of EIA acts globally and locally and is explained over a timeline in Fig. 5a. In Fig. 5b [9, 10].

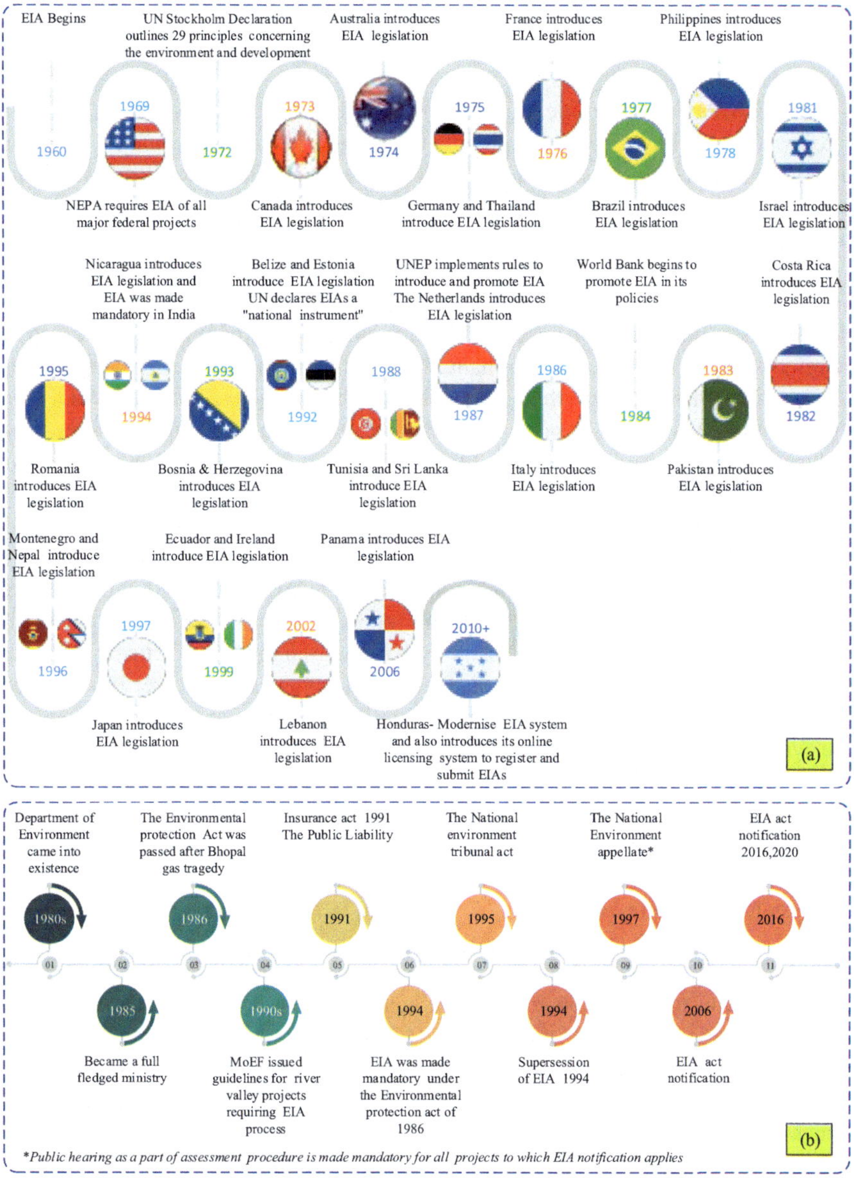

Fig. 5 The evolution of EIA: (**a**) A global perspective and (**b**) India's response to environmental change

The EIA Act 2006 notification replaced the EIA Act 1994, where additions and deletions were done to the existing information where more power was lived with the local authorities under which these industrial clusters come under. This notification gave rise to the existing industries, and several ancillary industries started more production orientation than without a limit. Living power to the local authority regime through the 74th and 73rd amendment left these grey parts in the metropolitan areas more polluted [11].

The EIA Act 2016 notification further added building environmental auditors into the sector with 1.5–3 lakh sqm area under the state authority and >3 lakh sqm under the union government for building permissions. These notifications mainly concentrated on the urban township's energy, water auditing, and professional pollutant impact assessment procedures in the industrial clusters [12].

One perspective holds that living in a safe environment is a fundamental right. When there is a right to equality for different resources across the nation, it means that there should be equality in terms of various means without any caste, creed, or religious bias. For instance, a merchant in the leather tanning industry who is trying to prepare leather goods for the public at a lower cost may live in an industrial plant vicinity to lower the transportation cost and giving maximum hours to work. Industrial clusters used to be located far from the city back in the 1980s, but in the current times, urban outgrowth had taken over the area, and the homes that were previously away to the industries now cover the industrial vicinities, if the industry's standard chimney height is 15 m as per Gaussian plume calculation which is based on wind speed, direction, and temperature but in the current context people moved to urban areas in search of equal income opportunities and occupied the buffer spaces around these industrial clusters thus putting larger mass of the immigrated population. The local government planned and designated areas over time to meet the needs of all parties involved in the land market. The area's unique characteristics are evident before any market is established, and the environmental legalities aspect is given minimal weight in the development plan preparation process. When starting a project, the factors considered are primarily mechanical, and even industrial laws and other matters are never scoped upon in SIA (Social Impact Assessment). The process of industrial permit works as a simple EIA documentation is provided if the industry is taking more land.

Let us discuss this issue in detail the industries are categorized majorly into two types, which are category A and category B, where category B is again divided into more two subcategories, category B1, and category B2, where only these category B industries require EIA documentation [13]. The fact that SIA is not included in any of the EIA regulations throughout the entire process demonstrates the lack of connection between the issues raised now and the plans created for the people, which are unable to address the problems of the people living around these industries.

The prime law that came into existence after industrialization during the start of the eighteenth century was the law of torts as shown in Fig. 6, which first started in England and then was passed throughout the world. Law of torts comes into the case when the duty of person A affects the rights of the other person B where the responsibility would be of any sort where the work of person A would carry the wrongful omission which would cause damage to the other person B and the person A will be

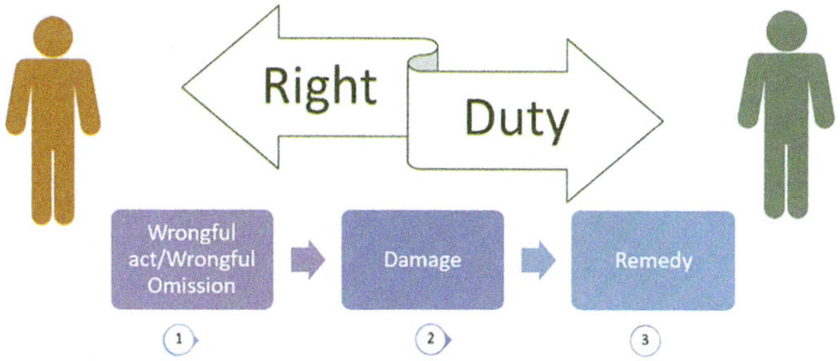

Fig. 6 Exploration of Law of Torts and its origins

in a position to pay the remedy to the person B. This was given by the cases as discussed by the (i) Absolute liability and (ii) Strict liability and there are few exemptions to the same.

2.1.1 Absolute Liability

In simple words, the rule of absolute liability can be defined as the rule of strict liability minus the exceptions. In India, the rule of absolute liability evolved in the case of MC Mehta v Union of India (1984–1999), subject of Bhopal Gas Tragedy case. Public Liability Insurance Act in the year 1991, let the principle of Absolute liability be explained by the following instances in which there used to be only strict liability during start of the law of torts but over the period of time defendant started supporting to his views by supporting that people had come into their premises which makes the liability void or null some times. Though in the case of the Bhopal gas tragedy, the gas leaked from the industrial premises, and the defendant shifted the blame onto the permitted authority for negligence in excess chemical storage. In the subsequent years, especially after 1986, it was proven that the precautionary measures followed by Union Carbide India Limited, the plaintiff in this case, were inadequate compared to industry standards abroad. The consequences of such accidents continue to be visible even 25 years after the incident [1, 14, 15]. A reconnaissance survey was conducted at the affected vicinity next to union carbide site location, resulting in over 16,000 deaths in the year 1986. The settlement condition is studied and the identifications were made as depicted in Fig. 7. The questions asked regarding the settlement and land rights, where the respondents when asked about "Why are they not willing to settle down somewhere else? as they still suffer because of the 1986 effect. They said "Getting the space in which they are currently living, will cost their 10 years of savings and they said as the industry is closed down they don't have any problem with this area still knowing it is going to affect them" though the soil concentrations showed very high alarming values within this vicinity.

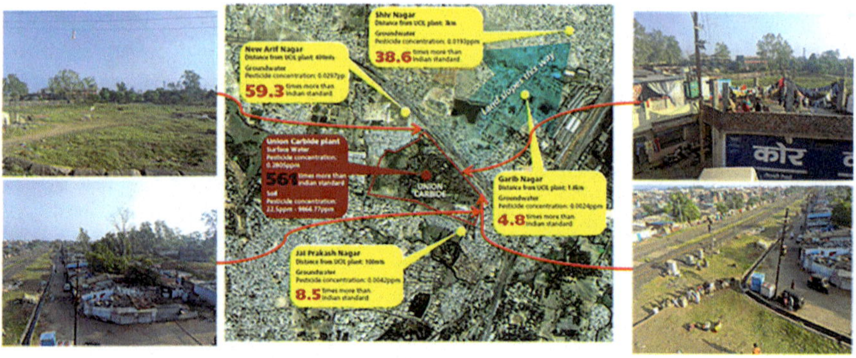

Fig. 7 Bhopal gas tragedy impact area. (Source: Author)

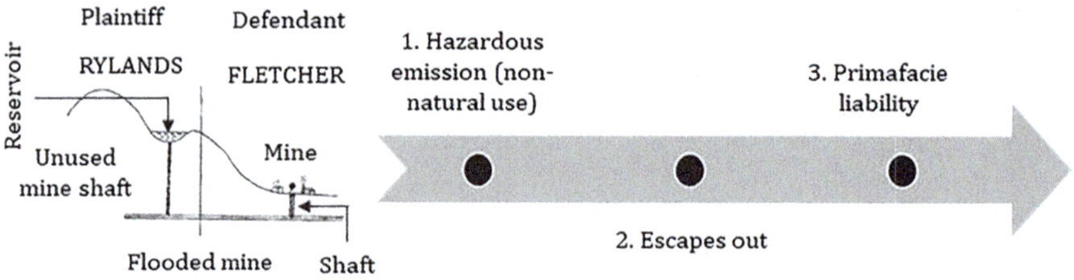

Fig. 8 The genesis of strict liability: analysing the landmark case of Rylands Vs Fletcher and its impact on hazardous release management

2.1.2 Strict Liability

The principle of strict liability evolved in the case of Rylands v Fletcher in 1868. The principle of strict liability states that any person who keeps hazardous substances on his premises will be held responsible if such substances escape the premises and cause any damage. In the case of Rylands and Fletcher, Rylands had a mill which needed the production of energy, where he stored the water within his vicinity but during heavy rain, the water seeped to the downward side where Fletcher had his cattle and agricultural land. The sudden flooding created the huge animal and agricultural loss, where the defendant charges the plaintiff in this case where any non-natural use emission when escapes out from the plaintiff vicinity. The case law is explained in Fig. 8; it becomes the prima facia liability where the defendant can sue the plaintiff, in any case. This is also called no-fault liability, where the following cases are the exceptions in the strict liability.

- Plaintiffs Own Fault (Pointing Vs Noakes (1849) 2QB 281)
- Act of God (Nicolas Vs Marsland)
- Mutual Benefit (Voluntary Not Fit Injuria)
- Act of Stranger
- Statutory Authority

Overall, the strict liability states that the amount of damage caused is equal to the amount of production done by the industry.

Case 1 The case of Mystery Gas Spreads Panic v. The Director, A.P. Pollution Control Board, Hyderabad and Others. (Sudden Panic in the surrounding settlements caused the high court to take the su-mo-to where the defendant was proven guilty).

Case 2 V.S. Damodaran Nair v. State of Kerala (To understand fully the extent to which the Pollution is predominant, NEERI (National Environmental Engineering Research Institute) was initiated into the study and recommendations were strictly followed).

Case 3 P.C. Cherian v. State of Kerala (The defendant runs small-scale rubber industry, and the soot coming out from this plant is causing a nuisance. In the nearby church vicinity where the defendant stated that carbon does not cause any health hazard to which the local court sued the defendant in view of the community health).

Case 4 Rukhiya Beevi v. State of Kerala (woman who had failed to gain consent in respect of an industrial plant. She contended that she did not require consent as she was not causing any pollution to be emitted. The Kerala State Pollution Control Board conducted an inspection of her plant and categorized it as the most polluting red category and the word Industrial plant is redefined because of the same case as small-scale industries can also be hazardous).

2.2 Air Pollution Modelling

Air pollution modelling is classified differently based on the purpose and method used, where the modelling follows sheer mathematical functions in creating a 3D or a 2D models of the pollutant concentrations. The model uses average emission data, wind velocity, and other time-dependent parameters to prepare an air quality model [16]. The mathematical models use different functions to show the concentration heat maps. The mathematical modelling includes different functions where the parameters change based on the different gas combinations, vertical temperatures which decide the flow direction of the gas and wind velocity fluctuations. These parameters are further altered in the urban areas due to the presence of the gradient surfaces [17, 18] which changes the gradient velocity vertically where the unevenness in the urban fabric gives rise to more grey spots, creating lesser velocity gradient spaces throughout the urban areas.

The gradient wind changes within the boundaries of the city where the velocity gradient as per Table 1 shows there is less discussion about the following case studies independently and methods used the presence in the external atmosphere of one or more contaminants (pollutants) or combinations thereof in such quantities and of such duration as may be or may cause injury to human health, plant or animal life

Table 1 Basics of air pollution and legislation (Authors compilation from NOAA, URDPFI 2016)

Level	Vertical scale	Time scale	Level of action	Map scale
Local	Height of stack	Hours	Municipal scale/sub-regional	1:2000
Urban	Lower kilometres	Days	District scale/regional	1:50,000
Regional	Troposphere	Months	State/national scale	1:100,000
Continental	Stratosphere	Years	National /international scale	1:150,000
Global	Atmosphere	Decades	International scale	1:250,000

or property (materials), or which unreasonably interfere with the comfortable enjoyment of life or property or the conduct of business [19].

3 Conclusion

In conclusion, the nuanced interaction between industry, urbanization, and environmental repercussions in India reveals a complex interplay of advantages and difficulties. Unquestionably, the development of many industrial clusters has fuelled economic expansion, created job possibilities, and significantly increased the country's GDP. Nonetheless, given the ongoing difficulties and the recent industrial mishaps in India and worldwide, a more coordinated and aggressive strategy is required to address the detrimental effects of industrial activity.

The dual nature of industrialization is highlighted by the socioeconomic division observed in urban environments, particularly in metropolitan regions. While it generates employment, it exacerbates issues such as pollution, greenhouse gas emissions, and hazards to public health. A major barrier to sustainable development is the absence of a standardized framework for calculating and mitigating these negative impacts.

The fact that India is the most polluted nation in the world draws attention to the serious issue of indoor pollution with more than 388 industrial clusters existing and more are coming up in and around major cities across the subcontinent which highlighted necessity of conducting in-depth studies on this problem and actively participating in international conventions that address the negative health effects of industrialization. As the EIA 2020 suggested the use of air pollution modelling in its documentation process for extension of industrial permits and on the other side nation increasingly urbanizes, practical approaches and policies are badly needed to assess and mitigate the negative social and environmental repercussions of industrial growth.

The reality that India needs the basic understanding of concept of liabilities and need a clear definition of plaintiff (The one who brings the case against damages) and defendant (The person who is being sued) as due to problem of excessive urbanization people encroach into these spaces and industries become the plantiff supported by the urban local body and making the problem more critical in larger scale.

References

1. Mishra, P. K., et al. (2009). Bhopal gas tragedy: Review of clinical and experimental findings after 25 years. *International Journal of Occupational Medicine and Environmental Health, 22*, 193–202.
2. Panjala, P., Gumma, M. K., Mesapam, S., Shukla, A. K., & Pignatta, G. (2025). Geospatial analysis of crop residue burn areas and their dates for emission mitigation strategies. *Sustainability, 17*(6), 2508. https://doi.org/10.3390/su17062508
3. Shrikant.umrikar. *List of industrial centres in India*. Wikipedia.
4. Drisse, M.-N. B., Goldizen, F., & Adair-Rohani, H. (2017). *Atlas on children's health and the environment*.
5. Gupta, A., & Shukla, A. K. (2024). Optimal approaches in global warming mitigation and adaptation strategies at city scale. *Discover Sustainability, 5*(1), 272. https://doi.org/10.1007/s43621-024-00497-8
6. Guttikunda, S. K., Nishadh, K. A., & Jawahar, P. (2019). Air pollution knowledge assessments (APnA) for 20 Indian cities. *Urban Climate, 27*, 124–141.
7. Niti. (2014). Chapter 2. Five year plans and SSIs in Indian economy.
8. EIA 2020 Draft. 804 (2020).
9. Eia, T., & Manual, G. (2009). *Technical EIA guidance manual*.
10. IISD. (2016). *Environmental impact assessment training manual*.
11. Act, P., Act, A. & Acts, P. (1905). *The Bengal Smoke-Nuisances Act, 1905*.
12. Sand, S., et al. (2013). *Draft EIA/EMP Report*.
13. Gupta, A., De, B., Shukla, A. K., & Pignatta, G. (2024). Vulnerability assessment of a highly populated megacity to ambient thermal stress. *Sustainability, 16*(8), 3395. https://doi.org/10.3390/su16083395
14. Sriramachari, S. (2004). The Bhopal gas tragedy: An environmental disaster. *Current Science, 86*, 905–920.
15. Malla, T. M., Senthilkumar, C. S., Sharma, N. C., & Ganesh, N. (2011). Chromosome instability among Bhopal gas tragedy survivors. *American-Eurasian Journal of Toxicological Sciences, 3*, 245–249.
16. Pullen, J., Boris, J. P., Young, T., Patnaik, G., & Iselin, J. (2005). A comparison of contaminant plume statistics from a Gaussian puff and urban CFD model for two large cities. *Atmospheric Environment, 39*, 1049–1068.
17. Venkatram, A., Isakov, V., Pankratz, D., & Yuan, J. (2005). Relating plume spread to meteorology in urban areas. *Atmospheric Environment, 39*, 371–380.
18. Tominaga, Y., & Stathopoulos, T. (2013). CFD simulation of near-field pollutant dispersion in the urban environment: A review of current modeling techniques. *Atmospheric Environment, 79*, 716–730.
19. Summary, E., et al. (1996). Health effects of outdoor air pollution. Committee of the Environmental and Occupational Health Assembly of the American Thoracic Society. *American Journal of Respiratory and Critical Care Medicine, 153*, 3–50.

Applications of Remote Sensing and GIS Techniques in Monitoring and Assessment of Land Surface Temperature (LST)

Gowri Shenoy B. and Anoop Kumar Shukla

1 Introduction

Nearly 50% of the world's inhabitants have been living in cities over the last few years, as shown in Fig. 1. According to the UN Report 2009, urbanization will increase, with 590 million people staying in cities by 2030. "Urbanization is considered one of the important factors of change in land cover" [1–4]. The rise in "population growth has caused uncontrolled urban growth" [5–11] to provide shelter, i.e., sprawl, which will result in a few identified issues, such as unlicensed settlements, environmental degradation, "conversion of agricultural lands to buildings" [12], reduction of forest area and surface water bodies, and overall leads to a change in land cover and land use permanently [13, 14]. The rapid rise in population in recent years has resulted in urbanization faster than usual at the global level. Even if urbanization has many positive impacts on the growth of cities in developing countries, unplanned growth affects the quality of life and the overall environment [15, 16]. Biodiversity is affected by city expansion and the fall of green cover, which leads to variations in surface temperature. Unplanned urban extension will comprise the urban green cover to accommodate impervious surfaces like buildings, roads, and other infrastructure requirements. Expansion in urban areas will affect the geographical size of cities and the fabric of the urban regions. Cities expansion should protect natural resources to avoid the destruction of natural environments and preserve the ecosystem.

Figure 1 depicts the increase in urban population over the years. Graph mentions the rise in population in urban area in both percentage and numbers.

Gowri Shenoy B. · A. K. Shukla (✉)
Manipal School of Architecture and Planning, Manipal Academy of Higher Education, Manipal, Karnataka, India
e-mail: anoop.shukla@manipal.edu

S. Pathak et al. (eds.), *Geo-Data Revolution*, Advances in Geographic Information Science, https://doi.org/10.1007/978-3-031-94078-1_8

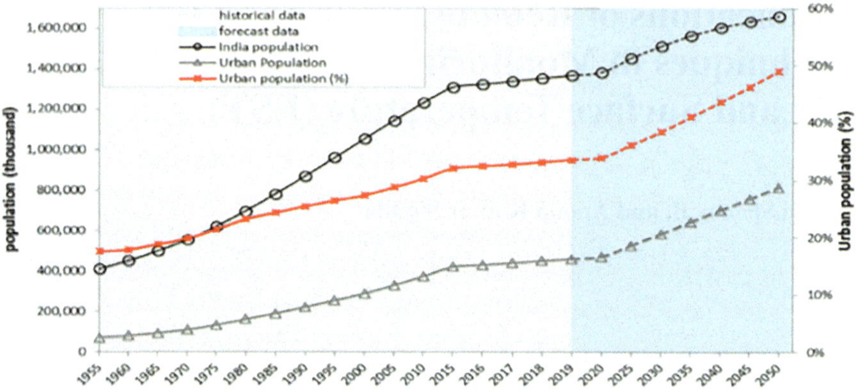

Fig. 1 Graph of rise in urban population. (Source: [15])

LST, i.e., land surface temperature, is a measure to monitor vegetation, buildings, and overall climate change [16]. "Monitoring variations in LST and analysing its relationship with LULC maps" [17] is required for a sustainable urban environment and to mitigate temperature variations from urban growth. There will be variation in temperature in clusters of urban areas compared to the outer periphery. Moreover, solar radiation, absorption, thermal capacity, and thermal conductivity will be higher in urban areas due to fewer green areas and more concrete areas. In line with this, it's essential to be monitored to understand the impact of urbanization and land cover changes in temperature variations at surface level. Variations in day and night temperature and temperature differences between clusters are due to the generation of urban heat. This temperature difference between clusters is termed as UHI, i.e., urban heat island. "UHI is also formed because of changes in LULC" [18] like the reduction of green areas and the land surface being transformed due to anthropogenic activities. UHI is defined as "a temperature increase in built-up areas compared to their rural surroundings caused by human activities" [19]. Manely, in 1953, termed the UHI effect. Due to rapid urban growth and climate change, urban heat island effects have increasingly impacted people's output and life. Urban heat islands are one of the major environmental concerns for society [20, 21].

UHIs created on surfaces quantify temperature at the land surface and are termed surface urban heat islands. Surface Urban Heat Island, i.e., SUHI, is "the temperature of the entire 3-D envelope of a surface" [21, 22]. SUHIs are usually present both at day and night [23]. However, their effect will be more during the daytime due to sun radiation and accumulation of heat at the city level. During the nighttime, the heat is released within the built-up areas due to the loss of cooling effect, as there will be less or a lack of green spaces within the neighborhoods.

Figure 2, the rise in population causes urbanization, which tends to change land use and transformations of new land surfaces. This will create differences in surface temperature and the creation of SUHI. Due to variations in sun intensity and land cover, there will be changes in urban heat island magnitude for different seasons. Surface heat islands are usually stronger and higher in summer. "SUHI Urban Heat

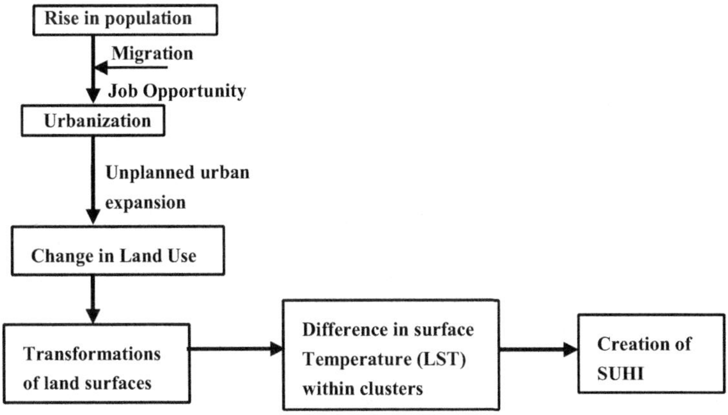

Fig. 2 Flowchart of the relationship between urbanization and surface urban heat islands. (Source: Author)

Island is measured by land surface temperature [23]" using remote sensing techniques [23].

The book chapter explores and demonstrates the effective utilization of RS and GIS tools in monitoring and assessing LST patterns and trends. The chapter also attempts to provide a broad understanding of LST dynamics influenced by urbanization and land use changes. The chapter explains the fundamentals of RS and GIS tools in assessing and monitoring the LST dynamics. Further, the role of RS tools and GIS, along with the scope has been elaborated with various datasets required for assessing the LST. This chapter will help to understand the importance of RS and GIS tools in the assessment of LST and formulate strategies for SUHI to enhance and protect the environment.

2 Fundamentals of RS and GIS Tools

GIS applications are more helpful for collecting data faster than other manual methods like surveying, observation tools, etc. GIS is referred to as an advanced tool compared to traditional tools for its exclusive descriptions and resources [23]. Unlike traditional tools like observational surveys, interviews, etc., GIS effectively combines the maps with graphs or images [23]. Hence, spatial integration will help users derive fine clarity from its data since it will be referred to as geographic reference. GIS survey provides efficient data collective tools compared to traditional methods. Any geographical changes will be recorded in GIS very precisely. GIS will help users to analyze data spatially more precisely than traditional tools.

GIS is a very effective surveying tool in decision-making, restoration projects, and handling spatial data effectively with precise, complex analysis [23]. However, mapping urban areas and urban expansion, i.e., an increase of impervious surfaces, reduction in green cover, and other analyses using remote sensing and geographic information systems [23] is effective and extensively used worldwide. In recent years, a combination of remote sensing, geographic information systems, and global positioning systems can be considered effective tools for "mapping urban areas, modelling urban growth, and assessing land use/land cover changes" [24]. Hence, GIS can be a useful analytical tool in urban planning to address urban issues [25].

RS, a modern, helpful technology tool, is used to prepare satellite images digitally through large quantities of raw data and information and made accessible to researchers [26, 27]. Spatial and temporal studies and their analysis using traditional methods of surveying will be time-consuming, and not all data will be available [28]. Hence, remote sensing will be an advanced tool for the collection of data in a faster mode. For any study related to urban planning, it's very significant to realize and analyze the changes and patterns of a city's growth due to urbanization. Moreover, urban growth "can be effectively analysed using Geographic information systems, a remote sensing tool along with available ground data" [29–32]. Mapping of land cover data has usually been done by remote sensing tools for many years [33].

3 Role of RS and GIS Tools in the Assessment of LST

RS tools like GIS play a role in measuring and monitoring land surface temperature by providing relevant data for spatial and temporal variations [33] analysis. The RS approach can help to understand two aspects of LST: (i) the effects of change in land use on LST and (ii) the impact of LU change and variation in LST on the local environment. The RS approaches for assessing LST will signify the importance of understanding the implications of surface temperature changes at the meso, macro, and micro levels. The data collection for LST [33] through RS tools will allow precise assessment of UHI effects and land use changes. Hence, RS data will help to monitor LST at different scales.

Using a variety of sensors with different spatial and temporal resolutions, LST can be effectively monitored and assessed with its "variations along with urban heat islands and changes in land use and land cover" [35]. Normally, satellite sensors capture thermal data, which is helpful in calculating land surface temperature calculations. Moreover, GIS data, along with "land use and land cover indices, will be essential data" [35] to compute "the relationship between Land surface temperature and environmental factors" [36]. The traditional method, the ground meteorological method for measuring LST, will have limitations in accurately assessing the intensity of LST [37]. The satellite data will give precise information for analyzing the LST patterns. Moreover, there is a need for actual and accurate data to mitigate the variations of "LST due to land use cover changes in urban areas" [38]. The datasets from remote sensing tools will play a major role in assessing the LST [39]. Remote sensing datasets are essential in assessing the "variations in land surface temperature, its impact on land use, land cover dynamics" [23] and environmental factors. These datasets will help monitor and analyze seasonal temperature fluctuations, as mentioned in Table 1. The various datasets of RS are listed in Table 1, in which Landsat and MODIS are commonly used to study vegetation, climatic patterns, UHI, and LST variations are influencing the surface temperature.

However, Land Surface Temperature through Landsat or any other satellite data can be retrieved from the following algorithms as mentioned in Table 2. The above table explains how LST can be retrieved using RS tools with the help of algorithm mentioned for each quantified unit related to LST. LST can be retrieved by using parameters like spectral radiance (pixel value), brightness temperature, NDVI, and fractional vegetation (vegetation), through which surface emissivity data can be retrieved. However, LST is retrieved in the following sequence [47], as mentioned in the steps below:

1. Obtain the reflectance data for red and NIR (near-infrared) to compute NDVI. NDVI is the RS index to measure vegetation.
2. Calculate the fractional vegetation using the values of NDVI. Fractional vegetation FV is the surface area that quantifies the area of vegetation.
3. Next, estimate the surface emissivity using derived fractional vegetation cover through NDVI values. Surface emissivity is useful to understand the thermal radiation emitted to the surface.
4. Further, the brightness temperature will be derived from spectral radiance.
5. With reference to emissivity, brightness temperature will be converted to LST.

Table 1 List of sensors used in remote sensing

Sl. no	Types of sensors	Spatial resolution	Inferences	Reference
1	Landsat Thematic Mapper (TM) and Enhanced Thematic Mapper Plus (ETM+)	30 m	Monitors UHI, vegetation, and land cover changes, which influence surface temperature. Analyze the temporal trends in LST. Identifies environmental changes at local, regional, and even global levels	[34]
2	Moderate Resolution Imaging Spectroradiometer (MODIS)	250–1000 m	Widely used for studying UHIs, land cover changes, and their impacts on surface temperature. Monitors LST variations (diurnal and seasonal), vegetation, and climate patterns	
3	National Oceanic and Atmospheric Administration (NOAA) Advanced Very High-Resolution Radiometer (AVHRR)	1.1 km	Monitors LST changes at a global scale Used for studying climate patterns, vegetation dynamics, and urban heat islands Useful in assessing the impact of land cover changes on surface temperatures	
4	IKONOS and QuickBird	Less than 1 m	Studying LST variations at a local scale Mapping land cover changes microclimatic variations, urban heat islands, and the impact of land use changes on surface temperatures	

Table 2 Land surface temperature data retrieved using RS

Quantified unit	Algorithm	Source
Land Surface Temperature, LST	$LST = \frac{TB}{[1+(\lambda \times TB/\rho)\ln\varepsilon]}$ where TB is band 6 in Landsat ETM+, and band 10 in Landsat OLI/TIRS; λ is the wavelength of emitted radiance (11.5 μm for Landsat ETM+ band 6 and 10.8 μm for Landsat OLI/TIRS band 10); ρ = h × c/σ (1.438 × 10–2 mK), σ = the Boltzmann constant (1.38 × 10–23 J/K), h = Planck's constant (6.626 × 10–34 Js), and c = velocity of light (2.998 × 108 m/s); and ε is the land surface emissivity.	[40–42]
TB = Brightness temperature	$Tb = \frac{K_2}{\ln((K_1/L\sigma)+1)} - 273.15$	[41, 43, 44]
Spectral Radiance	$L\lambda$ = 0.0003342*DN + 0.1 where DN is digital number of pixels in sensor	[44–46]
Surface Emissivity, S€	S€ = 0.004 Fv + 0.986; where Fv is Fractional vegetation	[43, 45, 46]
Fractional Vegetation, FV	$Fv = \frac{NDVI - NDVI_{min}}{NDVI_{max} - NDVI_{min}}$	[43, 45, 46]
Normalized difference vegetation index, NDVI	NDVI = (NIR – R)/(NIR + R) where NIR = near-infrared from Landsat data	[47–49]

4 Scope of RS Dataset in the Assessment of LST

Most of the previous studies have been conducted regarding LST in developed countries or metropolitan cities. We learned that various spectral data, which could be long term and based on spatial resolution, can be obtained from Landsat, Sentinel, MODIS, and other satellites. In line with the listed datasets in Table 1, other remote sensing tools, such as SPOT satellites, airborne sensors, and radar imaging, can be used for land surface temperature (LST) assessment. SPOT data can be used for regional studies due to medium spatial resolution, whereas radar sensors can be used for weather monitoring since sensors can access clouds and haze. Most of the studies depend on free and open data sources [48].

The methodological framework (Fig. 3) used to assess LST is initiated with data acquired from satellite images. After three steps of image preprocessing, "the map can be prepared using the object-based image analysis (OBIA) method" [50–52]. In the OBIA approach, "there will be two types of map classification, i.e., Supervised classification and Unsupervised classification" [52] ."Object-based image analysis (OBIA) is used to prepare LST through Land use land cover (LULC) maps" [48]. The classification indices for the OBIA approach used are: (1) The normalized

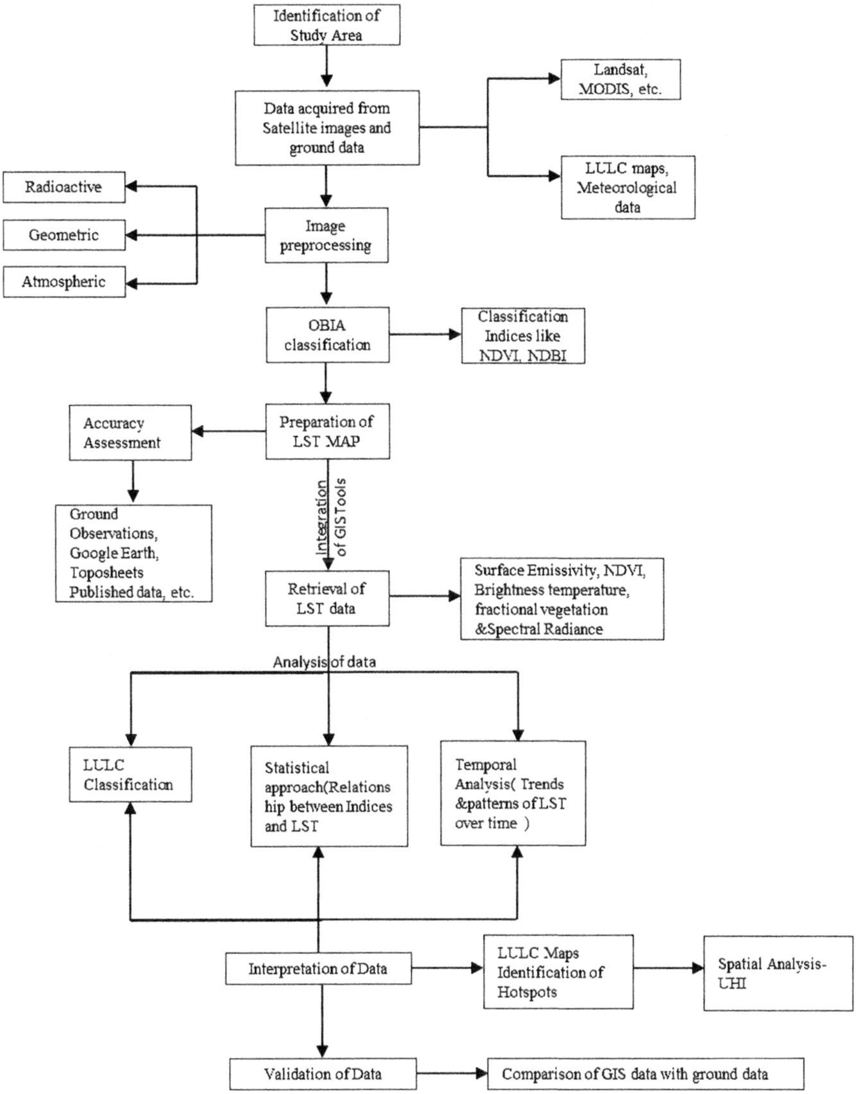

Fig. 3 Methodological framework used for assessment of LST using RS and GIS

Difference Vegetation Index (NDVI) assesses the quantity and quality of vegetation [48] in any identified area using remote sensing [53, 54] (2) "The Normalized Difference Built-Up Index (NDBI) is a remote sensing tool used to quantify the urban built-up area" [55]. Collection of data through satellite images can be through Landsat or MODIS satellite images. The data of LST is then retrieved through surface emissivity, fractional vegetation, etc., through remote sensing tools. Further, the data is analyzed, interpreted, and validated with the help of GIS applications.

Besides remote sensors, many other forms of GIS data can be used. Most of the datasets are collected from remote sensing; since they are derived directly without remote sensing tools, it can be classified differently. It is classified as spatial and nonspatial data. Spatial data consists of toposheets of local, regional, and global levels. In line with this, the spatial data also have digital elevation models, i.e., DEM, contour maps, land use maps, etc. Nonspatial data consists of "precipitation, temperature, wind speed, relative humidity, solar Radiation, etc., on daily, monthly, seasonal, and yearly data" [55], which can be collected from the local meteorological department.

5 Conclusion

In recent years, the urban environment has been affected due to urbanization. Remote sensing tools like GIS techniques will be essential to computing trends in Land Surface Temperature by identifying hot spots, evaluating the impacts of urbanization on surface temperature, and formulating mitigation strategies for urban heat island phenomena. The combinations of different satellite datasets can give an accurate and thorough understanding of changes in surface temperature and its dynamics with land use. Hence, this will help to formulate policies for mitigating climate change and environmental degradation.

Integrating data from multiple remote sensing platforms can formulate a holistic approach to understanding environmental changes and their impacts on surface temperature [55]. Monitoring the land surface temperature through remote sensing tools will be essential for urban planners and policymakers to formulate mitigation strategies effectively in city planning. As per the findings, the integration of remote sensing and GIS tools, combined with multiple data sources and open-source data source datasets like Google Earth Engine, significantly enhances the assessment of LST and achieves substantial environmental benefits.

References

1. Alsharif, A. A. A., Pradhan, B., Mansor, S., Zulhaidi, H., & Shafri, M. (2015, February). Urban expansion assessment by using remotely sensed data and the relative Shannon entropy model in GIS:: A case study of Tripoli, Libya. *Theoretical and Empirical Researches in Urban Management, 10*(1), 55–71. https://doi.org/10.2307/24873521
2. Deka, J., Tripathi, O. P., & Latif Khan, M. (2012, May). *Urban growth trend analysis using Shannon Entropy approach-a case study in North East India* (pp. 1072–1078). https://www.researchgate.net/publication/236623940
3. Weng, Q. (2001, January). A remote sensing? GIS evaluation of urban expansion and its impact on surface temperature in the Zhujiang Delta, China. *International Journal of Remote Sensing, 22*(10), 1999–2014. https://doi.org/10.1080/713860788

4. Gupta, A., De, B., Shukla, A. K., & Pignatta, G. (2024, April). Vulnerability assessment of a highly populated megacity to ambient thermal stress. *Sustainability (Switzerland), 16*(8). https://doi.org/10.3390/su16083395
5. Al-sharif, A. A. A., Pradhan, B., Shafri, H. Z. M., & Mansor, S. (2013). Spatio-temporal analysis of urban and population growths in tripoli using remotely sensed data and GIS. *Indian Journal of Science and Technology, 6*(8), 5134–5142. https://doi.org/10.17485/ijst/2013/v6i8.9
6. Grimm, N. B., Grove, J. M., Pickett, S. T. A., & Redman, C. L. (2008). Integrated approaches to long-term studies of urban ecological systems. In *Urban ecology: An international perspective on the interaction between humans and nature* (pp. 123–141). Springer. https://doi.org/10.1007/978-0-387-73412-5_8
7. Hedblom, M., & Söderström, B. (2008, January). Woodlands across Swedish urban gradients: Status, structure and management implications. *Landscape and Urban Planning, 84*(1), 62–73. https://doi.org/10.1016/j.landurbplan.2007.06.007
8. Yadav, N., Jianping Wu, J., Banerjee, A., Pathak, S., Garg, R. D., & Yao, S. (2024). Climate uncertainty and vulnerability of urban flooding associated with regional risk using multi-criteria analysis in western coast of India. *Environmental Research, 117962*, 117962.
9. Weng, Q., Liu, H., & Lu, D. (2007, June). Assessing the effects of land use and land cover patterns on thermal conditions using landscape metrics in city of Indianapolis, United States. *Urban Ecosystem, 10*(2), 203–219. https://doi.org/10.1007/s11252-007-0020-0
10. Al-Sharif, A. A., Abubakr, B. P., Mansor, S., & Shafri, H. (2015, February). Urban expansion assessment by using remotely sensed data and the relative Shannon entropy model in GIS: A case study of Tripoli. *Theoretical and Empirical Researches in Urban Management, 10*(1), 55–71.
11. Marzluff, J. M., et al. (2008). Urban ecology: An international perspective on the interaction between humans and nature, New York, USA
12. Jato-Espino, D., & Pathak, S. (2021). Geographic location system for identifying urban road sections sensitive to runoff accumulation. *Hydrology, 8*(2), 72.
13. Fang, J., Liu, M., Liu, W., Pathak, S., Li, S., Tang, X., Zhou, L., & Sun, F. (2020). Piloting a capital-based approach for characterizing and evaluating drivers of Island sustainability-an application in Chongming Island. *Journal of Cleaner Production, 261*, 121123.
14. Morefield, P. E., Fann, N., Grambsch, A., Raich, W., & Weaver, C. P. (2018, November). Heat-related health impacts under scenarios of climate and population change. *International Journal of Environmental Research and Public Health, 15*(11). https://doi.org/10.3390/ijerph15112438
15. Ramaiah, M., & Avtar, R. (2019, September 1). Urban green spaces and their need in cities of rapidly urbanizing India: A review. *Urban Science, 3*(3). https://doi.org/10.3390/urbansci3030094
16. Bagyaraj, M., et al. (2023, September). A study of urban heat Island effects using remote sensing and GIS techniques in Kancheepuram, Tamil Nadu, India. *Urban Climate, 51*, 101597. https://doi.org/10.1016/j.uclim.2023.101597
17. Ayanlade, A., Aigbiremolen, M. I., & Oladosu, O. R. (2021, December). Variations in urban land surface temperature intensity over four cities in different ecological zones. *Scientific Reports, 11*(1), 20537. https://doi.org/10.1038/s41598-021-99693-z
18. Mallick, J., Alsubih, M., Ahmed, M., Almesfer, M. K., & Ben Kahla, N. (2022, April). Assessing the spatiotemporal heterogeneity of terrestrial temperature as a proxy to microclimate and its relationship with urban hydro-biophysical parameters. *Frontiers in Ecology and Evolution, 10*. https://doi.org/10.3389/fevo.2022.878375
19. Bukovetz, J., Kochubovski, M., & Kaplan, G. (2023). *Urban heat island and green spaces in the city of Skopje: An environmental health approach combined with remote sensing data.* https://doi.org/10.21203/rs.3.rs-3193084/v1
20. Rose Amirtham, L., David Devadas, M., Rose, L. A., & Devadas, M. D. (2009). *Analysis of land surface temperature and land use/land cover types using remote sensing imagery-a case in Chennai city, India. Analysis of land surface temperature and land use/land cover types using remote sensing imagery-a case in Chennai city, India.* http://glcf.umiacs.umd.edu/index.shtml

21. Satheendran, S. S., Chandran, S. S., & Varghese, A. (2018, November). Space based spatio-temporal assessment of land surface temperature in Karunagappally municipality, a fast growing city in the western coast of India. *The International Archives of the Photogrammetry, Remote Sensing and Spatial Information Sciences, XLII–5*, 773–779. https://doi.org/10.5194/isprs-archives-xlii-5-773-2018
22. Mennecke, B. E., & Crossland, M. D. (1996). Geographic information systems: Applications and research opportunities for information systems researchers. In *Proceedings of the Annual Hawaii International Conference on System Sciences, IEEE Computer Society* (pp. 537–546). https://doi.org/10.1109/HICSS.1996.493249.
23. Rahman, A., Sen Roy, S., Talukdar, S., & Shahfahad (Eds.). (2023). Advancements in urban environmental studies. In *GIScience and geo-environmental modelling*. Springer. https://doi.org/10.1007/978-3-031-21587-2
24. Bhagyanagar, R., Kawal, B. M., Dwarakish, G. S., & Surathkal, S. (2012, September). Land use/land cover change and urban expansion during 1983–2008 in the coastal area of Dakshina Kannada district, South India. *Journal of Applied Remote Sensing, 6*(1), 063576–063571. https://doi.org/10.1117/1.jrs.6.063576
25. Sumandeep, Anju. (2017). *Identification of future urban development potential in Mangalore city based on Geographic Information System technology*. http://www.ijims.com
26. Al-shalabi, M., Billa, L., Pradhan, B., Mansor, S., & Al-Sharif, A. A. A. (2013, September). Modelling urban growth evolution and land-use changes using GIS based cellular automata and SLEUTH models: The case of Sana'a metropolitan city, Yemen. *Environment and Earth Science, 70*(1), 425–437. https://doi.org/10.1007/s12665-012-2137-6
27. Al-shalabi, M., Pradhan, B., Billa, L., Mansor, S., & Althuwaynee, O. F. (2013, June). Manifestation of remote sensing data in modeling urban sprawl using the SLEUTH model and brute force calibration: A case study of Sana'a city, Yemen. *Journal of the Indian Society of Remote Sensing, 41*(2), 405–416. https://doi.org/10.1007/s12524-012-0215-6
28. Al-sharif, A. A. A., & Pradhan, B. (2014, October). Monitoring and predicting land use change in Tripoli Metropolitan City using an integrated Markov chain and cellular automata models in GIS. *Arabian Journal of Geosciences, 7*(10), 4291–4301. https://doi.org/10.1007/s12517-013-1119-7
29. Li, W., Radke, J., Liu, D., & Gong, P. (2012). Measuring detailed urban vegetation with multisource high-resolution remote sensing imagery for environmental design and planning. *Environment and Planning. B, Planning & Design, 39*(3), 566–585. https://doi.org/10.1068/b37135
30. Moskal, L. M., Styers, D. M., & Halabisky, M. (2011, October). Monitoring urban tree cover using object-based image analysis and public domain remotely sensed data. *Remote Sensing, 3*(10), 2243–2262. https://doi.org/10.3390/rs3102243
31. Zhou, W., Troy, A., & Grove, M. (2006). Measuring Urban parcel lawn greenness by using an object-oriented classification approach. In *International Geoscience and Remote Sensing Symposium (IGARSS)* (pp. 2693–2696). https://doi.org/10.1109/IGARSS.2006.695
32. Shekhar, S., & Aryal, J. (2019, December). Role of geospatial technology in understanding urban green space of Kalaburagi city for sustainable planning. *Urban Forestry & Urban Greening, 46*, 126450. https://doi.org/10.1016/j.ufug.2019.126450
33. Khan, A. (2022). *Global urban heat Island mitigation*. Elsevier.
34. Ayanlade, A. (2017, July 03). Remote sensing approaches for land use and land surface temperature assessment: A review of methods. *International Journal of Image and Data Fusion, 8*(3), 188–210. https://doi.org/10.1080/19479832.2017.1299802
35. Sharma, A., & Vashishtha, D. (2023, April). Spatio-temporal assessment of land use land cover changes and their impact on variations of land surface temperature in Aligarh municipality. *Journal of the Indian Society of Remote Sensing, 51*(4), 799–827. https://doi.org/10.1007/s12524-022-01652-2
36. Sinha, S., Sharma, L. K., & Nathawat, M. S. (2015, December). Improved Land-use/land-cover classification of semi-arid deciduous forest landscape using thermal remote sensing. *Egyptian*

Journal of Remote Sensing and Space Science, 18(2), 217–233. https://doi.org/10.1016/j.ejrs.2015.09.005

37. Xin, J., Yang, J., Sun, D., Han, T., Song, C., & Shi, Z. (2022, August). Seasonal differences in land surface temperature under different land use/land cover types from the perspective of different climate zones. *Land (Basel), 11*(8). https://doi.org/10.3390/land11081122
38. Roy, P. S., et al. (2022, August 01). Anthropogenic land use and land cover changes—A review on its environmental consequences and climate change. *Journal of the Indian Society of Remote Sensing, 50*(8), 1615–1640. https://doi.org/10.1007/s12524-022-01569-w
39. Al Kafy, A., et al. (2021, January). Remote sensing approach to simulate the land use/land cover and seasonal land surface temperature change using machine learning algorithms in a fastest-growing megacity of Bangladesh. *Remote Sensing Application, 21*, 100463. https://doi.org/10.1016/j.rsase.2020.100463
40. Wang, R., Hou, H., Murayama, Y., & Derdouri, A. (2020, February). Spatiotemporal analysis of land use/cover patterns and their relationship with land surface temperature in Nanjing, China. *Remote Sens (Basel), 12*(3). https://doi.org/10.3390/rs12030440
41. Liu, W., Ji, C., Zhong, J., Jiang, X., & Zheng, Z. (2007). Temporal characteristics of the Beijing urban heat Island. *Theoretical and Applied Climatology, 87*(1–4), 213–221. https://doi.org/10.1007/s00704-005-0192-6
42. Cristóbal, J., Jiménez-Muñoz, J. C., Prakash, A., Mattar, C., Skoković, D., & Sobrino, J. A. (2018, March). An improved single-channel method to retrieve land surface temperature from the Landsat-8 thermal band. *Remote Sensing, 10*(3). https://doi.org/10.3390/rs10030431
43. Guha, S., Govil, H., Dey, A., & Gill, N. (2018, January). Analytical study of land surface temperature with NDVI and NDBI using Landsat 8 OLI and TIRS data in Florence and Naples city, Italy. *European Journal of Remote Sensing, 51*(1), 667–678. https://doi.org/10.1080/22797254.2018.1474494
44. Siddique, M. A., Wang, Y., Xu, N., Ullah, N., & Zeng, P. (2021, November). The spatiotemporal implications of urbanization for urban heat islands in Beijing: A predictive approach based on ca–Markov modeling (2004–2050). *Remote Sensing, 13*(22). https://doi.org/10.3390/rs13224697
45. Ullah, N., Siddique, M. A., Ding, M., Grigoryan, S., Zhang, T., & Hu, Y. (2022, April). Spatiotemporal impact of urbanization on urban heat island and urban thermal field variance index of Tianjin city, China. *Buildings, 12*(4). https://doi.org/10.3390/buildings12040399
46. Macarof, P., & Statescu, F. (2017, November). Comparison of NDBI and NDVI as indicators of surface urban heat Island effect in Landsat 8 imagery: A case study of Iasi. *Present Environment and Sustainable Development, 11*(2), 141–150. https://doi.org/10.1515/pesd-2017-0032
47. Ullah, N., et al. (2023, February). The impact of urbanization on urban heat island: Predictive approach using Google earth engine and CA-Markov modelling (2005–2050) of Tianjin city, China. *International Journal of Environmental Research and Public Health, 20*(3). https://doi.org/10.3390/ijerph20032642
48. Pandey, M., Gupta, A. K., & Oliveto, G. Disaster resilience and green growth. In A. K. Gupta, S. V. R. K. Prabhakar, A. Surjan (Series Eds.), *River, sediment and hydrological extremes: Causes, impacts and management*.
49. Kumar Shukla, A., et al. (2018, September). Population growth, land use and land cover transformations, and water quality nexus in the Upper Ganga River basin. *Hydrology and Earth Systems Science, 22*(9), 4745–4770. https://doi.org/10.5194/hess-22-4745-2018
50. Shukla, A. K., Ojha, C. S. P., Garg, R. D., Shukla, S., & Pal, L. (2020, October). Influence of spatial urbanization on hydrological components of the upper ganga river basin, India. *Journal of Hazardous, Toxic, and Radioactive Waste, 24*(4). https://doi.org/10.1061/(asce)hz.2153-5515.0000508
51. Wu, C., et al. (2019, December). Understanding the relationship between urban blue infrastructure and land surface temperature. *Science of the Total Environment, 694*, 133742. https://doi.org/10.1016/j.scitotenv.2019.133742

52. Huang, S., Tang, L., Hupy, J. P., Wang, Y., & Shao, G. (2021, February 01). A commentary review on the use of normalized difference vegetation index (NDVI) in the era of popular remote sensing. *Journal of Forestry Research, 32*(1). https://doi.org/10.1007/s11676-020-01155-1
53. Pandee,·C. B., Moharirr,·K., & Ahmeddelbeltagii, S. Q. editors. *Climate climate change impacts on natural resources, ecosystems and agricultural systems*. Springer
54. Allawi, M. F., Jaafar, O., Mohamad Hamzah, F., Abdullah, S. M. S., & El-shafie, A. (2018, May 01). Review on applications of artificial intelligence methods for dam and reservoir-hydro-environment models. *Environmental Science and Pollution Research, 25*(14), 13446–13469. https://doi.org/10.1007/s11356-018-1867-8
55. Nega, W., & Balew, A. (2022, June 1). The relationship between land use land cover and land surface temperature using remote sensing: systematic reviews of studies globally over the past 5 years. *Environmental Science and Pollution Research, 29*(28), 42493–42508. https://doi.org/10.1007/s11356-022-19997-z

Crop Yield Estimation in Indiana Using a Semi-Physical Model

Khushal Sen, Vishal Gautam, and Shray Pathak

1 Introduction

Early yield forecast is one of the determining aspects for efficient crop decision-making and to boost farmer's profit. Accurate crop yield estimation is essential for sustainable farming practices and food security for a region. Traditional yield prediction models typically fall into two categories: statistical models and mechanistic models. The first order relies on historical yield trends but lack the adaptability to changing climatic conditions. Mechanistic models provide detailed crop physiology but require extensive computational resources and input data for processing [1]. To bridge the gap between these approaches, this study employs a semi-physical model, integrating both remote sensing and meteorological data for enhanced yield estimation accuracy while maintaining computational efficiency. A semi-physical model is an integrated approach by combining crop canopy parameters (NDVI), physiological parameters (Photosynthetically Active Radiation and Light Use Efficiency), physical parameter (harvest index), and stress parameters (Land Surface Wetness Index and Temperature Stress) and are extensively used to predict crop yield and to establish a relation between crop growth and productivity [2–4].

Furthermore, the availability of diverse geospatial data in the modern era from various Earth observation programmes, including NASA's LANDSAT and ESA's Sentinel missions, has been proven to be highly effective in crop yield prediction. In the following study, MODIS satellite data from 2019 to 2024 are used alongside meteorological records to develop a regional crop yield estimation framework for

K. Sen (✉)
Department of Geography, Panjab University, Chandigarh, India

V. Gautam · S. Pathak
Department of Civil Engineering, Indian Institute of Technology Ropar, Rupnagar, Punjab, India
e-mail: 2023cem1016@iitrpr.ac.in; shray.pathak@iitrpr.ac.in

S. Pathak et al. (eds.), *Geo-Data Revolution*, Advances in Geographic Information Science, https://doi.org/10.1007/978-3-031-94078-1_9

Indiana. Indiana located in the Midwest United States has a diverse landscape which has a significant effect on crop yields. The state is favourable for large-scale agricultural operations due to the unique combination of plains, rolling hills, and river valleys. Although wheat and fodder crops are also grown, corn and soybeans are the most common crops. Indiana has warm summers and chilly winters due to its humid continental climate (Fig. 1). The region's agricultural operations are supported by the 1020–1270 mm (40–50 in.) of precipitation that falls there each year. Alfisols and Mollisols, the two most common soil types in Indiana are known for their fertility and capacity for holding onto moisture, making them ideal for the production of corn and soybeans [5].

The semi-physical model improves the overall prediction accuracy based on empirical information by taking account of the environmental variables available and combining them with the machine learning approaches for simplifies crop development equations [6]. Through efficient resource allocation this approach improves the prediction reliability for yield, enables precision farming, and offers upscaling capability for vast agricultural monitoring [7]. By leveraging real-time satellite and climate data, the semi-physical modelling approach provides a robust, adaptable, and computationally efficient tool for yield prediction. It will support agricultural stakeholders in decision-making contributing to sustainable agriculture and helps in mitigating climate change's effects on Indiana's crop output.

2 Dataset Used

Light Use Efficiency (LUE) One of the key concepts in crop growth modelling is LUE which indicates how well a plant transform absorbed photosynthetically active radiation (APAR) into biomass and crop growth. It is can be used in yield estimation

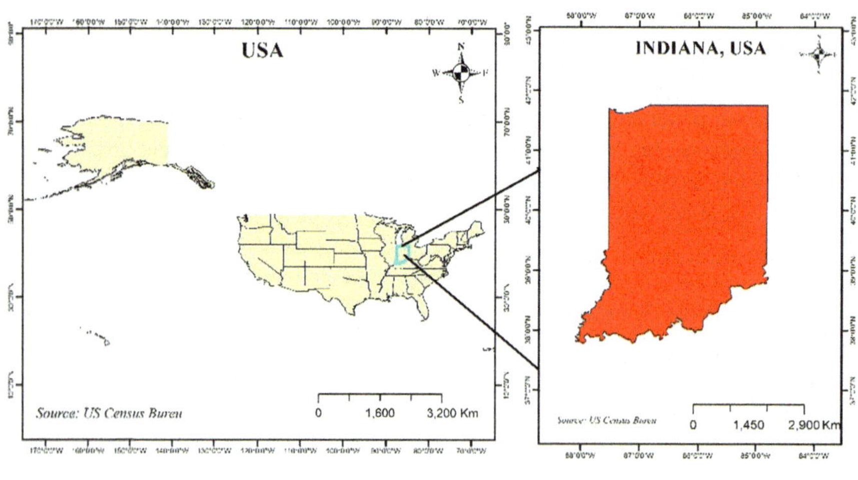

Fig. 1 Indiana, USA

models based on remote sensing notably when combined with vegetation indices like NDVI and EVI [8]. Incorporation of LUE into yield prediction models have been shown increased accuracy particularly when combined with remote sensing data and machine learning (ML) [9]. Environmental factors such as temperature, water availability, and nutrient supply significantly influence LUE. Stress conditions such as drought or extreme temperatures reduce LUE, resulting in lower biomass accumulation and yield. LUE is calculated by using the Eq. 1 [10].

$$\varepsilon = \varepsilon_{\max} \times T_s \times W_s \times P_s \tag{1}$$

where

ε_{max} is the maximum light use efficiency (gC/MJ) and its value is 0.34 [11].
W_s is the water stress.
T_s is the temperature stress.
P_s is the vegetation growth.

Water Stress Water stress occurs either when the demand of water exceeds the available supply or due to access to poor water quality which restricts its usage leading to reduced crop yields and ecosystem degradation [12]. In this study, the water stress has been calculated between a linear combination of the NIR (0.78–0.89 μm) and SWIR (1.58–1.75 μm) spectral bands known as the Land Surface Water Index (LSWI) [13, 14]. Higher LSWI more availability of water content indicating healthy vegetation and lower LSWI suggests poor moisture content which further indicates dry vegetation. The ratio explains the moisture stress in the crops as SWIR reflectance increases as the vegetation water content decreases indicating draught conditions or excessive evapotranspiration [13, 15]. It is calculated using the equation [16].

$$\text{LSWI} = \frac{\text{NIR} + \text{SWIR}}{\text{NIR} - \text{SWIR}} \tag{2}$$

From LSWI, water-stressed zone is identified. The value of LSWI ranges from −1 to 1.

$$W_s = \frac{(1 - \text{LSWI})}{(1 + \text{LSWI}_{\max})} \tag{3}$$

FPAR and PAR Fraction of Photosynthetically Active Radiation (FPAR) one of the crucial biophysical parameters available for the estimation of plant growth and crop yield, it represents the total available Photosynthetically Active Radiation absorbed by the vegetation for photosynthesis [17]. The equation developed by [18] has been used to quantify the fAPAR. The fAPAR is defined as the ratio of absorbed PAR (APAR) to incident PAR (0 < fAPAR < 1). It ranges from 0 to 1, where PAR is cumulative intercepted photosynthetically active radiation. Photosynthetically

Active Radiation (PAR) is the portion of light spectrum (400–700 nm) which plant uses for photosynthesis process. The intensity and accumulated quality of PAR significantly affect the vegetative growth by regulating photosynthesis, biomass, and development process for leaf canopy structure. In general, FPAR has a strong linear relationship with NDVI, and it is derived either by using vegetation-specific regression constants of such relationship [19, 20] or linear scaling based on NDVI [21]. In the present study, we used MODIS fAPAR MOD15A2H was acquired for the whole season at a temporal resolution of 8 days and at a 500 m spatial resolution which provides dynamic monitoring for both the soybean and corn. In an extensive study carried out in Midwest US found out that MODIS FPAR explained 85% of variance in the biomass production of maize giving more reliable results as compared to traditional meteorological-based methods [22]. Moreover, MODIS-derived FPAR has been proved to be a reliable parameter across different ecosystems in up-scaling Gross Primary Production (GPP) and improved Net Primary Product by 20–30% as compared to traditional vegetation indices [14].

Temperature Stress Temperature stress influences crop photosynthesis, transpiration, and metabolic efficiency. Tstress is a temperature scalar used in semi-physical models to adjust biomass accumulation based on daily temperature variations. It is estimated at each time step using the equation from equation [16].

$$T_s = \frac{(T - T_{\min})(T - T_{\max})}{(T - T_{\min})(T - T_{\max}) - (T - T_{\text{opt}})2} \tag{4}$$

where

T is the Mean temperature (°C).
$T_{\min}$ is the Minimum temperature (°C).
$T_{\max}$ is the Maximum temperature (°C).
T_{opt} is the Optimum temperature for photosynthesis (°C) and its value is 30 °C for corn and soybeans.

MODIS-LST (Land Surface Temperature) dataset is commonly used to estimate Tstress for large-scale yield modelling [23]. Tstress is used with (LUE) models to adjust Gross Primary Productivity (GPP) under heat stress [21]. It is used for early warning model for heatwaves for maize and wheat, assessment of thermal stress, etc.

Vegetative Growth Vegetative growth refers to the plant development primarily focused on the production of leaves, stems, and roots rather than reproductive parts [24]. This stage is crucial for establishing plant structure, biomass production, and optimization of photosynthetic activity [14].

$$P_{s1} = 1 \tag{5}$$

$$P_{s2} = \frac{1 + \text{LSWI}}{2} \tag{6}$$

Calculation of Crop Yield In particular, NPP Net Primary Production (NPP) is defined as the difference of what amount of carbon plants take in through photosynthesis and what they are using for their own energy needs. In simpler terms, what's left over to help plants grow and build biomass. This makes NPP a key factor in understanding of monitoring and evaluating crop growth conditions and biomass accumulation [25].

Yield, on the other hand, refers to the total harvested portion of a crop. It is usually measured in Ton/hac which the total weight per area. Studies have shown a strong agronomic relationship between the NPP and the yield established [25, 26].

Net primary productivity is calculated as:

$$\text{NPP} = \sum \text{PAR} \times \text{fAPAR} \times \varepsilon \times \Delta t \tag{7}$$

Now final yield can be calculated as:

$$\text{Yield} = a \times \sum \text{NPP} + b \tag{8}$$

3 Results and Discussion

To check the accuracy of the model, we used the RMSE to measure the averaged deviation between the observed and predicted values, with lower RMSE indicates better model accuracy and R^2 value to measures how well a model predicts crop yields, with values closer to 1 indicating greater performance [27].

For corn yield performance, the semi-physical model showed a strong predictive performance for corn yield estimation by achieving an R^2 value of 0.76 and an RMSE of 0.8319 ton/ha (Fig. 2). The high R^2 value indicates a strong correlation between the predicted and observed values. The reduced RMSE also shows the model's accuracy in showing less error in the predictions. For Soybean yield estimation, the model demonstrated moderate predictive performance with an R^2 value of 0.57 and an RMSE of 1.4278 t/ha. But the lower R^2 value indicates that only 57% of the variability in actual soybean yield is explained by the model suggesting a weaker correlation as compare to corn. Also the higher RMSE also indicates greater potential challenges in accurately modelling soybean yield and prediction errors.

The lower predictive capability for soybean yield may be due to the complex physiological responses of soybean plants to environmental variations (Fig. 3).

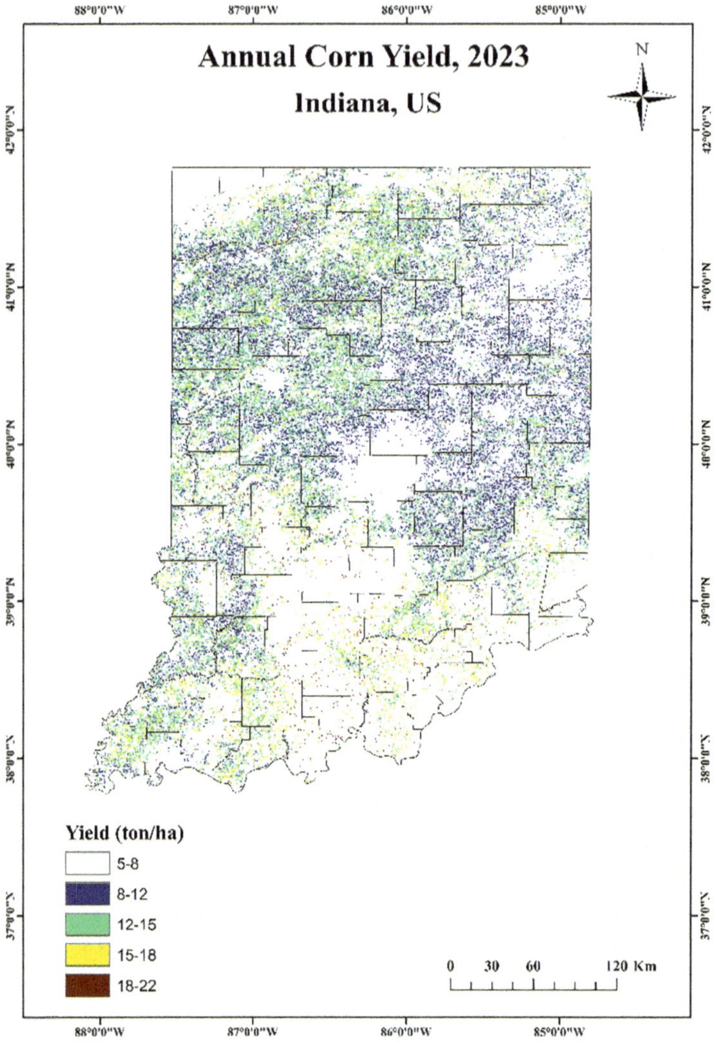

Fig. 2 Annual corn yield of 2023 using semi-physical model

Unlike corn, soybean growth is more sensitive to water availability, temperature fluctuations, and reproductive phase stresses.

Additionally, variations in management practices, soil fertility, and crop rotations may introduce inconsistencies that are not fully captured by the model. Incorporating additional crop-specific parameters, such as evapotranspiration, nitrogen uptake, or stress indices, may improve the model's accuracy [28].

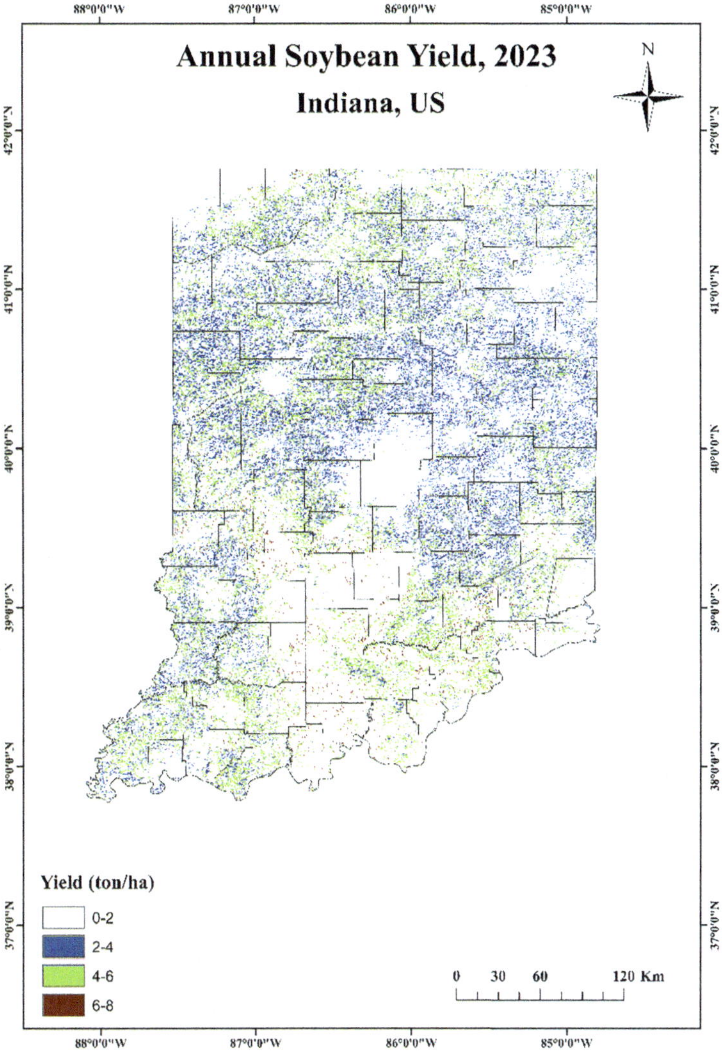

Fig. 3 Annual soybean yield of 2023 using semi-physical model

4 Conclusion

This study employed a semi-physical model integrating remote sensing and meteorological parameters for estimating corn and soybean yields in Indiana, USA. The results showed that the model performed better for corn (R^2 value 0.76 and RMSE = 0.8319 ton/ha) compared to soybean yield estimation (R^2 value 0.57 and

RMSE = 1.4278 ton/ha). The higher predictive accuracy for corn suggests that the model effectively captures crop growth patterns influenced by key environmental variables. On the other side for the lower correlation for soybean yield prediction highlights the need for further improvements, possibly due to the crop's complex physiological response to environmental stressors and management practices.

Despite setting aside these differences semi-physical model provides a computationally efficient and scalable approach for crop yield prediction. Future improvements can be made by focusing on model calibrations, incorporating other crop stress factors and machine learning approaches to improve the prediction accuracy. The study underscores the potential of geospatial technologies and semi-physical modelling in precision agriculture contributing to better resource management, improved yield prediction and enhanced food security strategies for the future.

References

1. Lobell, D. B., & Burke, M. B. (2010). On the use of statistical models to predict crop yield responses to climate change. *Agricultural and Forest Meteorology, 150*(11), 1443–1452.
2. Doraiswamy, P. C., Moulin, S., Cook, P. W., & Stern, A. (2003). Crop yield assessment from remote sensing. *Photogrammetric Engineering & Remote Sensing, 69*(6), 665–674.
3. Prasad, N. R., Patel, N. R., & Danodia, A. (2021). Crop yield prediction in cot-ton for regional level using random forest approach. *Spatial Information Research, 29*, 195–206.
4. Sakamoto, T., Gitelson, A. A., & Arkebauer, T. J. (2013). MODIS-based corn grain yield estimation model incorporating crop phenology information. *Remote Sensing of Environment, 131*, 215–231.
5. Homoya, M. A., Abrell, D. B., Aldrich, J. A., & Post, T. W. (1984, January). The natural regions of Indiana. *Proceedings of the Indiana Academy of Science, 94*, 245–268.
6. White, J. W., Hoogenboom, G., Kimball, B. A., & Wall, G. W. (2018). Methodologies for simulating impacts of climate change on crop production. *Field Crops Research, 219*, 1–18.
7. Liakos, K. G., Busato, P., Moshou, D., Pearson, S., & Bochtis, D. (2018). Machine learning in agriculture: A review. *Sensors, 18*(8), 2674.
8. Van Klompenburg, T., Kassahun, A., & Catal, C. (2020). Crop yield prediction using machine learning: A systematic literature review. *Computers and Electronics in Agriculture, 177*, 105709.
9. Lagergren, F., Eklundh, L., Grelle, A., Lundblad, M., Mölder, M., Lankreijer, H., & Lindroth, A. (2005). Net primary production and light use efficiency in a mixed coniferous forest in Sweden. *Plant, Cell & Environment, 28*(3), 412–423.
10. Na, S., Hong, S., Kim, Y., & Lee, K. (2014). Estimation of corn and soybean yields based on MODIS data and CASA model in Iowa and Illinois, USA. *Korean Journal of Soil Science and Fertilizer, 47*(2), 92–99.
11. Potter, C. S., Randerson, J. T., Field, C. B., Matson, P. A., Vitousek, P. M., Mooney, H. A., & Klooster, S. A. (1993). Terrestrial ecosystem production: A process model based on global satellite and surface data. *Global Biogeochemical Cycles, 7*(4), 811–841.
12. FAO. (2012). *Crop water information: Drought and water stress.*
13. Xiao, X., Boles, S., Frolking, S., Salas, W., Moore Iii, B., Li, C., Li, H., & Zhao, R. (2002). Observation of flooding and rice transplanting of paddy rice fields at the site to landscape scales in China using VEGETATION sensor data. *International Journal of Remote Sensing, 23*(15), 3009–3022.

14. Xiao, X., Zhang, Q., Braswell, B., Urbanski, S., Boles, S., Wofsy, S., Berrien Moore, I. I. I., & Ojima, D. (2004). Modeling gross primary production of temperate deciduous broadleaf forest using satellite images and climate data. *Remote Sensing of Environment, 91*(2), 256–270.
15. Jurgens, C. (1997). The modified normalized difference vegetation index (mNDVI) a new index to determine frost damages in agriculture based on Landsat TM data. *International Journal of Remote Sensing, 18*(17), 3583–3594.
16. Jha, M. N., Kumar, A., Dubey, S., & Pandey, A. (2022). Yield estimation of rice crop using semi-physical approach and remotely sensed data. In *Geospatial technologies for resources planning and management* (pp. 331–349). Springer.
17. Myneni, R. B., & Williams, D. L. (1994). On the relationship between FPAR and NDVI. *Remote Sensing of Environment, 49*(3), 200–211.
18. Monteith, J. L. (1977). Climate and the efficiency of crop production in Britain. *Philosophical Transactions of the Royal Society of London B, Biological Sciences, 281*(980), 277–294.
19. Potter, C., Klooster, S., Myneni, R. B., & Genovese, V. (2004). Terrestrial carbon sinks predicted from MODIS satellite data and ecosystem modeling. *Earth Observer, 16*(2), 15–20.
20. Prince, S. D., & Goward, S. N. (1995). Global primary production: A remote sensing approach. *Journal of Biogeography, 22*(4/5), 815–835.
21. Sellers, P. J., Randall, D. A., Collatz, G. J., Berry, J. A., & Field, C. B. (1996). A revised land surface parameterization (SiB2) for atmospheric GCMs. *Journal of Climate, 9*(4), 676–705.
22. Zhang, L., Guo, X., Li, J., & Zhao, W. (2020). Estimating crop gross primary production using fraction of photosynthetically active radiation. *Science of the Total Environment, 734*, 139436.
23. Li, L., Wang, B., Feng, P., Wang, H., He, Q., Wang, Y., , De Li Liu, Yi Li, Jianqiang He, Hao Feng, Guijun Yang & Yu, Q. (2021). Crop yield forecasting and associated optimum lead time analysis based on multi-source environmental data across China. Agricultural and Forest Meteorology, 308, 108558.
24. Taiz, L., Zeiger, E., Møller, I. M., & Murphy, A. (2015). *Plant physiology and development.* Sinauer Associates.
25. Wang, P. J., Zhou, Y. Y., Huo, Z. G., Han, L. J., Qiu, J. X., Tan, Y., & Liu, D. (2018). Monitoring growth condition of spring maize in Northeast China using a process-based model. *International Journal of Applied Earth Observation and Geoinformation, 66*, 27–36. https://doi.org/10.1016/j.jag.2017.11.001
26. Yao, F. M., Tang, Y. J., Wang, P. J., & Zhang, J. H. (2015). Estimation of maize yield by using a process-based model and remote sensing data in the Northeast China Plain. *Physics and Chemistry of the Earth, 87–88*, 142–152. https://doi.org/10.1016/j.pce.2015.08.010
27. Chai, T., & Draxler, R. R. (2014). Root mean square error (RMSE) or mean absolute error (MAE)? Arguments against avoiding RMSE in the literature. *Geoscientific Model Development, 7*(3), 1247–1250. https://doi.org/10.5194/gmd-7-1247-2014
28. Kutner, M. H., Nachtsheim, C. J., Neter, J., & Li, W. (2004). *Applied linear statistical models.* McGraw-Hill Irwin.

Geospatial Analysis for Assessment of Change Detection in Land Use Land Cover Pattern in Puranpur Block of Pilibhit District of Uttar Pradesh

Mamta Shukla, Pallavi Singh, and Sudhakar Shukla

1 Introduction

Land use and land cover studies show a vital role in compassioning how anthropological actions and natural practices form the Earth's surface. These studies deliver valuable visions into the spatial distribution of land assets, the changing aspects of landscapes over time, and the connections between human cultures and surroundings. Through different patterns of land use (in what way land is developed and exploited by people for innumerable tenacities like expansion, cultivation, or conservation) land cover (the physical as well as biotic cover of the Terrain's surface, comprising flora, aquatic forms, plus constructed buildings), investigators can evaluate conservational changes, monitor ecosystem health, and inform sustainable development practices. This introduction sets the stage for discovering the intricacies and importance of land use and land cover studies addressing contemporary environmental challenges in addition to planning for an irrepressible future. Land use and land cover studies stay precarious owing to the district's miscellaneous terrestrial, environmental, and socio-economic landscapes. Here's a condition that elucidates the prominence and application of such studies.

Pilibhit District well-known as one of the most conspicuous district is full of cosmic biodiversities. Over the past few decades, this area has witnessed significant population growth, industrial extension, and urban growth. To comprehend the

M. Shukla (✉)
Integral University, Lucknow, Uttar Pradesh, India

P. Singh
Remote Sensing Applications Centre, Lucknow, Uttar Pradesh, India

S. Shukla
School of Geoinformatics, Remote Sensing Applications Centre, Lucknow, Uttar Pradesh, India

S. Pathak et al. (eds.), *Geo-Data Revolution*, Advances in Geographic Information Science, https://doi.org/10.1007/978-3-031-94078-1_10

impact of these changes on the environment and plan sustainable development, comprehensive cover studies are essential.

Remote sensing and Geographic Information Systems (GIS) persist operative approaches aimed at gaining accurate plus suitable data for the three-dimensional dissemination of variations in land use and land cover over considerable areas [1–4]. The majority of the research done in the former and current day by administrations as well as associations worldwide has focused on the implementation of LULC modifications. The stretchy environment accessible via GIS marks it promising to collect, supply, demonstrate, examine, and evaluate the ordinal facts required for land change detection [5–7]. Geospatial intelligence's most valuable data resource is remote sensing imagery. GIS technology and satellite imagery are employed to map the extension of urban areas. Researchers evaluate changes in land cover from natural or agronomic land to built-up areas like residential, commercial, and industrial zones. This helps urban designers to contemplate where development is taking place and its implications for agriculture and natural habitats. It is helpful in assessing diverse agricultural landscape of District Pilibhit. Studies of land use trail various changes in agricultural land, including modifications from traditional crops to cash crops or the transformation of agricultural land to non-agricultural practices. This information enlightens policies directed at improving food security and maintainable agricultural practices.

To various forest ecosystems, the district is a real home, which is crucial for carbon sequestration and conservation of biodiversity. Satellite data is used to monitor fluctuations in forest cover, identifying deforestation, afforestation efforts, and forest's dreadful conditions. These insights guide forest management strategies and conservation enterprises to mitigate climate change impacts and protect wildlife habitats. Land use studies also play a significant role in managing water resources. They consider changes in, wetlands, waterbodies, and watershed areas affected by expansion, agriculture, and industrial activities. This information supports in flood risk assessment, water shed management, and certifying sustainable supply of water for rural and urban communities. Extensive scale infrastructure projects, like dams or highways, involve environmental impact assessments (EIAs) that embrace analysis of land cover and land use. These studies appraised how future developments will vary the biodiversity, natural landscape and local communities, controlling decisions to curtail environmental destruction and encourage ecological development.

These studies are essential for informing and advising strategies, urban development, natural resource management, and environmental conservation exertions. They deliver an ample understanding of how human actions and natural practices interact with the landscape, critical for safeguarding a balance between expansion and environmental sustainability in a promptly changing world.

2 Study Area

Pilibhit District is located at India-Nepal border at the feet of Himalayas. Pilibhit District is one of the most noticeable districts of this state including seven blocks covering an area of 1712 km^2. Out of seven blocks Puranpur is the most conspicuous block. In last few years, this northerly area is facing changing environmental conditions in a more rapid way. The fluctuating ecological conditions in this part has given rise to altering biodiversities.

The Terai area is located in the northerly portion of the area casing the northerly portions of Pilibhit as well as Puranpur. Various rivulets by means of fluctuating pathways in addition to forested area are the key features of Terai. River Sharda fixtures on the borderline from north-west to south-east course. Besides this numerous rivulets trench the domain for a small space. Geographically, the region fits to Dun and alluvial grates creations.

Land area of the district is mostly plain and fertile. The whole gross seeded area is 63% of the total geographic area which is 2372.24 km^2, accessible for cultivation; maximum of the zone is cast off as dual harvested zone. Typically, the terrestrial assets stay minor in extent. 92.49% of whole part is plough land, of which 98.28% of entire plough land is watered.

Innumerable natural plus climatic deviations, developmental actions and several other anthropogenic doings like deforestation, industrial activities, and unplanned urbanization along with encroachment pose extortions to ecosystems, human health, and overall terrestrial sustainability. Due to exhausting natural assets such as woodlands and waterbodies, the pronounced loss in the ecological conditions has been perceived. Geologically, the district has many features. It is like 'Terai area' in the north-west and looks like with common features of Rohilkhand while the Southern portion is parallel to the zones of Shahjahanpur and Bareilly. The eastern sector is similar to the forest area of Kheri and the north of Awadh. The district is basically interconnected by several brooks flowing mainly in the southern direction. There are various changeable trenches and depressions in the inner level. Geographically, the region belongs to Alluvium (recent) creations, which falls under Terai agro climatic conditions. Location map of the study as shown in Fig. 1.

3 Objectives of the Study

The major objectives of the study are here as under:

- To identify and categorize land use and land cover pattern through satellite-borne data and GIS investigations for different time periods.
- To analyse changes over the time and to monitor fluctuations through remote sensing data and GIS.
- To study and evaluate the impact of these land use pattern on the environment and the ecosystems.

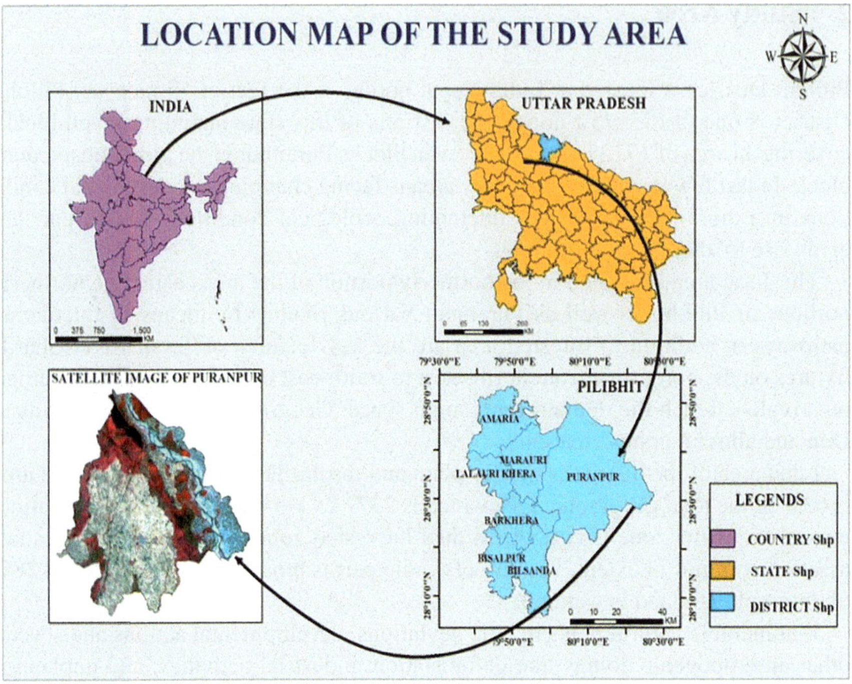

Fig. 1 Location map of the study area

4 Dataset Used

To achieve the objectives following data collections and material are used:

- Survey of India (SOI) Toposheet
- Satellite Data
- Field Data
- Ancillary Data

4.1 SOI Toposheets

A toposheet is a sheet which is well-known as topographic sheet; basically display information regarding railways, roads, canals, rivers, settlement, etc. of a study area. Conferring to the practices, topographic sheets are accessible at diverse scales.

The key database of the study area has been engendered on a scale of 1:50,000 with the help of Survey of India topographic maps. Other than field investigations,

Table 1 Shows a total number of toposheets used

S. no.	Toposheets number
1.	62d/1
2	62d/2
3	62d/3
4.	62d/6
5.	62d/7
6.	53P/14
7.	53P/15

the Survey of India topographic sheets are very valuable to represent the foremost and elementary information of tectonic elements of this landscape or terrain. The topographic maps deliver provincial geomorphological location and other substantial evidence like significant zones, transportation network, etc. of the zone. On a scale of 1:50,000, a total of seven toposheets are used casing the entire study area shown in Table 1.

4.2 Satellite Data

- **Landsat Data:** Visual data from the USGS EarthExplorer in the form of Landsat-8 Series having resolution of 30 m were used. Landsat 8 was launched on 2013 which has two sensors. This digital data is used in disaster response, agriculture, vegetation, land use planning, and vegetation and monitoring of water resources. This information is convenient in having hazed or cloud-free sights for the universal continent. With a spatial resolution of 30 m spectral bands; band1, band2, band3, band4, band5, band6, and band7 are used.

4.3 ArcGIS Software

A topographical information system for working with maps and geographic evidence is well-known as ESRI's ArcGIS. It is used for generating and using maps, assembling topographical data, examining mapped data, distributing and ascertaining geographic data, by means of maps and geographic data in a series of solicitations, and handling geographic data in a database. ArcGIS delivers suitable tools and facilities for mapping and spatial analysis so as to discover data and share location-based considerations. To study and examine the temporal assessment through geospatial techniques, here ArcGIS 10.8 Software is used.

4.4 Field Data

Field inquiries show a crucial part in supporting the tool of remote sensing. Different kinds of field data such as geomorphological features, surface waterbodies position, *X*, *Y* coordinates, information have been collected to formulate the final product.

5 Methodology

Methodology is the assortment and provision of tools and datasets used to get anticipated results of any problem. In the current study, the methodology has been alienated into three portions. It provides with the collection of basic information of the study area. After procurement of relevant information of the study area, preparation of various thematic maps is another significant task. In the current study, the maps and figures have been prepared with the help of SOI (Survey of India) topographic maps and satellite images. Detailed methodology of the present study area is shown in (Fig. 2) as flow chart diagram.

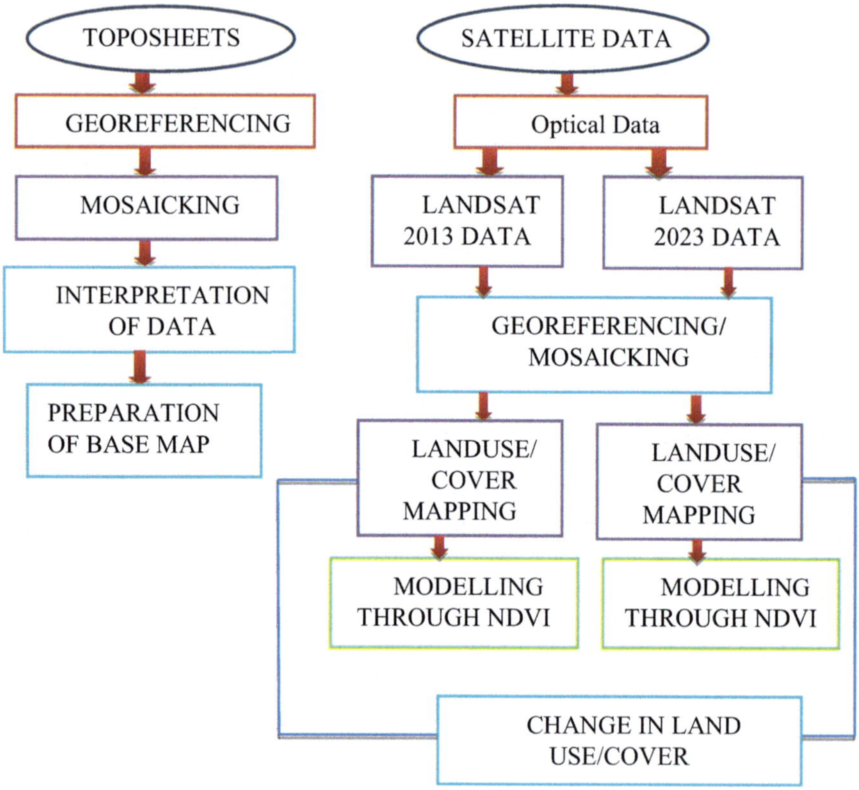

Fig. 2 Flow diagram of methodology

6 Results and Analysis

6.1 Studies Related to Land Resources

A supervised classification technique has been found suitable to create land use/land cover maps in the study area for the year 2013 and 2023 using ArcGIS. In this classification, training sets have been assigned to different categories on the basis of spectral signatures. Different classes' categories are summarized in Tables 2 and 3 for the year 2013 and 2023. For understanding the decadal difference in different land use/land cover classes, maps at Figs. 3 and 4 are also displayed.

Among the most valuable natural resources is land. Local geological, hydrological, ecological, and climatic change are significantly influenced by changes in land use and land cover brought on by urbanization. It causes the neighbouring rural areas to undergo incredible socio-economic change, which eventually leads to urban encroachment [8]. Coming towards examining at the land use pattern, the analysis presented in Table 3 shows the surprising facts that even after having the availability of very fertile land, only 46.93% of land with an area of 69,795.94 ha are of agricultural land and crop land; urban areas cover an area of 9581.51 ha with 6.44% of coverage. However, the forest area is also very small accounting for 5305.11 ha covering 35.84% of total area. The area under cultivable waste, pasture/grazing, orchards, and fallows is only 2.96% covering an area of 4398.80 ha. The area covered by waterbodies is 11,641.03 ha. Rivers, reservoirs, and canals included under waterbodies, and they occupy about 7.83% of the area in 2013 and 6.92% in 2023 [9].

Table 2 Showing different land use/land cover patterns with their area and percentage (%) of the year 2013

S. no.	Parameter	Sum of area (ha)	Percentage (%)
1.	Agricultural field and crop land	69,795.94	46.93
2.	Built-up	9581.51	6.44
3.	Fallow land	4398.80	2.96
4.	Forest	5305.11	35.84
5.	Waterbodies	11,641.03	7.83
	Grand total	**148,722.39**	**100**

Table 3 Showing different land use/land cover patterns with their area and percentage (%) of the year 2023

S. no.	Parameter	Sum of area (ha)	Percentage (%)
1.	Agricultural field and crop land	76,862.80	51.68
2.	Built-up	10,191.98	6.85
3.	Fallow land	6109.73	4.12
4.	Forest	45,263.75	30.43
5.	Waterbodies	10,294.13	6.92
	Grand total	**148,722.39**	**100**

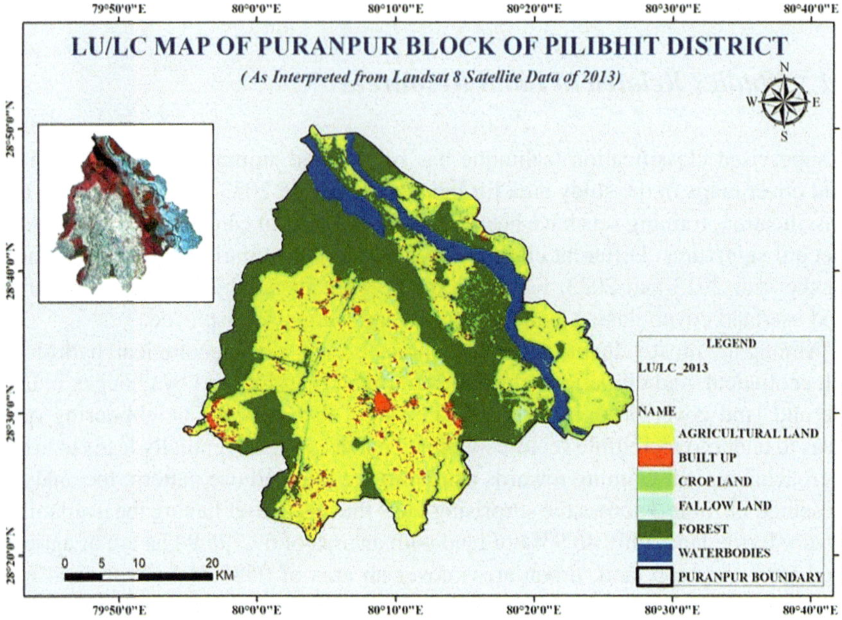

Fig. 3 Showing land use/land cover pattern in the study area (2013)

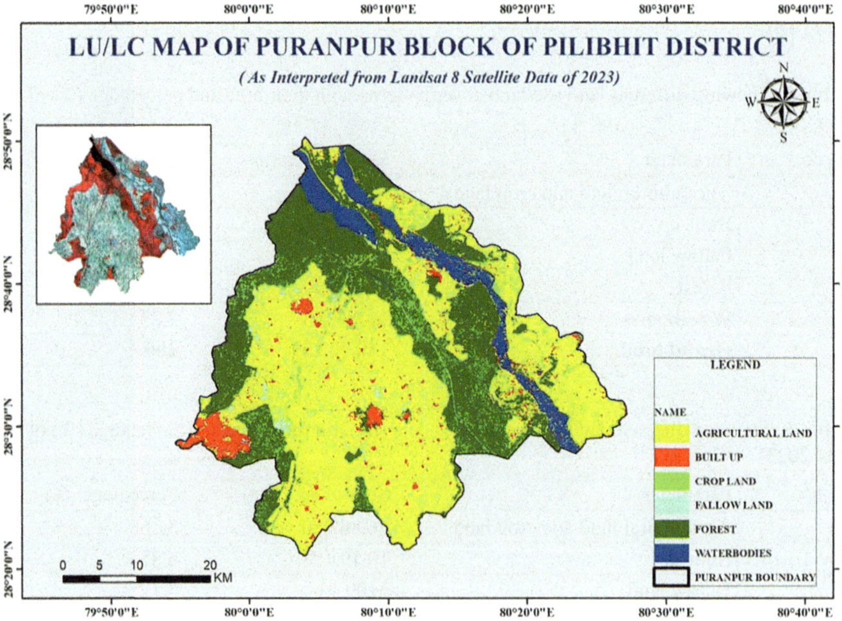

Fig. 4 Showing land use/land cover pattern in the study area (2023)

By examining both the tables, the land use pattern shows that agricultural land and crop land in the year 2023 is increased by covering an area of 76,862.80 with 51.68%. By various anthropogenic activities, urban areas have also got expanded covering an area of 10,191.98 ha with 6.85%. On the other hand because of continuous deforestation and urbanization, the forest area remains only 30.43% and waterbodies with an area of 6.92% only.

Thus, by comparative study of both years it has been observed that agricultural land has been increasing as well as various anthropogenic activities and encroachment by human results in an increase of urbanization causing a decrease in forest cover so as waterbodies from the year 2013 to 2023.

6.2 Change Detection of Land Use and Land Cover

During this decadal variation, Table 4 shows the conditions of land transformations.

Change detection between 2013 and 2023 in land use/land cover is depicted in below mentioned Fig. 5. Recognizing LULC on digital images that exhibit changes in interesting aspects over a period of more than two ages is the objective of the change detection technique [10]. Cutting of forests, loss of biodiversity, global warming, and upsurge in various natural disasters as overflowing or flooding are mainly the outcomes of fluctuations in land use and land cover, together natural and human-caused [11–13]. These ecological concerns are repeatedly associated to swings in LULC. As a result, evidence on LULC modifications can be a vivacious component of environmental monitoring and future planning decisions [14, 15].

6.3 Normalized Difference Vegetation Index

NDVI-based study has also been conducted to study the health of the vegetation for 2013 and 2023. Normalized Difference Vegetation Index (NDVI) studies in Pilibhit district are valuable for evaluating and assessing vegetation health, land cover variations, and dynamics of ecosystem. NDVI is calculated using satellite data and measures the difference between near-infrared (NIR) and red light reflected by vegetation. Higher NDVI values indicate healthier vegetation, while lower values may indicate scarce vegetation or areas without vegetation. In Fig. 6 given below, Normalized Difference Vegetation Index in Puranpur Block of Pilibhit of year 2013 shows value from −0.1053 to 0.48518 which is calculated through satellite imagery. The formula to calculate NDVI is: **NDVI = (Band 5 – Band 4)/(Band 5 + Band 4).**

According to this formula, Band 5 indicates NIR and Band 4 Red Band. This shows rationing of different bands which is a part of image transformation technique. In NDVI, a value varies from −1 to +1. Higher the value of NDVI reflects

Table 4 Change detection between 2013 and 2023 in land use/land cover

Row labels	Sum of area change
Agricultural field—agricultural field	45,951.70
Agricultural field—built-up	4604.87
Agricultural field—crop land	2537.14
Agricultural field—fallow land	3904.12
Agricultural field—forest	1752.09
Agricultural field—waterbodies	1236.50
Built-up—agricultural field	5384.86
Built-up— built-up	1947.43
Built-up—crop land	624.79
Built-up—fallow land	333.77
Built-up—forest	260.60
Built–up—waterbodies	99.84
Crop land—agricultural field	5079.38
Crop land—built-up	989.51
Crop land—crop land	518.30
Crop land—fallow land	308.38
Crop land—forest	542.33
Crop land—waterbodies	284.68
Fallow land—agricultural field	3071.19
Fallow land—built-up	197.46
Fallow land—crop land	86.09
Fallow land—fallow land	441.64
Fallow land—Forest	2.711
Forest—agricultural field	5946.28
Forest—built-up	764.82
Forest—crop land	1642.45
Forest—fallow land	142.62
Forest—forest	44,040.27
Forest—waterbodies	1615.11
Waterbodies—agricultural field	4321.22
Waterbodies— built-up	292.88
Waterbodies—crop land	379.86
Waterbodies—fallow land	17.40
Waterbodies—forest	783.86
Waterbodies—waterbodies	8616.06
Grand total	**148,722.4**

high near-infrared (NIR), means dense greenery. Some potential areas of focus for NDVI studies in Pilibhit district could include**:** Monitoring Forest Cover: Assessing changes in forest cover over time using NDVI can help monitor deforestation, reforestation, and forest health in Pilibhit Tiger Reserve and other forested areas in the district.

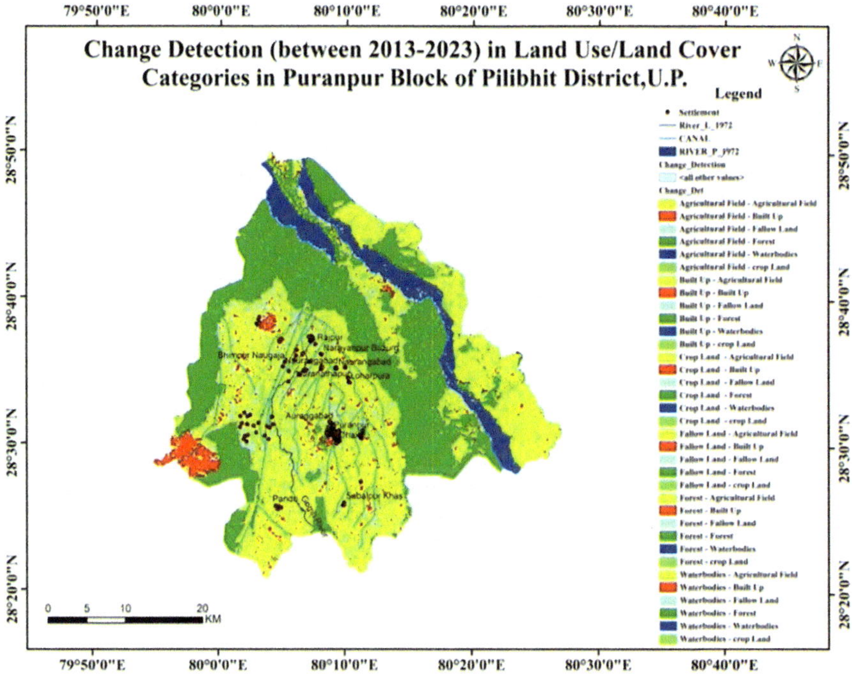

Fig. 5 Shows change detection in land use/land cover categories in Puranpur Block of Pilibhit, District U.P. (2013–2023)

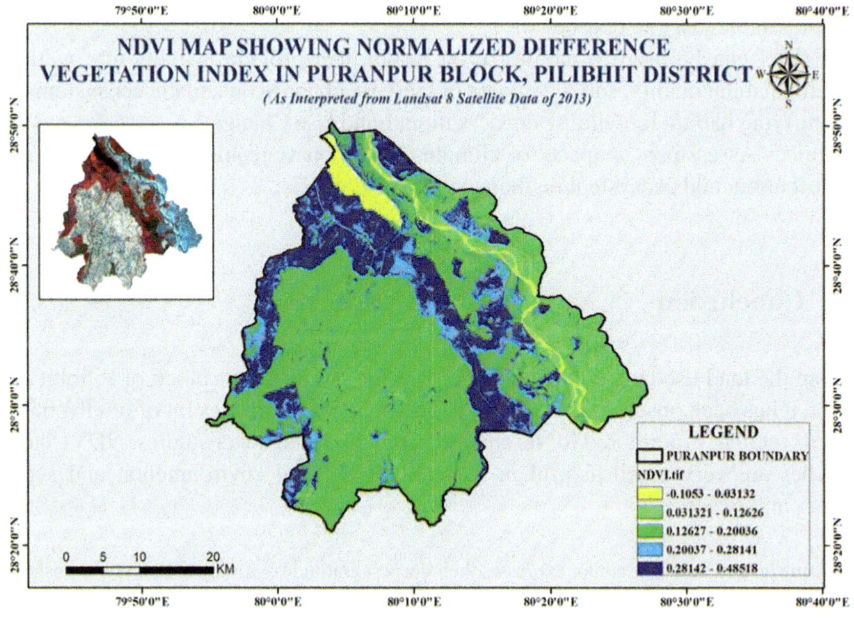

Fig. 6 Shows NDVI in study area during 2013

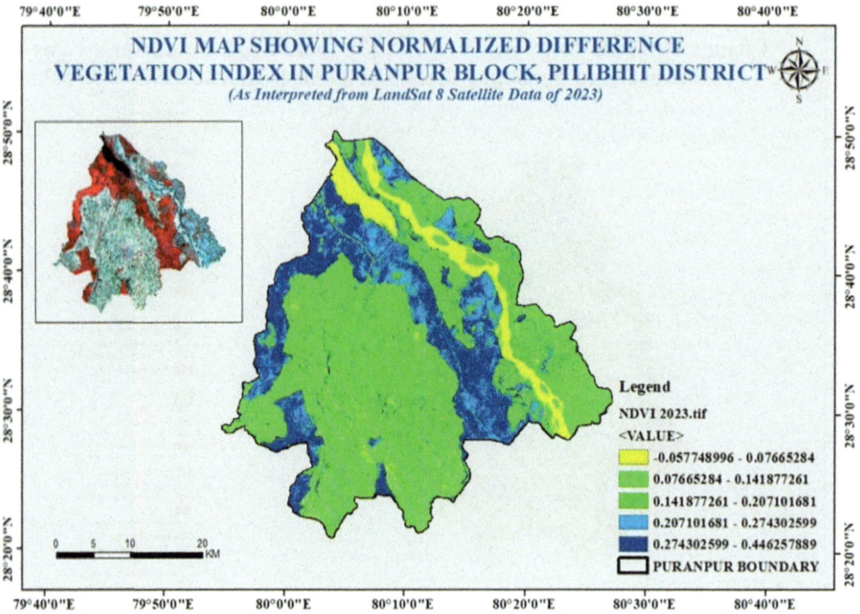

Fig. 7 Shows NDVI in study area during 2023

Normalized Difference Vegetation Index of year 2023 shows Fig. 7 value from −0.05774 to 0.44625 which is deliberated through satellite imagery. Continuous decrease in the NDVI values from 2013 to 2023 reflects the poor health of the vegetation in the last one decade.

NDVI can be used to monitor crop health, can provide insights into wetland health, habitat quality, and the impact of land use changes on aquatic ecosystems, in identifying habitat for wildlife conservation, Land Use Change Analysis, Ecosystem Services Assessment, impacts of climate change on vegetation dynamics, species distribution, and ecosystem resilience in Pilibhit district.

7 Conclusion

From the land use/land cover studies conducted for Puranpur block of Pilibhit district, it has been observed that the study area which experiences lot of biodiversities needs regularly monitored for its changing land use/land cover studies. NDVI-based studies are very much helpful in the context of local environmental and socio-economic factors.

Acknowledgement The author expresses their deepest gratitude to survey of India for toposheets, USGS EarthExplorer for providing Landsat satellite data. I would like to extend my heartfelt thanks to Dr. Sudhakar Shukla Scientist—SE Head, SGID Remote Sensing Applications Centre U.P., Lucknow, for their invaluable insights and support throughout this chapter.

References

1. Carlson, T. N., & Azofeifa, S. G. A. (1999). Satellite remote sensing of land use changes in and around San José, Costa Rica. *Remote Sensing of Environment, 70*, 247–256.
2. Guerschman, J. P., Paruelo, J. M., Bela, C. D., Giallorenzi, M. C., & Pacin, F. (2003). Land cover classification in the Argentine Pampas using multi-temporal Landsat TM data. *International Journal of Remote Sensing, 24*, 3381–3402.
3. Kumar, S. S., Pandey, M., & Shukla, A. K. (2024). Spatio-temporal analysis of riverbank changes using remote sensing and geographic information system. *Physics and Chemistry of the Earth, Parts A/B/C, 136*, 103692. https://doi.org/10.1016/j.pce.2024.103692
4. Mishra, P., Jena, D., Thakur, R. R., Chand, S., Javed, B., & Shukla, A. K. (2024). Peri-urban floodscapes: Identifying and analyzing flood risk areas in North Bhubaneswar in Eastern India. *Water, 16*(21), 3019. https://doi.org/10.3390/w16213019
5. Demers, M. N. (2005). *Fundamentals of geographic information systems*. Wiley.
6. Yomralıoğlu, T. (2000). *Coğrafi Bilgi Sistemleri Temel Kavramlar ve Uygulamalar*. Seçil Ofset.
7. Wu, Q., Li, H. Q., Wang, R. S., Paulussen, J., He, H., Wang, M., Wang, B. H., & Wang, Z. (2006). Monitoring and predicting land use change in Beijing using remote sensing and GIS. *Landscape and Urban Planning, 78*, 322–333.
8. Sakthivel, M., Priyadharshini, J., & Vijayakumar, S. (2021). Accuracy assessment of land use land cover classification using GIS and remote sensing approach: A case study of Tiruchirappalli District, Tamil Nadu, India. *The International Journal of Analytical and Experimental Modal Analysis, 13*, 861–875.
9. Nayak, D., Shukla, A. K., & Devi, N. R. (2024). Decadal changes in land use and land cover: Impacts and their influence on urban ecosystem services. *AQUA—Water Infrastructure, Ecosystems and Society, 73*(1), 57–72. https://doi.org/10.2166/aqua.2024.211
10. Shukla, A. K., Ojha, C. S. P., Mijic, A., Buytaert, W., Pathak, S., Garg, R. D., & Shukla, S. (2018). Population growth, land use and land cover transformations, and water quality nexus in the Upper Ganga River basin. *Hydrology and Earth System Sciences, 22*(9), 4745–4770. https://doi.org/10.5194/hess-22-4745-2018
11. Deeksha, Shukla, A. K., & Rama Devi, N. (2023). Impact of land use and land cover in water resources. In *River conservation and water resource management* (pp. 217–231). Springer. https://doi.org/10.1007/978-981-99-2605-3_11
12. Mas, J. F., Velazquez, A., Gallegos, J. R. D., Saucedo, R. M., Alcantare, C., Bocco, G., Castro, R., Fernandez, T., & Vega, A. P. (2004). Assessing land use/cover changes: A nationwide multidate spatial database for Mexico. *International Journal of Applied Earth Observation and Geoinformation, 5*, 249–261.
13. Tillihal, S. B., & Shukla, A. K. (2023, July). Use of NDVI and PCA to analyze land use and land cover change in Upper Cauvery River Basin (UCRB), India. In *IGARSS 2023-2023 IEEE International Geoscience and Remote Sensing Symposium* (pp. 2973–2976). IEEE.
14. Fan, F., Weng, Q., & Wang, Y. (2007). Land use land cover change in Guangzhou, China, from 1998 to 2003, based on Landsat TM/ETM+ imagery. *Sensors, 7*, 1323–1342.
15. Prenzel, B. (2004). Remote sensing-based quantification of land-cover and land-use change for planning. *Progress in Planning, 61*, 281–299.

Implementing and Validating a Remote Sensing-Driven Crop Yield Estimation Model in Illinois, USA

Vishal Gautam, Shivanshu Giri, and Shray Pathak

1 Introduction

Crop yield acting an important role in the economic and agricultural progress of a country hence accurately estimating crop yield before the harvesting period is imperative for informed management and policy-making decisions. Over the past many years, there has been a high demand for farm products and high rates of profits. This has led to a change in traditional agricultural practices and requirements [1].

Crop yield prediction refers to quantifying crop yield through using meteorological data, modeling methods, and environmental and soil conditions.

Over the years, many studies have been conducted to optimize crop yield prediction by establishing a relationship between crop growth and environmental factors to propose a simulation of growth by crop growth model [2–9].

A crop growth model is a mathematical equation or a series of mathematical equations that quantifies crops' behavior during their growth by simulation of the inner growth process and the relationship between crop growth and environmental conditions. With the introduction of accurate prediction models using machine learning and mathematical optimization techniques efforts have been made to integrate remote sensing data and surface parameters with models predicting crop

V. Gautam (✉) · S. Pathak
Department of Civil Engineering, Indian Institute of Technology Ropar, Rupnagar, Punjab, India
e-mail: 2023cem1016@iitrpr.ac.in; shray.pathak@iitrpr.ac.in

S. Giri
Department of Civil Engineering, Indian Institute of Technology Roorkee, Roorkee, Uttarakhand, India
e-mail: shivanshu_g@ce.iitr.ac.in

S. Pathak et al. (eds.), *Geo-Data Revolution*, Advances in Geographic Information Science, https://doi.org/10.1007/978-3-031-94078-1_11

growth and a more accurate estimation of crop yield by training the model with large historical datasets [10–12].

In our research, we utilized a crop yield estimate model trained on a 19-year corn and soybean crops dataset in Iowa [10]. We then adapted this model to predict corn and soybean yields specifically for the state of Illinois.

The prediction model employs a non-linear piecewise linear regression methodology with breakpoints, implemented using the Quasi-Newton method [13–15]. The model consists of a two-piece Empirical equation with the yield as the dependent variable and other parameters (NDVI, Rainfall, Land Surface Temperature, and Soil Moisture) as independent variables. Using the Non-linear approach coefficients of this Empirical equation are calculated with the historical 19-year yield dataset of corn and soybean from the state of Iowa [10] and remote sensing data used to calculate NDVI, land surface temperature, soil moisture, and rainfall.

The prediction model equation utilizes parameters such as the Normalized Difference Vegetation Index (NDVI), daily rainfall, land surface temperature, and soil moisture. Illinois was selected for this study because it is part of the Corn Belt, along with Iowa, and has corn and soybean as its primary crops.

Illinois is a major agricultural state in the United States, significantly contributing to national crop production. According to the US Department of Agriculture, Illinois accounts for 15% of the country's corn production (Fig. 1) and 13% of its soybean production (Fig. 2) [16]. These statistics highlight Illinois' vital role in the national agricultural sector, particularly in the cultivation of these key crops, which are essential for food, feed, and various industrial applications.

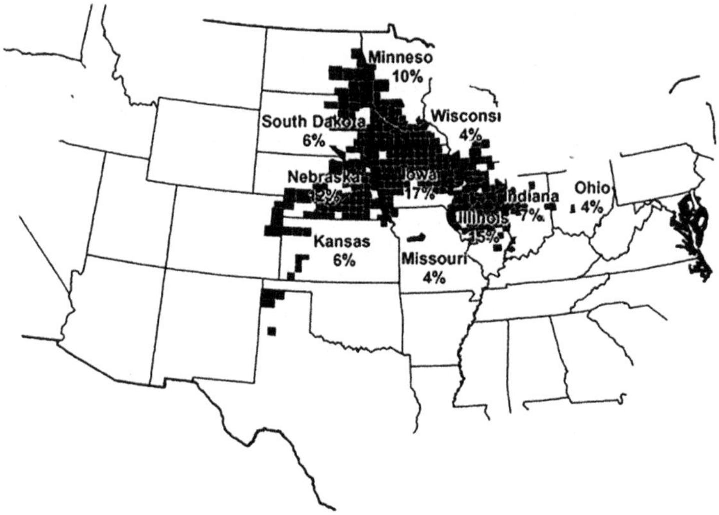

Fig. 1 United States: corn production

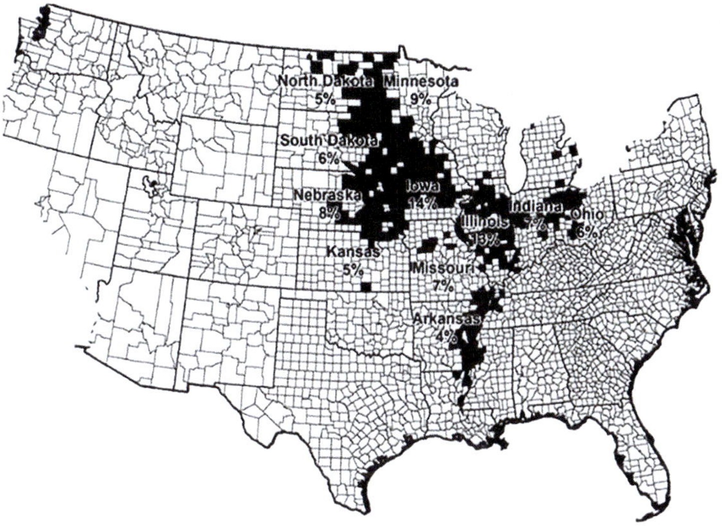

Fig. 2 United States: soybean production

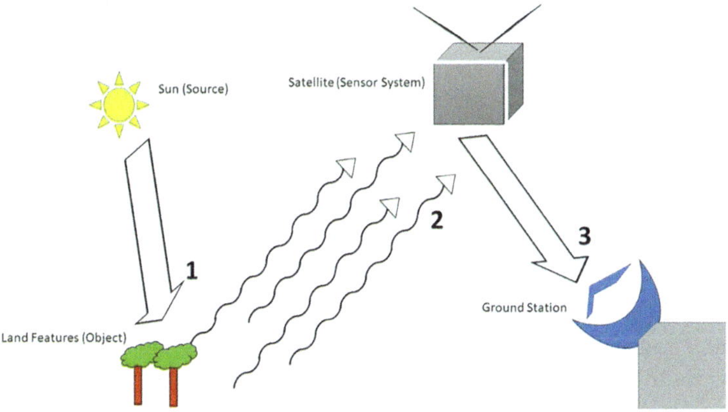

Fig. 3 RS using passive sensor system

2 Remote Sensing Data: Fundamentals

2.1 Remote Sensing

It is a process of collecting and monitoring the physical characteristics of a region without actually coming in contact with them physically. These physical characteristics or information are based on radiation emitted or radiated at a distance (usually an aircraft or satellite) from the area or object under observation. If the source of measured energy is the sun, the process is called passive remote sensing (Fig. 3) and

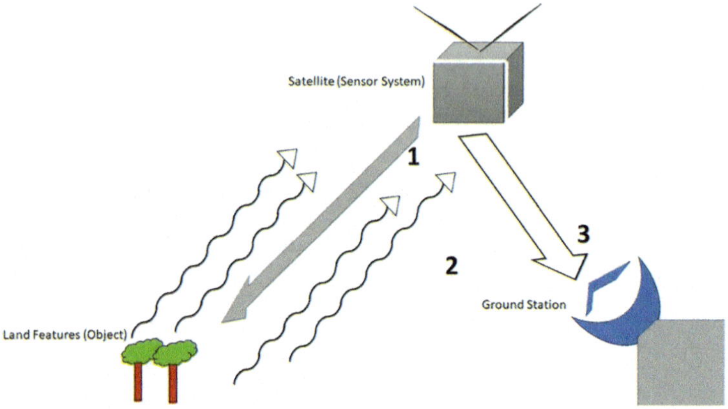

Fig. 4 Remote sensing using an active sensor system

Table 1 Satellite imagery applications

Parameter	Dataset used	Source
NDVI	Landsat 8–9	https://earthexplorer.usgs.gov/
Soil moisture	Daily 4-in. (10 cm) average soil moisture (water fraction by volume)	https://warm.isws.illinois.edu/warm/soiltemp/displaymap.asp?from=sl&data=sm10cm#toptitlebar
Land surface temperature	Landsat 8–9	https://earthexplorer.usgs.gov/
Precipitation	Daily precipitation dataset—Illinois state water survey	https://warm.isws.illinois.edu/warm/soiltemp/displaymap.asp?from=wx&data=prec

if the source of measured energy is a sensor such as radar the process is called active remote sensing (Fig. 4).

Remote sensing uses special cameras on a satellite or airplane to take images of areas on Earth's surface. This data in the form of an image helps in determining the nature and composition of the surface of earth and atmosphere. The captured image data can be used to analyze and assess changes on the surface at different points in time.

3 Data and Methodology

Table 1 below lists the datasets used along with their respective applications. The Landsat 8 dataset was used to calculate the Normalized Difference Vegetation Index (NDVI). Additionally, the daily 4-in. (10 cm) average soil moisture (water fraction by volume) was converted to millimeters, and land surface temperature was measured in Kelvin, along with daily rainfall in millimeters.

3.1 NDVI

Remote sensing data has been instrumental in supporting various research initiatives focused on creating indices like the Normalized Difference Vegetation Index (NDVI), Vegetation Condition Index (VCI), and Temperature Condition Index (TCI). These indices help in vegetation monitoring, drought detection, and evaluating the effects of weather on vegetation which can help in vegetation health and monitoring.

NDVI is a metric that quantifies vegetation greenness, used to assess vegetation density and track changes in plant health over time. Dense and healthy vegetation absorbs increased amounts of red and blue light and reflects higher levels of near-infrared (NIR) and green light compared to other wavelengths. NDVI is calculated using satellite imagery with the help of a satellite's spectrometer and radiometric sensors which stores surface reflectance values of both red and near-infrared wavelengths. Figure 5 presents the spatial distribution of Normalized Difference Vegetation Index (NDVI) values throughout the study area in Illinois.

NDVI is calculated by dividing the Red (R) value by the near-infrared (NIR) value, using the formula:

$$\mathrm{NDVI} = \left(\mathrm{NIR} - \mathrm{R}\right) / \left(\mathrm{NIR} + \mathrm{R}\right) \tag{1}$$

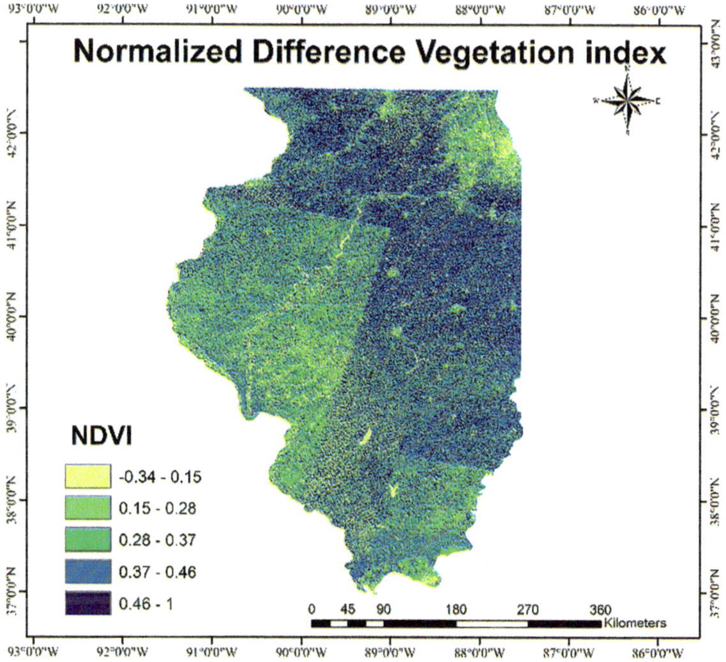

Fig. 5 NDVI of Illinois

NDVI values range from −1 to +1. A negative value indicates water region and cloud and values that are positive and near zero indicate bare soil with insignificant vegetation or even an urbanized area. A higher positive value of NDVI ranging from 0.5–1.0 indicates sparse vegetation while a value above 0.5 indicates dense vegetation.

3.2 *Soil Moisture*

Soil moisture refers to the total amount of water, both in liquid and vapor forms, present in unsaturated soil. It is crucial for predicting droughts and floods, monitoring agriculture, and managing groundwater. Soil moisture, a critical component of the land surface, plays an essential role in controlling the heat and water exchange between the ground and the atmosphere. Soil water impacts ground structure and thickness, as well as the temperature and heat capacity of the ground. Factors that affect soil moisture include soil texture, organic matter content, soil density, temperature, salinity, and depth. In this study, we obtained soil moisture from the range of 4.22–30.58 mm (Fig. 6) [17].

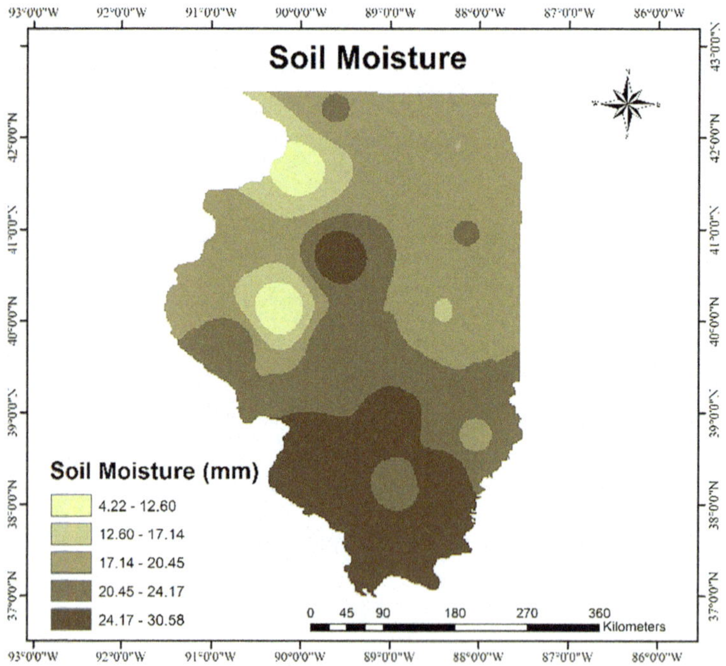

Fig. 6 Soil moisture (mm) of Illinois

3.3 *Land Surface Temperature*

Land surface temperature (LST) is a key surface parameter that plays a significant role in the interaction between surface energy and water balance within terrestrial and atmospheric systems. It quantifies the average radiative skin temperature of the Earth's land surface as observed within the instrument's field of view. Land surface temperature is distinct from atmospheric temperature [18]. The former refers to the temperature of the ground, while the latter refers to the temperature of the air roughly 2 m above the ground. In this study, we calculated the LST from Landsat 8–9 satellite imagery from the various steps. Figure 7 shows the calculated LST of Illinois for August to September of 2023.

3.4 *Precipitation*

Precipitation occurs when water falls from clouds in the form of rain, freezing rain, sleet, snow, or hail. Precipitation is the primary method through which atmospheric water comes back down to Earth's surface. The majority of precipitation falls as

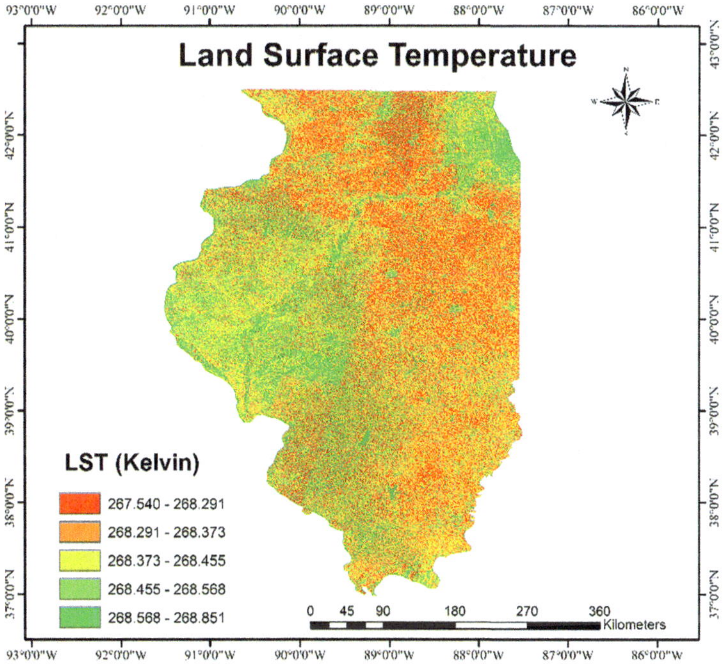

Fig. 7 Land surface temperature (kelvin) of Illinois

rain. Measuring precipitation entails determining whether water reaches the horizontal ground or its projected plane, and it is quantified in terms of water depth or the equivalent depth of solid precipitation. Typically, precipitation is measured in millimeters (mm). Different types of rain gauges, including non-recording cylindrical containers, recording weighing devices, float-based instruments, and tipping-bucket gauges, have traditionally been used to measure liquid precipitation at specific locations [19]. Another method involves radar technology, which measures the strength of echo returns from backscattered signals. This various rain gauge types, such as non-recording cylindrical containers, recording weighing devices, float-based instruments, and tipping-bucket gauges, have historically been utilized to measure liquid precipitation. Each of the gauge's measures rainfall at a specific location. An alternate way to measure precipitation involves utilizing radar technology. The radar measures the power of the echo returns that are backscattered. The return power is employed in the computation of a reflectivity factor, Z, which is subsequently utilized in a formula known as the "Z-R relationship" for rainfall rate determination. In Illinois, the rainfall we obtain is from 0.27 to 2.76 mm (Fig. 8) [20].

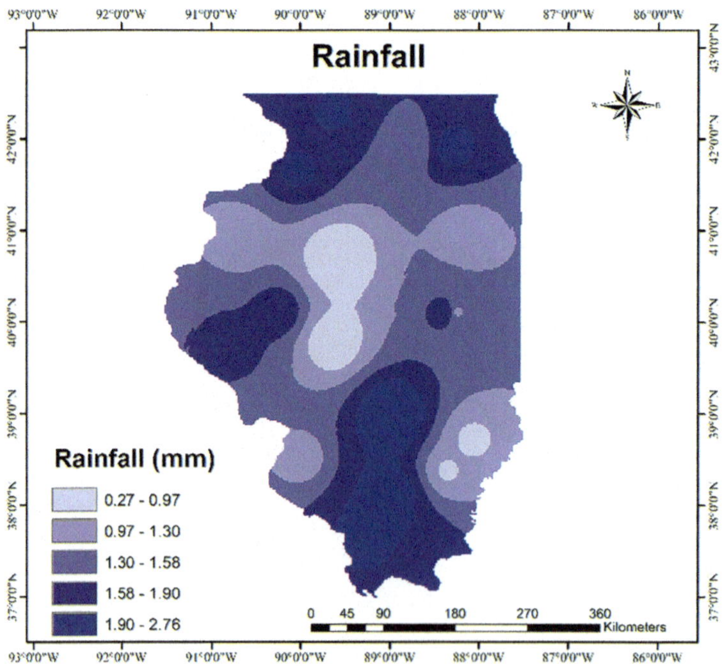

Fig. 8 Rainfall (mm) of Illinois

3.5 Model

We adapted the crop yield estimation model originally developed by Prasad et al. [10] for Iowa to the context of Illinois. This model employs remote sensing data (NDVI) and surface parameters (temperature, soil moisture, rainfall) as inputs.

$$\text{Crop yield} = \left(C_2 + \left(A_1 \times \text{NDVI}\right) + \left(A_2 \times \text{SM}\right) + \left(A_3 \times \text{ST}\right) + \left(A_4 \times \text{RF}\right)\right)$$

Data for these parameters were obtained for the growing season of 2023, specifically from August to September. This period was selected due to higher NDVI values compared to other months in the growing season. The adapted crop model comprises two equations based on crop yield and a breakpoint, which determines if the yield should be less or more than a certain value. Both crop yield model equations were tested, and the following equation was found to be more accurate for estimating crop yield in this location. The flowchart of the methodology is represented in Fig. 9.

3.6 Flowchart of the Methodology

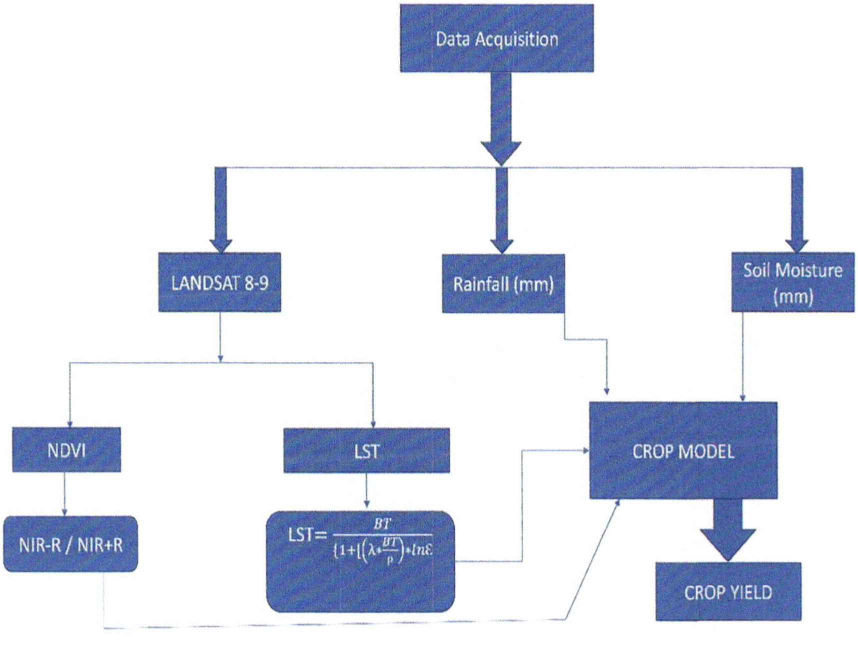

Fig. 9 Flowchart of the methodology

4 Results and Discussion

The study adapted the crop yield estimation model, initially designed for Iowa, to the conditions of Illinois, using the original coefficients without the need for recalibration. This model incorporates remote sensing data (NDVI) along with surface variables such as temperature, soil moisture, and rainfall to estimate crop yields.

The model predicted an average corn yield of 280 bushels per acre (BU/acre) across Illinois, which falls within the typical range of corn yields in the state, varying between 238 and 314 BU/acre (Fig. 10) depending on growing conditions. For soybeans, the model forecasted an average yield of 50 BU/acre statewide, consistent with historical average soybean yields in Illinois ranging from 45 to 61 BU/acre (Fig. 11).

The model's performance, as demonstrated by these yield projections, suggests that the coefficients originally developed for Iowa can be effectively applied to Illinois conditions, yielding satisfactory results without requiring recalibration.

The successful application of the Iowa-based crop yield estimation model to Illinois, using its original coefficients, highlights the model's robustness and adaptability across similar agricultural regions in the Midwest. This finding suggests a consistent relationship between remote sensing data, surface parameters, and crop yields in these two major agricultural states. Despite potential variations in local

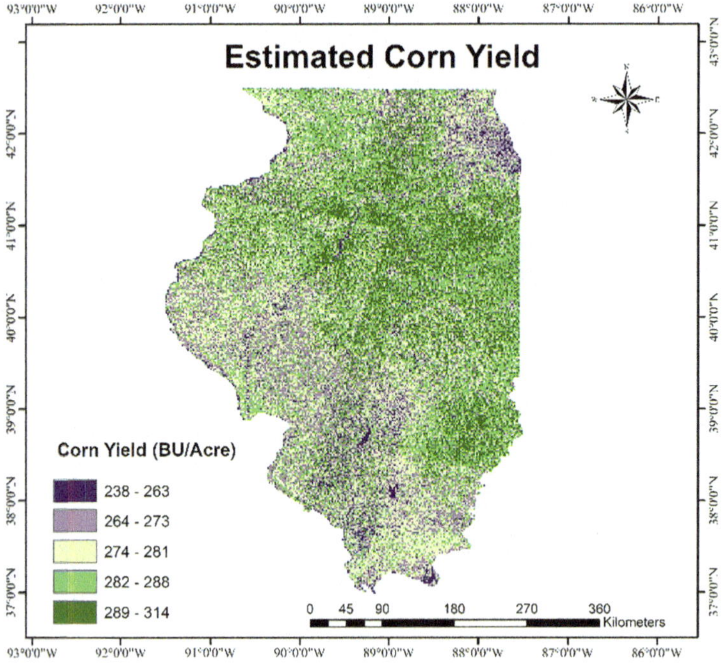

Fig. 10 Estimated corn yield of Illinois, 2023

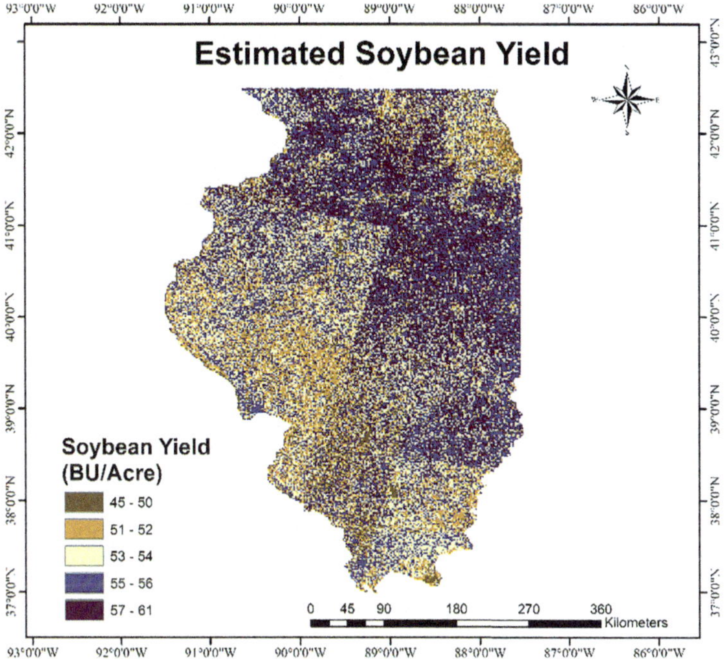

Fig. 11 Estimated soybean yield of Illinois, 2023

conditions between Iowa and Illinois, the model effectively captures the primary factors influencing corn yields.

5 Conclusion

This study effectively adapted and applied a crop yield prediction model, initially developed for Iowa, to the state of Illinois by employing remote sensing data and surface parameters. Employing the same coefficients as the Iowa study, the model demonstrated satisfactory performance in projecting corn and soybean yields across Illinois. While the results are promising, further research is necessary to validate the model's performance across multiple growing seasons and varying weather conditions. Additionally, exploring the potential for fine-tuning the model with Illinois-specific data could enhance its accuracy and reliability. This study advances the field of agricultural remote sensing by demonstrating the transferability of crop yield models between states. It lays the groundwork for future research into regional-scale crop yield estimation and offers a valuable tool for agricultural planners, policymakers, and stakeholders in Illinois and potentially other Midwestern states. The successful implementation of this model paves the way for more efficient crop yield forecasting, which could assist in agricultural decision-making, resource allocation, and food security planning on a regional scale. Future research should aim to expand the model's applicability to other crops and regions, as well as integrate additional data sources to enhance its accuracy and robustness.

Moreover, this research underscores the importance of leveraging advanced technologies in precision agriculture to improve decision-making processes, optimize resource allocation, and ultimately enhance food security. Future work will focus on refining the model by incorporating additional variables, such as soil health indicators and pest pressure data, to further improve its predictive capabilities. Expanding the model's applicability to other regions and crop types will also be a priority, ensuring broader utility and adaptability.

Thus, the successful implementation and validation of this remote sensing-driven crop yield estimation model mark a significant step forward in agricultural technology, offering a promising tool for modern farming practices in Illinois and beyond. The advancements achieved in this study pave the way for more efficient, sustainable, and data-driven agricultural management strategies in the future.

References

1. Tilman, D. (1999, May 25). Global environmental impacts of agricultural expansion: The need for sustainable and efficient practices. *Proceedings of the National Academy of Sciences of the United States of America, 96*(11), 5995–6000. https://doi.org/10.1073/pnas.96.11.5995
2. Boogaard, B. K., Bock, B. B., Oosting, S. J., Wiskerke, J. S. C., & Van Der Zijpp, A. J. (2010). Social acceptance of dairy farming: The ambivalence between the two faces of modernity. *Journal of Agricultural and Environmental Ethics, 24*(3), 259–282. https://doi.org/10.1007/s10806-010-9256-4
3. Brisson, N., Gary, C., Justes, E., Roche, R., Mary, B., Ripoche, D., Zimmer, D., Sierra, J., Bertuzzi, P., Burger, P., Bussière, F., Cabidoche, Y., Cellier, P., Debaeke, P., Gaudillère, J., Hénault, C., Maraux, F., Seguin, B., & Sinoquet, H. (2003). An overview of the crop model stics. *European Journal of Agronomy, 18*(3–4), 309–332. https://doi.org/10.1016/s1161-0301(02)00110-7
4. Panjala, P., Gumma, M. K., Mesapam, S., Shukla, A. K., & Pignatta, G. (2025). Geospatial analysis of crop residue burn areas and their dates for emission mitigation strategies. *Sustainability, 17*(6), 2508. https://doi.org/10.3390/su17062508
5. Jones, J. W., et al. (2007). The DSSAT cropping system model. *European Journal of Agronomy, 18*(3–4), 235–265.
6. Keating, B. A., et al. (2003). An overview of APSIM, a model designed for farming systems simulation. *European Journal of Agronomy, 18*(3–4), 267–288.
7. Nendel, C., et al. (2011). The MONICA model: Testing predictability for crop growth, soil moisture and nitrogen dynamics. *Ecological Modelling., 222*(9), 1614–1625.
8. Stöckle, C. O., et al. (2003). CropSyst, a cropping systems simulation model. *European Journal of Agronomy., 18*, 289–307.
9. Steduto, P, et al. (2009). AquaCrop—The FAO crop model to simulate yield response to water: I. Concepts and underlying principles. *Agronomy Journal*. https://doi.org/10.2134/agronj2008.0139s (wiley.com).
10. Prasad, A., Chai, L., Singh, R., & Kafatos, M. (2006). Crop yield estimation model for Iowa using remote sensing and surface parameters. *International Journal of Applied Earth Observation and Geoinformation., 8*, 26–33. https://doi.org/10.1016/j.jag.2005.06.002

11. Cen, H., & Wan, L. (2023). Crop yield estimation and prediction. In Q. Zhang (Ed.), *Encyclopedia of smart agriculture technologies*. Springer. https://doi.org/10.1007/978-3-030-89123-7_48-1
12. Manfré, L. A., Hirata, E., Silva, J. B., Shinohara, E. J., Giannotti, M. A., Larocca, A. P. C., & Quintanilha, J. A. (2012). An analysis of geospatial technologies for risk and natural disaster management. *ISPRS International Journal of Geo-Information, 1*(2), 166–185.
13. Raes, D., Steduto, P., Hsiao, T. C., & Fereres, E. (2009). AquaCrop—The FAO crop model to simulate yield response to water: II. Main algorithms and software description. *Agronomy Journal, 101*(3), 438–447. https://doi.org/10.2134/agronj2008.0140s
14. Belegundu, A. D., & Chandrupatla, T. R. (1999). *Optimization concepts and applications in engineering*. Pearson Education Asia.
15. Setiono, R., Kheng, L. W., & Zurada, J. M. (2002). Extraction of rules from artificial neural networks for nonlinear regression. *IEEE Transactions on Neural Networks, 13*(3), 564–577.
16. USDA/NASS QuickStats Ad-hoc Query Tool. (n.d.). https://quickstats.nass.usda.gov/#CD348873-C175-39C8-A139-B9339046116F
17. Illinois Climate Network Station Map: Soils, Illinois State Water Survey. (n.d.). https://warm.isws.illinois.edu/warm/soil/
18. Avdan, U., & Jovanovska, G. (2016). Algorithm for automated mapping of land surface temperature using LANDSAT 8 satellite data. *Journal of Sensors, 2016*, 1–8. https://doi.org/10.1155/2016/1480307
19. Shukla, A. K., Ojha, C. S. P., Garg, R. D., Shukla, S., & Pal, L. (2020). Influence of spatial urbanization on hydrological components of the upper Ganga River basin, India. *Journal of Hazardous, Toxic, and Radioactive Waste, 24*(4). https://doi.org/10.1061/(ASCE)HZ.2153-5515.0000508. ASCE.
20. Illinois Climate Network: Weather Data, Illinois State Water Survey. (n.d.). https://warm.isws.illinois.edu/warm/weather/

Analysing Landslide Hazard Propensity in Munnar Panchayat and Surroundings, Idukki District, Kerala

Prashasti Bhattacharyya, Sudhakar Shukla, Pallavi Singh, Shubhanita Dasgupta Chakrabarty, and Mousom Banerjee

1 Introduction

Landslides involves the movement of boulders, rock beds, debris, or soilmass down a slope and can involve a variety of mechanisms such as falling, toppling, slumping, flowing, and sliding. Landslides are one of the quasi-natural hazards that affect at least 15% of our country's land area specially in mountainous terrain. Landslides have had disastrous consequences and in 2005, over 500 lives were lost in India due to landslides. Landslides represent one of the most devastating natural hazards in mountainous regions worldwide, causing significant loss of life, damage to infrastructure, and economic disruption. The Western Ghats of India, particularly the high-altitude regions of Kerala, have witnessed an alarming increase in landslide frequency and intensity over the past several decades. Abraham and Kuriakose (2021) developed a multi-hazard vulnerability assessment framework specifically for the Western Ghats, incorporating landslide susceptibility as a key component. Their findings revealed that approximately 24% of Munnar Panchayat falls under "very high" susceptibility zones, with anthropogenic modifications significantly amplifying natural vulnerability [1]. Among these vulnerable areas, Munnar Panchayat and its surroundings in Idukki District stand out as a critical hotspot for

P. Bhattacharyya (✉) · M. Banerjee
Department of Geography, Sarsuna College, Kolkata, India

S. Shukla
School of Geo-Informatics, Remote Sensing Applications Centre—Uttar Pradesh, Lucknow, India

P. Singh
Remote Sensing Applications Centre—Uttar Pradesh, Lucknow, India

S. D. Chakrabarty
Deptartment of Geography, Bhairab Ganguly College, Kolkata, India

S. Pathak et al. (eds.), *Geo-Data Revolution*, Advances in Geographic Information Science, https://doi.org/10.1007/978-3-031-94078-1_12

landslide occurrences due to their unique geographical, geological, and anthropogenic characteristics. The study of landslides has drawn worldwide attention mainly due to natural and increasing awareness of the socioeconomic impact of landslides, as well as the increasing pressure of land use on the mountain environment [2]. The mountain environment, which is already a fragile zone as monsoon rain brings in its wake many devastating landslides in the region. Though intense and prolonged rainfall can be considered as a triggering factor, the other causes include geological, morphological, physical, and human factor. Slope, land use, vegetation density, anthropogenic elements, availability of unconsolidated sliding material, etc. activities, slope instability, and mass movements of highly crushed and deformed rocks, becomes highly vulnerable to slope failure when anthropogenic factors disturb the stability of the region. In order to reduce the damage caused by landslide initiations and reactivations, a landslide susceptibility map is really needed [3]. Landslide is a recurring disaster often occurring in conjunction with and triggered by other disasters like very heavy rains and further causing associated disasters like flood in the Western Ghats.

In the last several decades, the idea of landslide susceptibility and landslide susceptibility assessment have been introduced, and a number of methodologies have been created to assess the destructive potential of landslides and the processes that are linked to them. The ultimate goal is to develop a strategy that will enable timely implementation of suitable management actions to lower the risk of future landslides. It is helpful to have an LSZ (Landslide Susceptibility Zonation) map created that shows the area's ranking according to the actual and/or potential threat of future slides in order to prioritise the area for hazard mitigation activities. Identification of the factors causing slope failure, their methodical mapping, and an assessment of their respective contributions are necessary for any approach to LSZ.

Human activities have significantly exacerbated landslide risks in the Munnar region. The historical transformation of natural vegetation into tea plantations during the colonial period altered the region's hydrological regime and slope stability conditions. In recent decades, rapid urbanisation, unplanned construction on steep slopes, road network expansion, deforestation, and tourism development have further disturbed the delicate environmental balance. These anthropogenic interventions have modified natural drainage patterns, increased soil erosion, and created new stress configurations on vulnerable slopes, thereby intensifying landslide susceptibility throughout the area.

The scientific assessment of landslide susceptibility has become an essential component of disaster risk reduction strategies worldwide. Landslide susceptibility analysis involves the spatial evaluation of areas prone to landslides based on the correlation between past landslide occurrences and various causative factors. This spatial prediction provides critical information for land-use planning, infrastructure development, emergency preparedness, and implementation of appropriate mitigation measures. Scientists conducted a comprehensive analysis of the 2018 and 2020 Kerala landslide disasters, identifying unprecedented rainfall intensity combined with land-use changes as primary triggers. Their study employed an integrated approach using high-resolution satellite imagery and terrain modelling to identify critical zones in Idukki district that exhibited similar characteristics to previous failure sites [4].

The extraction of groundwater through bore wells and direct pumping at large scale has worsened the groundwater levels in the region which in turn adds to the drying of the river.

2 Study Area

The study area on Nilgiri range is located in the Idukki district of the south-western Indian state of Kerala with the famous hill station Munnar at the centre of the study area (as shown Fig. 1). The latitudinal and longitudinal extent is 10°0′2154″N to 10°15′50″N & 76°54′27″ E to 77°11′15″E. The area has a minimum height of 332 m (approximate) and maximum height is 2673 m (Relative Relief of the study area is 2341 m) and total area is 773.218 km^2. Munnar covers an area of 186.98 km^2 (approximate) and have 21 wards (shown in Fig. 2). Physiographical Munnar and its surroundings has foothill, valley, convex slope, concave slope, ridge crest, escarpment, and hilltop.

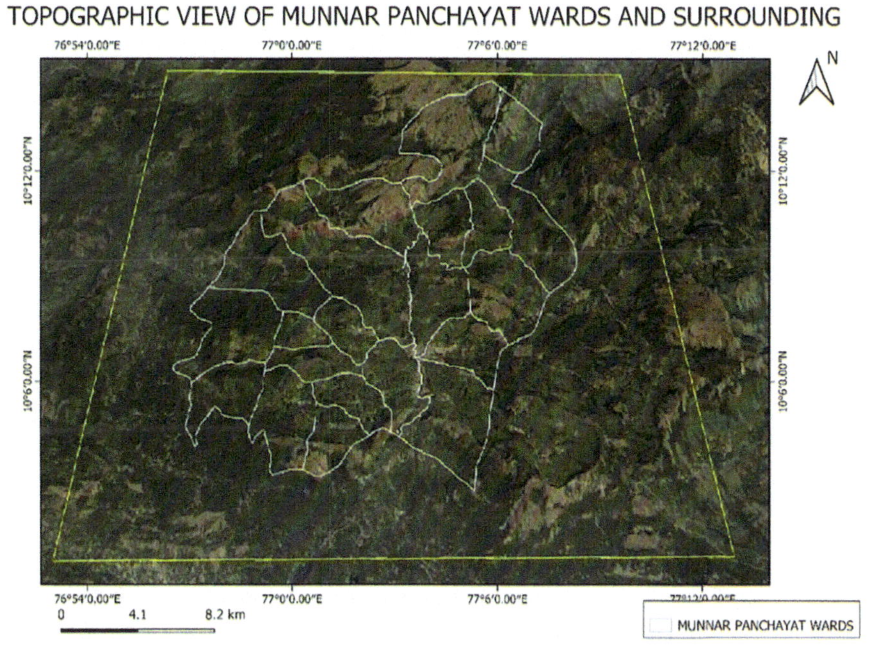

Fig. 1 Topographic view of Munnar and surroundings

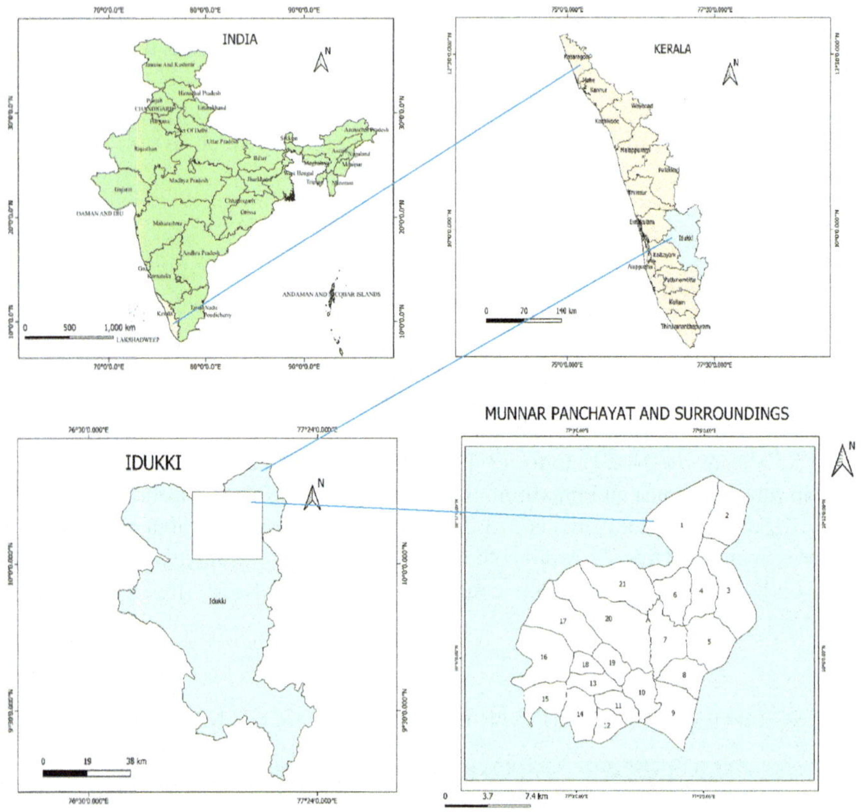

Fig. 2 Location map of Munnar Panchayat and surroundings

3 Literature Review

Literature review for the study can be divided into concept development regarding landslide susceptibility studies done so far and an information-based appraisal of the study area's natural and cultural features, land-use pattern, and tea plantation growth. Landslide hazard zonation (LHZ) is the classification of a land surface into homogeneous areas or domains and their ranking according to degrees of actual/ potential hazard caused by mass movement [5]. One important knowledge-building stage that is necessary for catastrophe mitigation, thorough landslip assessment management, and additional resilience building is landslip hazard zonation mapping. Although the landslip vulnerability assessment study's scope is multidisciplinary and based on spatiotemporal data that can be readily obtained via remote sensing, it is still lacking comprehensive empirical validation [6]. Geospatial techniques have made it possible to identify and evaluate landslides in even remote parts of mountain ranges, such as the Himalayas, and to identify landslide-vulnerable zones based on the main causative factors [7]. Numerous academics have attempted to map and analyse the causes of landslides using geo-informatics or GIS-based multi-criteria analysis [8]. Understanding the process, hazard analysis, and forecast are all part of the landslide analysis technique, which aims to minimise the upcoming dam damages. In order to go on to the overlay exercise of vulnerability map construction, studies entail the creation of landslip inventories, the collection and mapping of pertinent factor information, exploratory statistical analysis, the selection of final factor maps produced by weights, and the creation of a criteria tree hierarchy. Four approaches are preferred for conducting landslip susceptibility studies: deterministic, inventory-based probabilistic, heuristic, and statistical methodologies [9]. Heuristic analysis includes the multi-criteria decision analysis approach, which includes the ordered weighted average (OWA), weighted linear combination (WLC), and analytic hierarchy process (AHP) [10]. Heuristic approaches avoid the generalisation that is frequently a feature of data-driven approaches and enable site-specific and landslip type-specific assessments of the causative elements in addition to the combination of experts' knowledge and experience [11]. State Kerala is a witness of landslides and flood occurrences in Munnar and its surroundings of the district of Idukki.

4 Past Landslide Occurrences

Past landslide scars are identified by Google Earth and also based on a previous field survey by our college group. These landslide scars are indicators of areas which are vulnerable to landslide. Studying details of past landslide scar gives information to validate vulnerability assessment of landslide of the particular place. Detail study of landslide scar shows anthropogenic factors which are the cause of landslide, commonly depicting the locations along roads, de-vegetated points, and tea gardens. Past landslide scar location map (Fig. 3) shows the distribution of landslides in study area.

5 Aim of the Study

This research investigates the landslip hazard zonation maps of Munnar Panchayat wards and the surrounding area in order to determine the elements that contribute to the area's vulnerability to slope failure, with a focus on the natural and man-made effects in this terrain. The main objective is to evaluate preconditioning spatial factors based on image analysis and previous survey data integration on a GIS platform to prepare landslide vulnerability and susceptibility maps that can be probabilistically used to predict where landslides may occur in future.

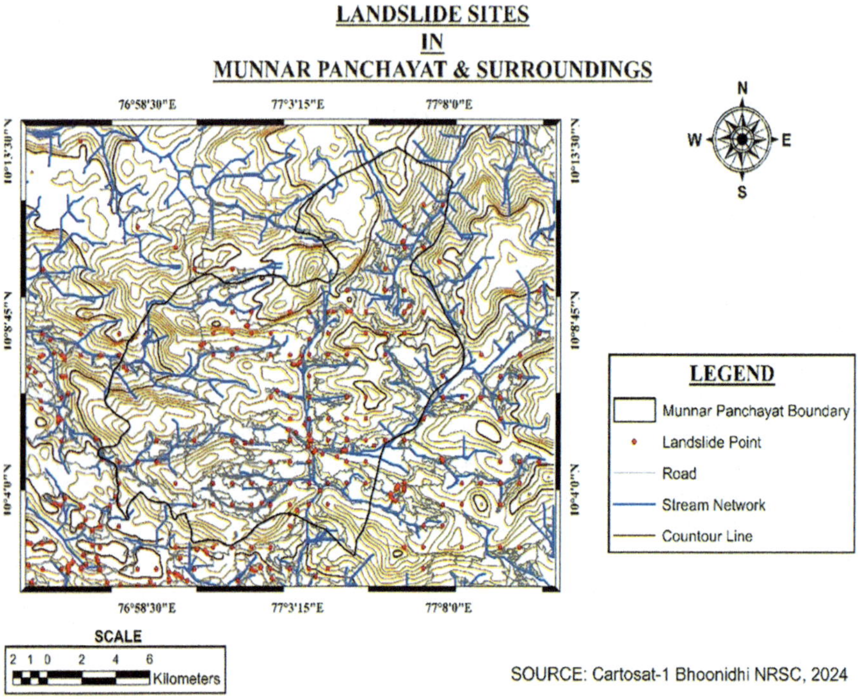

Fig. 3 Past landslide scar locations of Munnar and surroundings

6 Material and Methodology

In this study, geographic information system technology is used in combination with two methods: the analytical hierarchy process (AHP) approach and a method based on the expertise and experience of experts. Through the use of an AHP-weighted information content approach, the vulnerability and susceptibility analysis of Munnar Panchayat wards and the surrounding area is evaluated. The AHP approach is designed to boost the validation and acceptance of the landslip danger map.

In order to accomplish the goal, a comprehensive landslip inventory map of the research region has been created by digitising high-resolution satellite imagery and verifying the ground truth using GPS data from previous field studies. In addition, maps of the study area's landslip causal factor parameters have been created using a variety of sources, including satellite imagery, auxiliary data, the Digital Elevation Model (DEM), published reports and literature, and previous fieldwork. The research area's landslip susceptibility map is then created, and a landslip vulnerability and susceptibility assessment is conducted to determine which places are more vulnerable to these types of disasters so that appropriate action can be taken in the future.

The landslide susceptibility factors and their corresponding landslide susceptibility values are identified for the Landslide Vulnerability Mapping based on the intrinsic characteristics of the study region. The total number of factors is 100, and the factor values are divided depending on importance (Table 1). Depending on the data that is available, the factors might change. It is simpler to compute slope, relative relief, drainage density, roughness, dissection indices, and landform using topographical sheets and remote sensing image analysis. Geological map analysis was sufficient for lithological features. This is followed by the creation of a map for each factor and the identification of the classes it contains. Subclass-wise values or weights were then assigned keeping in mind that their sum must be equal to the corresponding LSV (Table 2). The assigning of subclass weights will also depend on significance of the classes on landslide susceptibility of the area concerned.

A detailed methodology flowchart (shown in Fig. 4).

Numerous landslip causative factor values were inferred using a digital elevation model (DEM) of the research area that was based on elevation data from the Advanced Spaceborne Thermal Emission and Reflection Radiometer (ASTER) (30 m resolution). Elevation, slope, aspect, and drainage to a certain degree are these elements. Using the ASTER DEM data, the drainage aspect was confirmed after being deciphered from SOI topographical sheets. From ASTER DEM data, contour elevation, slope, and aspect maps were created using the ArcGIS platform's spatial analysis capabilities, and they were verified by collating and validating topographical sheets. The Geological Survey of India provided geological parameters such as lithology, and LandSat provided information on lineament.

The factors' maps have zones organised in a grid-like framework. Furthermore, any grid may contain micro zonal fluctuations, single values, or point features. For this objective, taking into account micro zonal changes is more particular. Each grid

Table 1 Landslide vulnerability factors

Number	Factor	LSV
A	Slope	18
B	Geology	5
C	Lineament	8
D	Aspect	6
E	Relative relief	6
F	Dissection index	5
G	Ruggedness index	5
H	Drainage density	6
I	Distance to drainage	6
J	Landform	8
K	Vegetation quality	4
L	Land use and land cover	15
M	Distance to road	8

Table 2 Subclass wise breaking of LSV

Number	Factors	LSV	Subclass	Subclass value
A	Slope	18	>40.1	9
			30.1–40	5
			20.1–30	3
			10.1–20	0.5
			<10.1	0.5
B	Geology	5	Quartz vein and pegmatite	0.2
			Granite	1.5
			Granite gneiss	1.5
			Biotite gneiss	0.6
			Hornblende-biotite gneiss	0.6
			Calc granulite	0.6
C	Lineament	8	Very low and low	0.2
			Moderate	2
			High	2.5
			Very high	3.5
D	Aspect	6	Flat	0
			North	0
			North east	0.2
			East	0.2
			South east	1.2
			South	1.5
			South west	1.5
			West	1.2
			North west	0.2
E	Relative relief	6	Below 360 m	0.2
			360.01–535 m	0.6
			535.01–885 m	1.4
			885.01–1235 m	1.8
			Above 1234 m	2
F	Dissection index	5	<0.30	0.2
			0.30–0.40	0.3
			0.40–0.50	0.6
			0.50–0.60	0.9
			0.60–0.070	1.2
			>0.70	1.8
G	Ruggedness index	5	<0.6	0.2
			0.6–0.9	0.3
			0.9–1.2	0.6
			1.2–1.5	0.9
			1.5–1.8	1.2
			>1.8	1.8

(continued)

Table 2 (continued)

Number	Factors	LSV	Subclass	Subclass value
H	Drainage density	6	<2	0.2
			2–4 km	0.3
			4–6 km	1
			6–8 km	2
			>8	2.5
I	Distance from drainage	6	Within 100 m	3
			100–200 m	2
			200–300 m	1
			Above 300 m	0
J	Landform	8	Ridge crest and hilltop	0.3
			Escarpment and convex slope	3
			Foothill and valley	0.5
			Landslide scar	2.2
			Concave slope	2
K	Vegetation quality	4	Healthy	0
			Moderate	0
			Low	1.5
			Degraded	2
L	Land use and land cover	15	Settlement	4
			Forest and vegetation	0
			Grassland, open space, field	0.5
			Tea plantation	3
			Agriculture	1
			Rocky slope with partial grass cove	6.5
M	Distance from road	8	Within 0.5 km	4
			0.5–1 km	3
			1–1.5 km	0.8
			Above 1.5 km	0.2

is factor-wise assigned to the relevant subclass LSV if there is only one average value for each grid. The area percentages with the appropriate subclass LSVs are multiplied by the proportion of area under each subclass in a grid if micro zones are examined for each grid. The sum of the responses is then used to determine each grid's total subclass values for each factor. The gross LSV factor is applied to the overall value.

A landslide vulnerability map is then prepared with classes selected based on total LSV and two landslide susceptibility maps prepared based on two-way measurement of LSI.

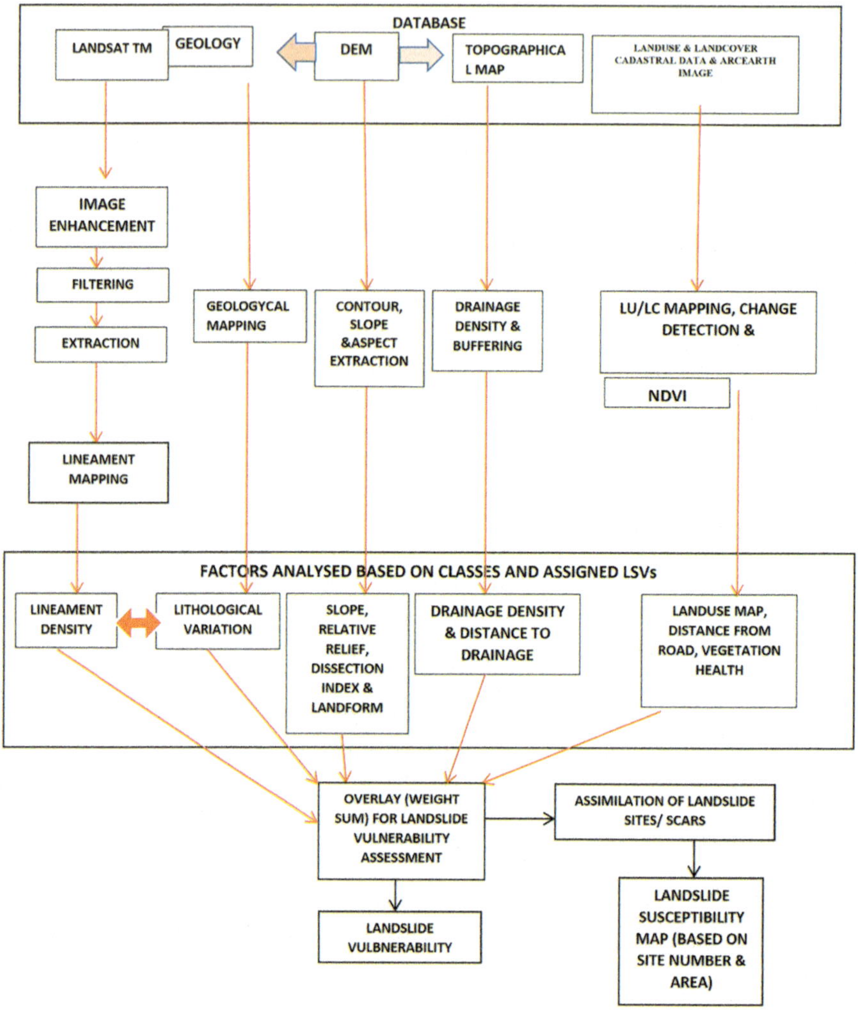

Fig. 4 Methodological process of landslide vulnerability and susceptibility mapping

7 Result and Discussion

7.1 Landform

Landform refers to features like hilltop, foothill, plain land, etc. A landform is a very important factor for landslide because it gives the idea of where debris is laying. If it is on the foothill, it is less vulnerable, but if it is on the convex slope or hill sides it become very much vulnerable. For landslide vulnerability assessment, it is

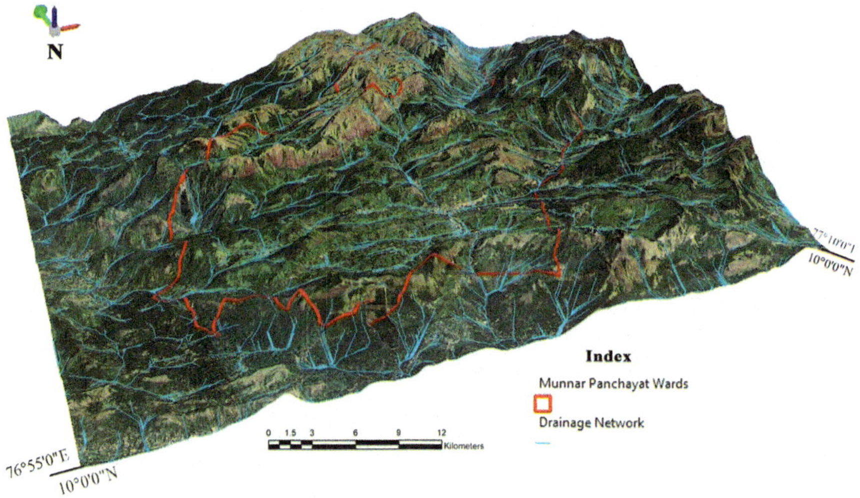

Fig. 5 3D topographic view of Munnar Panchayat and surroundings. (Source: USGS, USA)

an important parameter collaborating also with drainage pattern and stability of slope material.

The study area is divided into five types of landform features: ridge crest/hilltop, escarpment and convex slope, foothill and valley, landslide scar, concave slope. Northern part of Munnar is under ridge crest and central part of Munnar is foothill or valley. Landslides scar in the northern and southern parts of study area are situated on escarpment (Fig. 5).

7.2 Slope

Slope is the important factor for landslide. Slope is directly proportionate to landslide. An area under moderate slope is less vulnerable but area under high or very high slope is very much vulnerable because debris on high or very high slope become very much unstable and any triggering factor like rainfall led to landslide.

Study area (Fig. 6) is belongs to very low (below 10.1) to very high (>40.1°). We have divided the area into five classes and according to this class division most of the areas like north-western, western, south-western, and eastern parts are under very low to low. Southern, south-western, northern, and upper parts of Munnar belongs to moderate (20.1°) to very high (>40.1°) slope area.

Landslide scars are founding in northern, north-western, southern, south-western and central part of Munnar. This landslide places belongs to high (30.1°) to very high (>40.1°). Debris and loosely packed soil along with heavy rainfall act as triggering factors of landslides. Area under high to very high slope also increases run off, and gravitational pull increases the vulnerability for landslide.

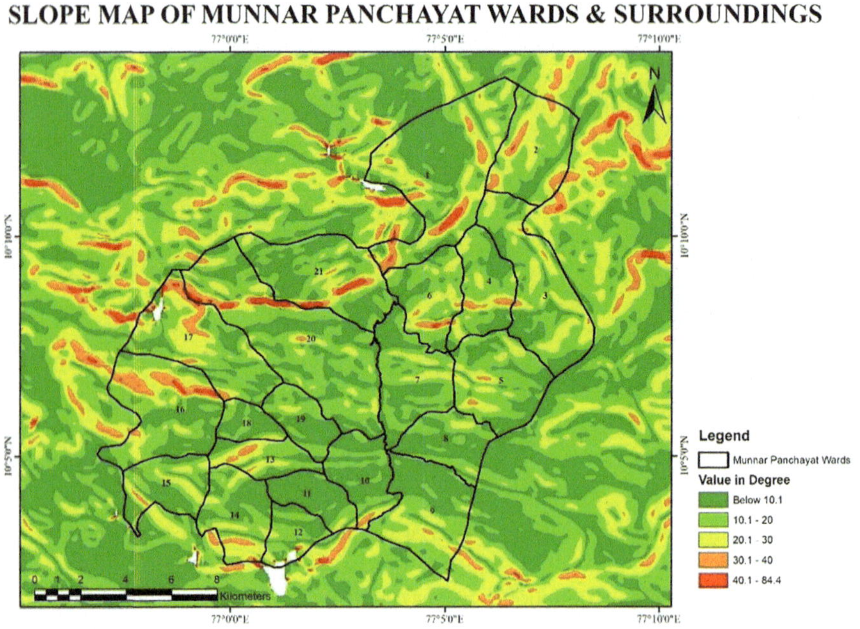

Fig. 6 Slope map of Munnar Panchayat wards and surroundings. (Source: USGS, USA)

7.3 *Geology*

Geology is the internal structure of rock beds; it also describes the rock bed types. From geological map, identifying the bedding planes, we can understand which areas are prone to rupture. Some rock beds are very much landslide prone and the structure of rock beds make them more vulnerable. This type of rock beds when highly triggered by geographical factors results in debris fall or mass fall.

The study area (Fig. 7) is under two geological time series Proterozoic and Archaean. Proterozoic series has two types of rock beds Quartz vein and pegmatite, Granite. The other series or older series is Archaean where four types of rock beds are found biotite gneiss, hornblende-biotite gneiss, calc granite, and granite gneiss.

Structurally, three types of features are found anticline, syncline, and lineament. Geomorphologically, foothill and mountainous area is dominating the study area. Granite gneiss and hornblende-biotite gneiss covers most of the study area. Biotite gneiss in south-western part and granite in south-eastern and eastern parts. Most of Munnar is under hornblende-biotite gneiss, granite gneiss (older series), and granite (younger series). The landslide scars are in granite gneiss (south, south east, northern parts of Munnar) and hornblende-biotite gneiss (central part of Munnar) rock beds. Both are older series rock beds. Some landslide scars are found on granite (central, south-eastern part of Munnar) which is younger series rock beds. These three types of rock beds are strong foliated in nature. Some landslide scars are near

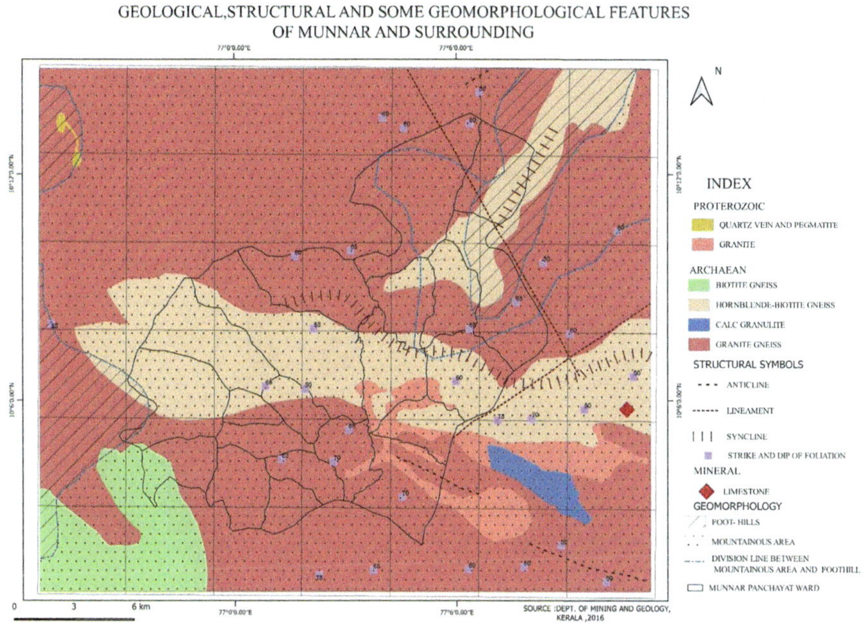

Fig. 7 Geological, structural, and some geomorphological features of Munnar and surrounding

and along lineaments. Landslide scar in lower part of Munnar is near 50 strike and dip foliation and central and northern part is near 65–70 strike and dip foliation which influence the landslide.

7.4 *Lineament Density*

The lineament indicates or depicts the weak areas where mass wasting is most likely to occur. Although the two do not precisely overlap, there is a positive correlation between structural discontinuities and lineaments.

The study area (Fig. 8) is mostly covered under very low-to-low lineament zones. Moderate to very high lineament is concentrated near main river channels. Central and southern and south-eastern part of Munnar and study area belongs to moderate to very high lineament density and most of the landslides in the southern and south eastern of belong to high to very high lineament density. Landslides in central part of Munnar belongs to very high lineament density and north-western parts of landslides belong to moderate lineament density. Lineament density is positively correlated to landslide occurrence. High lineament density increases the vulnerability for rock fall or debris flow and area becomes more vulnerable with the interplay of geomorphic agents like drainage channels and rainfall, etc.

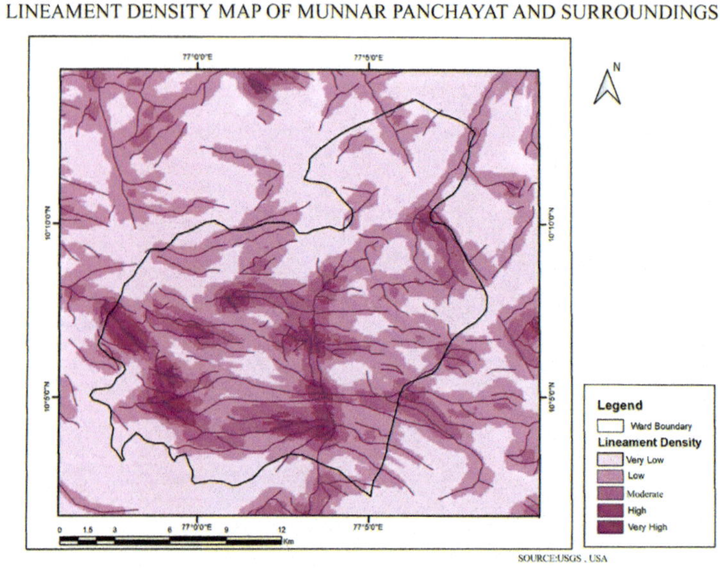

Fig. 8 Lineament density map of Munnar Panchayat and surroundings

7.5 Aspect Map

An aspect map shows simultaneously the direction and steepness of a terrain. The aspect map is generated from DEM (USGS).

The study area (Fig. 9) reveals that there is a combination of slopes from all directions in the region. Western, southern, south-western part of study area are the windward slopes. Most of the landslides are found in south, south-western and western, eastern parts, and these facets collaborate with highly steep slopes so that run-off moves swiftly during heavy rainfall events. Hence, these facets are very much vulnerable for landslides specially triggered by heavy rain. Loose soil and rock particles become unstable and falls. These aspect directions are given more emphasis and weightage is given as 6.

7.6 Relative Relief (RR)

Relative relief is the difference between maximum and minimum heights of a particular place. Relative relief is positively proportional to landslide. In a particular place where RR is higher the slope steepness becomes more, debris which are lying

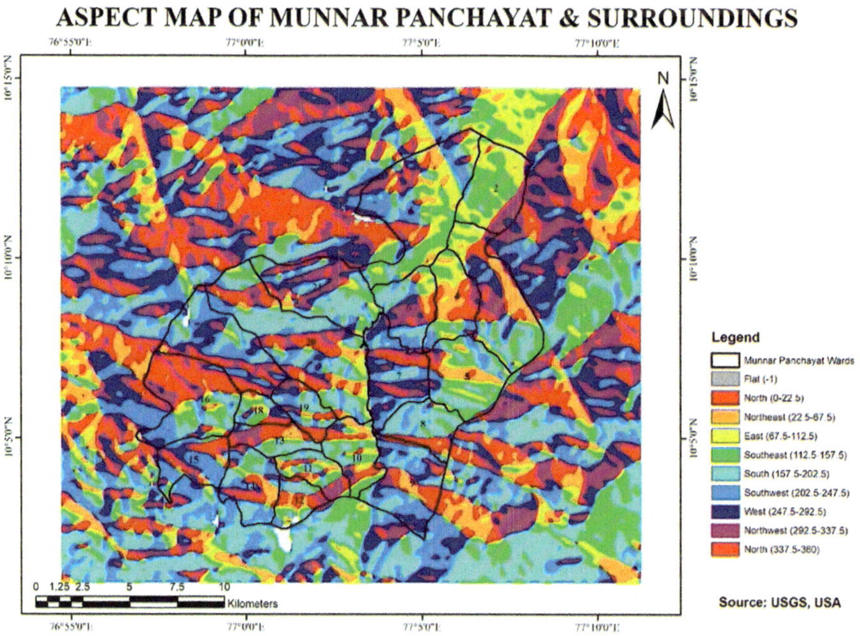

Fig. 9 Aspect map of Munnar and surroundings

on hill easily moves downward by gravitational pulls. Run off and other factors of landslide are very much activated in high RR values.

The study area (Fig. 10) has a diverse RR. Central part of Munnar has low RR of 200 m and northern, north-eastern, southern, western parts are dominated by high RR. The highest RR (1235 m) is in southern part of Munnar. The study area is classified into seven classes. Most of the area falls in the class 360–535 m to 710.01–885 m. Landslides are occurring in 885.01–1060 m and higher RR of above 1235 m. Landslides in southern part and north-western parts have occurred in high RR, greater than 885 m.

7.7 *Dissection Index (DI)*

Dissection Index reveals the roughness of topography, being the ratio between maximum relief and RR. Basically, it gives the idea of how much the area is degraded from its absolute relief.

The study area (Fig. 11) has diverse dissection value. Central part of Munnar has low dissection value where southern, western, and some parts of northern part of

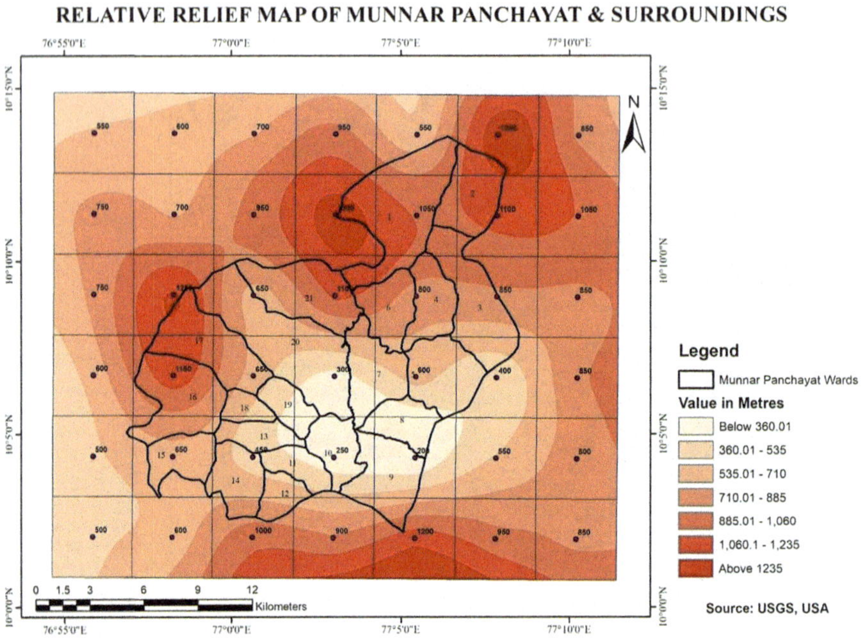

Fig. 10 Relative relief map of Munnar and surroundings

study area has high value of DI. The lowest dissection value is 0.17 in the central part and highest DI value is 0.74. The study area has been divided into seven classes. Most of the study area is under 0.30–0.40 class to 0.40–0.50. Landslides in the southern part of study area belong to 0.50–0.60 to above 0.70 class and the northern parts of study area is under 0.40–0.50 class. DI value indicates the steepness of the area, and how it affects the movement of debris and rocks downward. It increases the triggering factors of landslide.

7.8 *Ruggedness Index (RI)*

Ruggedness Index is the relation between Relative Relief and Drainage Density (DD). It indicates how the area is influenced together by RR and DD.

The study area (Fig. 12) has a lowest value of 0.23 in the Munnar Panchayat ward number 9 and 8 and high value of 2.36 in northern part of study area. According to the ruggedness values, the study area is classified into seven classes. Most of the study area belongs to 0.3–0.6 to 0.6–0.9 classes. Landslides in the northern and southern part of study area belongs to relatively high ruggedness value of more than 1 class. It also indicates that this area has been highly influenced by drainage

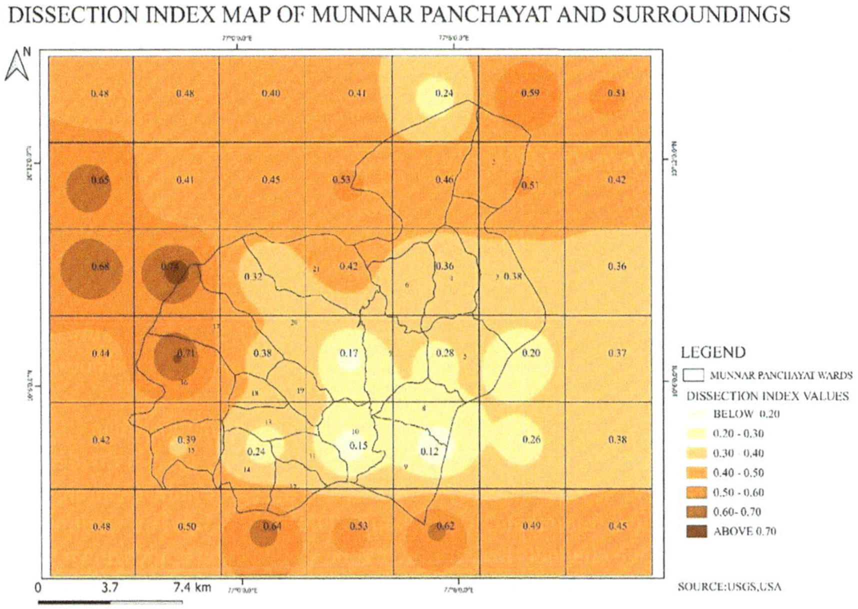

Fig. 11 Dissection index map of Munnar and surroundings

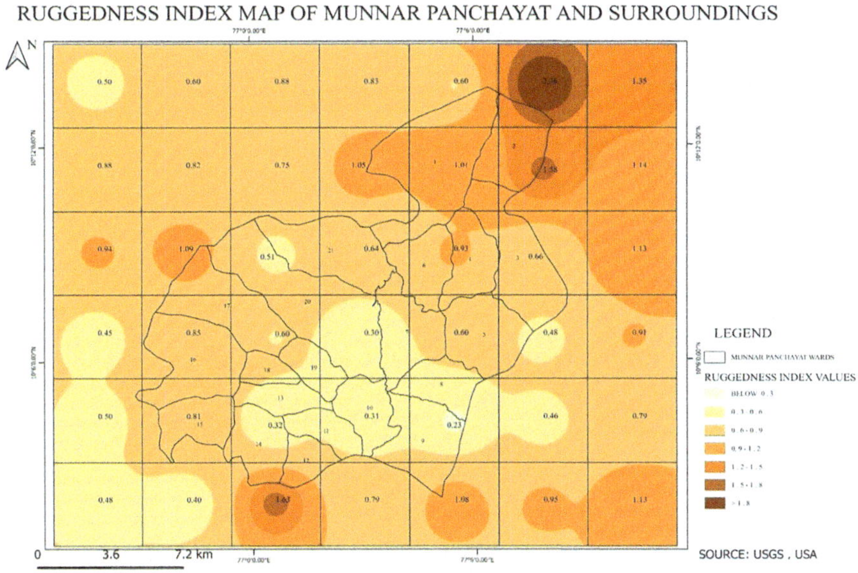

Fig. 12 Ruggedness index map of Munnar and surroundings

channels because of dominating high RR and hence assisting triggering factors for landslides. This factor is given an importance of weightage 5 in weightage table.

7.9 *Drainage Density (DD)*

Drainage density is one of the most important factors of landslide. Area under high drainage density has high vulnerability for landslides. Drainage density is positively proportionate to landslide, high drainage density affects the integrity of soil, adversely affect the weaker part of rock body and that's why during the rainy season with the DD and volume of surface water flow increase in channels and slopes, there is escalation of mass fall of rock and mud flow.

The study area (Fig. 13) has high DD, most of the area falls in more than 1.01 km/km^2 DD zone. The highest DD is more than 8/km^2. The study area is divided into nine classes. These classes are unevenly distributed in the study area. Central part of Munnar (Panchayat ward number 20,7) has high DD, near 8 km/km^2 attributed to the meeting of drainage channels here following the slope which comes down to this part from all directions. Drainage water congestion increasing many times in heavy storm events is one of the main reasons for landslide and flood in the central part of Munnar. The northern part of the study area where maximum places have drainage density more than 6 km/km^2 is very much landslide prone.

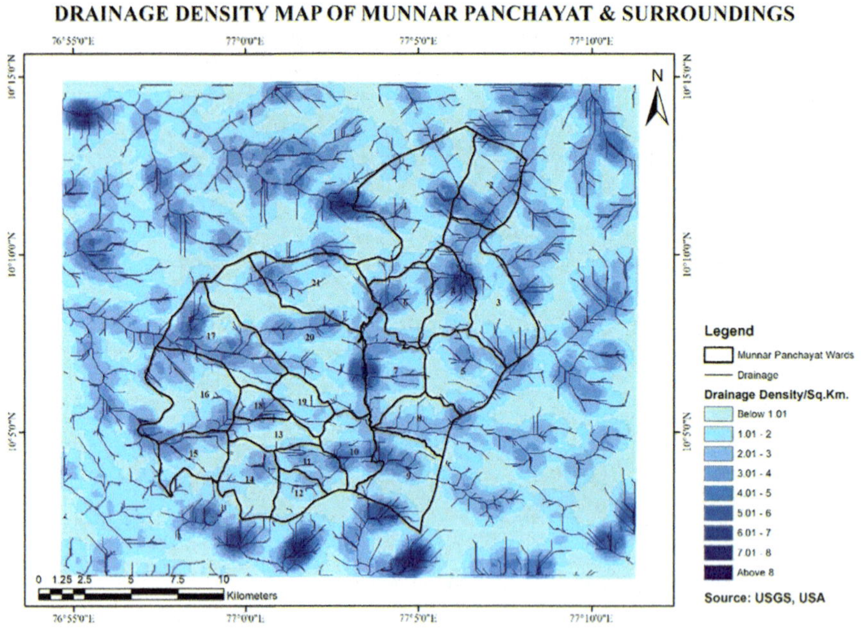

Fig. 13 Drainage density map of Munnar and surroundings

7.10 Drainage Buffer

Drainage buffer map depicts the areas in proximity to drainage channels. Landslide is very much influenced by the drainage channels and most of the previous landslides have been seen to occur in the nearest influence area of channels. The turbulence of water affects the loose particles of debris resting on slope.

The area (in Fig. 14) under the influence of 100 m from streams is most vulnerable for landslide. From the buffer map we found that with increasing distance from channels the landslide scars are decreasing. Some places of Munnar are free from stream influence, i.e. beyond buffer of 500 m.

7.11 Vegetation Quality

Vegetation quality based on NDVI (Normalised Difference Vegetation Index) is used to show the area under natural vegetation cover based on vegetation quality. Vegetation quality and landslide are negatively proportionate. Vegetation cover resists the soil from collapsing under direct rainfall impact.

The study area (Fig. 15) has been divided into four classes of NDVI-based vegetation cover quality zones. Most of the area is covered by moderate to healthy vegetation cover. The central part of Munnar has degraded vegetation, and other

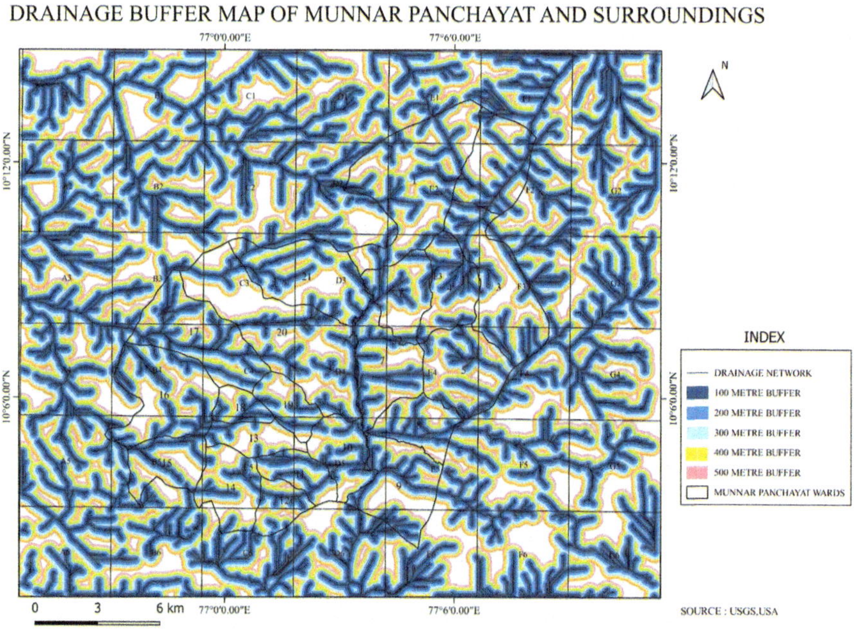

Fig. 14 Drainage buffer map of Munnar and surroundings

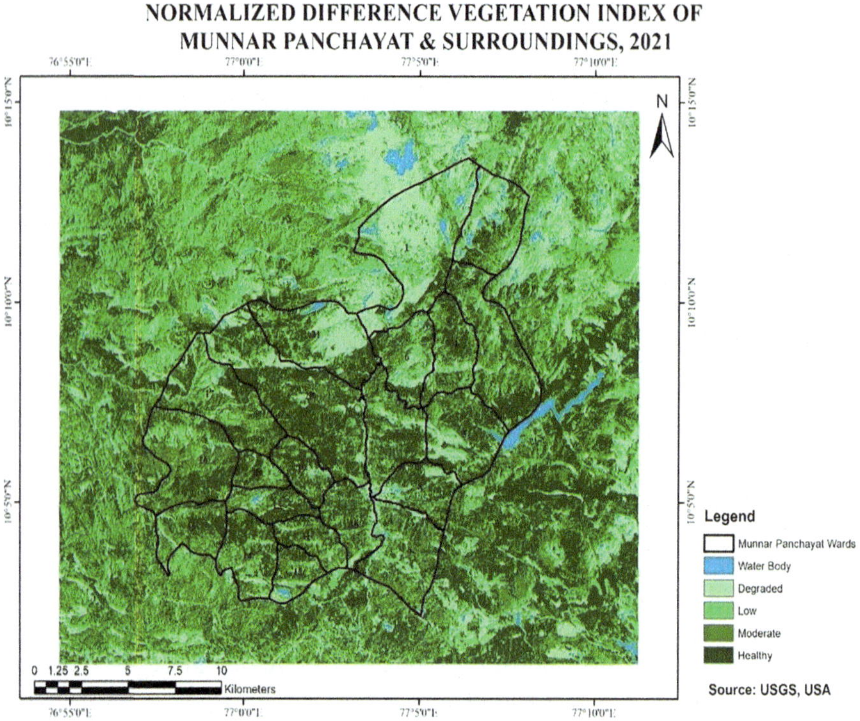

Fig. 15 NDVI of Munnar Panchayat and surroundings

parts like northern part has mostly degraded vegetation quality. Rest most of the entire area has combination of vegetation quality pockets. In this context, it needs to be mentioned that integral feature of Nilgiri vegetation is Shola forests interspersed with grasslands and in this specific area tea plantations.

7.12 *Land Use/Land Cover*

Land use/cover is an important factor for landslide vulnerability assessment because land use and land cover map indicate anthropogenic use of land. Unscientific use of land makes vegetation cover degraded, converts natural space to economic space.

The study area (Fig. 16) is mostly covered with forest and vegetation cover and grasslands. In north, north-western, western, and eastern parts, there is forest cover. The central part of Munnar where landslides are found is very much influenced by human interruption. This part is surrounded by the famous tea plantations of Munnar.

Anthropogenic factor is a major issue for landslide that's why LULC is given the second highest weightage. Cutting the slope for making roads and the debris resting on slopes makes slopes more unstable.

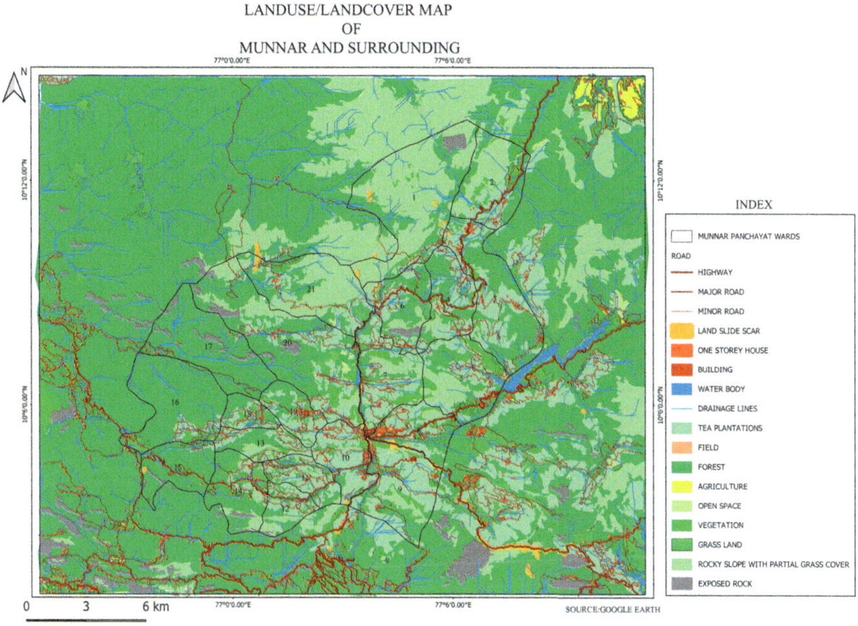

Fig. 16 Land use/land cover map of Munnar Panchayat and surroundings

The central part of Munnar is under extensive land use and in most places, there are degraded vegetation cover. Major, minor roads, highway are crossing in most places where slope debris are very much unstable. Most of the places are covered by houses. Area is very much urbanised and the population is maximum at that place. The area has school college, hospital, market, hotels, and all other amenities. From (Fig. 17) we can understand how much cultural landscape and life and property loss will happen if there is rise in occurrence of landslides.

7.13 Distance from Road

Road buffer map (Fig. 18) shows the influence of roads. In the hilly region where the main road or highway are built on the slopes of the hills, the stability of slope decreases. The debris which is lying on convex slope or mountain top becomes more vulnerable for downward movement. Cutting the slope and constructing roads or buildings area leads to degradation of the vegetation cover, drainage pattern, natural land use, slope stability, etc.

The study area has three highway and many major metalled roads, etc. The study area is divided into three zones based on road buffer, less than 0.5 km,1 km, 1.5 km. Most of the landslides are found in less than 0.5 km zone. Although much of the study area does not influence these three zones, those areas are mostly forest.

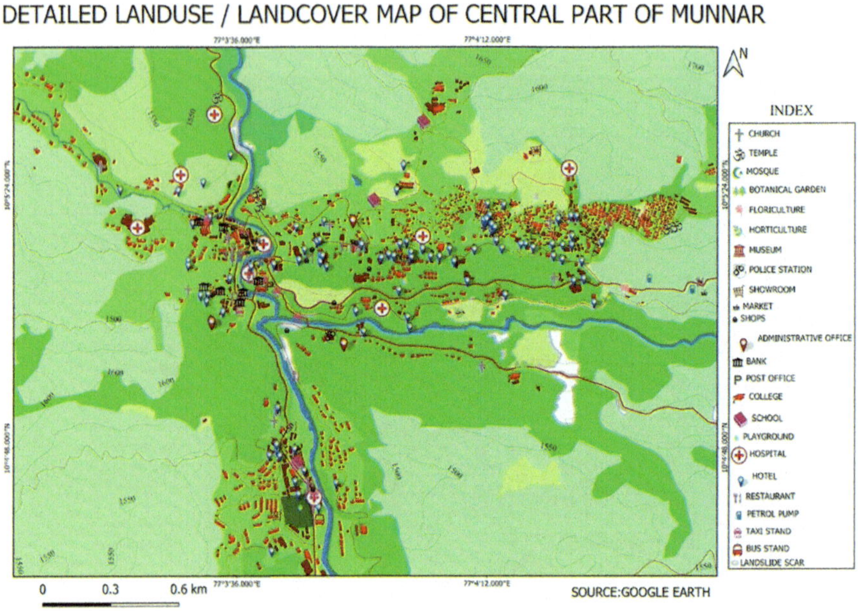

Fig. 17 Detailed land use/land cover map of central part of Munnar Panchayat

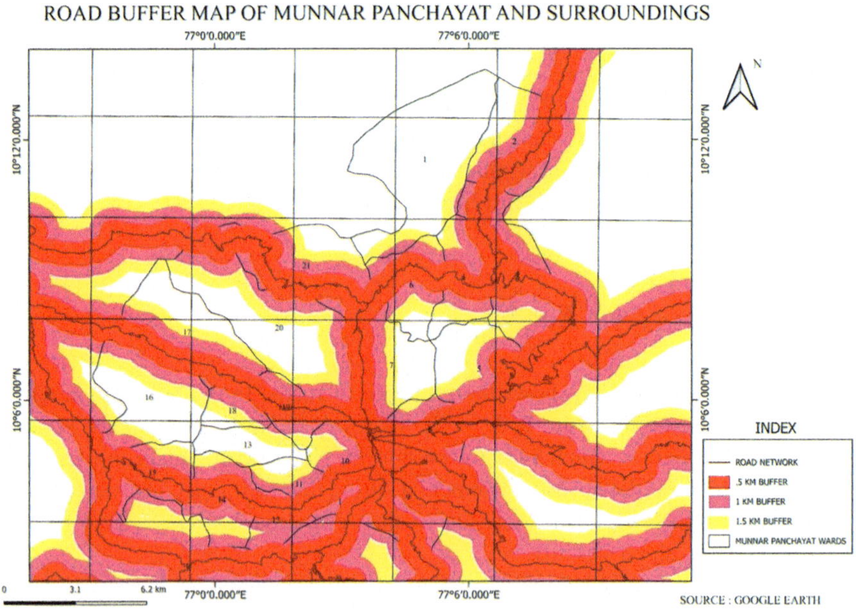

Fig. 18 Road buffer map of Munnar Panchayat and surroundings

Landslides in the southern part and the central part of Munnar and the northern part is near the highway and main roads, in less than 0.5 km buffer class.

7.14 *Major Findings*

The main aim of this study was to analyse the landslide susceptibility of Munnar and its surroundings.

With the help of the analyses of the landslide vulnerability and landslide susceptibility factors and the past landslide locations, the major results are

- Central part of Munnar has low relative relief, low slope, mostly foothill. Nevertheless, in this part drainage congestion, drainage buffer, road buffer, and anthropogenic factors like land use (built-up space and tea plantation) was important. With accelerated flow of discharge and sediments coming from higher parts and getting locked in this low area, there are many destructions caused in these parts to property like rads and buildings. Most of the landslides are on escarpments or convex slopes.
- Most of the landslide scars are very near (less than 0.5 km) to highway road and connecting roads.
- Landslides in the southern and north-eastern part of the study area are in very high Relative Relief, moderate to high (20.3–40° slope), degraded vegetation cover and more than 6 km/km^2 Drainage Density.
- Most of the past landslides are found in the south, south west, south-eastern facing slopes.

8 Landslide Vulnerability State

Landslide vulnerability map is composed on the basis of major landslide factors of slope, geology, lineament, aspect, Relative Relief, Dissection Index, Ruggedness Index, Drainage Density, distance from river, landform, vegetation quality, land use and land cover (LULC), distance from road and their subdivisions by giving the weightage. This weightage is given on the basis of their importance as a landslide aggravating factor. Slope, land use, distance from road, and landforms are given more emphasis. This LSV map (Fig. 19) has maximum value near 243 and minimum value of 151. North-western parts of study area and central part of Munnar (panchayat ward number 18, 13, 19) belongs to low landslide-vulnerable class. The LSV values are constantly increasing to northern and southern parts of study area. Munnar Panchayat ward numbers 1, 2 and some parts of 10, 11, 12, 9 have very high LSV value and many landslide scars have also been found in these parts. In southern parts of the study area, the maximum LSV value is found.

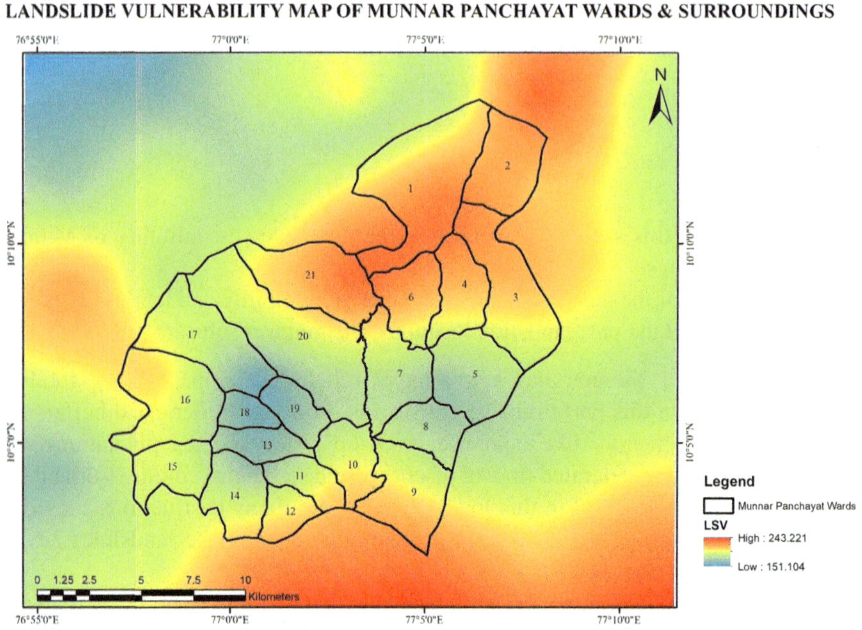

Fig. 19 Landslide vulnerability Map of Munnar Panchayat Wards and Surroundings

9 Landslide Susceptibility State

Landslide susceptibility index map has been prepared by integrating landslide vulnerability results with landslide occurrence data. The map is composed on the basis of landslide vulnerability value and considering (a) the area of landslide scars and (b) the number of landslide scars.

There are two types of calculation:

First one is on the basis of area of landslide occurrence, the maximum value of the study area is 3.2 and minimum value is 1.4. This map (Fig. 20) shows that the north-western parts of the study area and panchayat ward numbers 18, 13, 19, and 5 are in low susceptibility zones but the southern and some part of the north east like Munnar Panchayat ward number 21, 6, 2 and some part of 1 are in high to very high susceptibility zones.

The second map is on the basis on the number of landslides. In this type of calculation, number of landslides count in each grid was taken. Then the equation (number of landslide/area of the grid) + (LSV value/100) was followed. The maximum value is 62 and minimum value is 1. This landslide susceptibility map (Fig. 21) reveals that Munnar Panchayat wards 8, 9, 10 and southern part of the study area are in highly susceptibility zone where some parts of panchayat ward number 21, 1, and

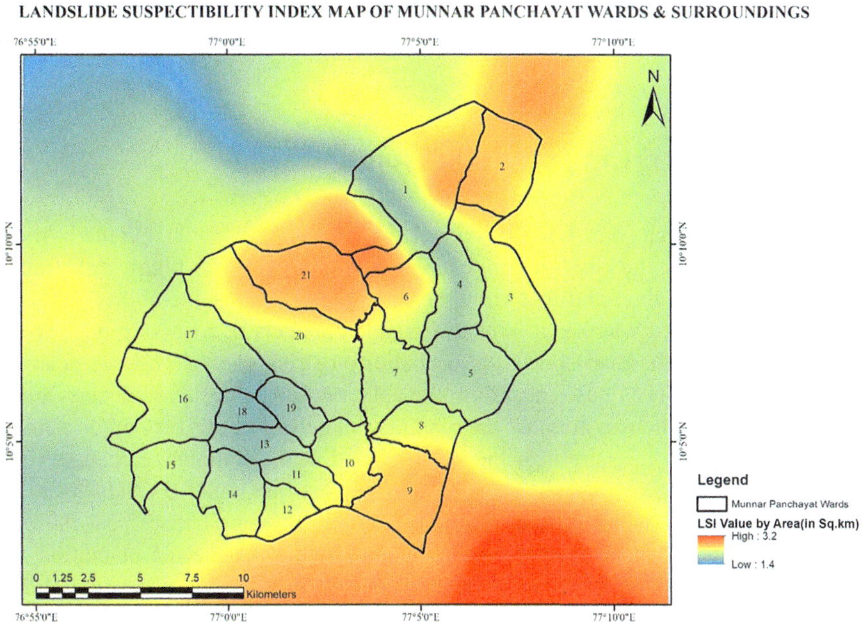

Fig. 20 Landslide Susceptibility Index Map of Munnar Panchayat and Surroundings (By the area of landslide scar)

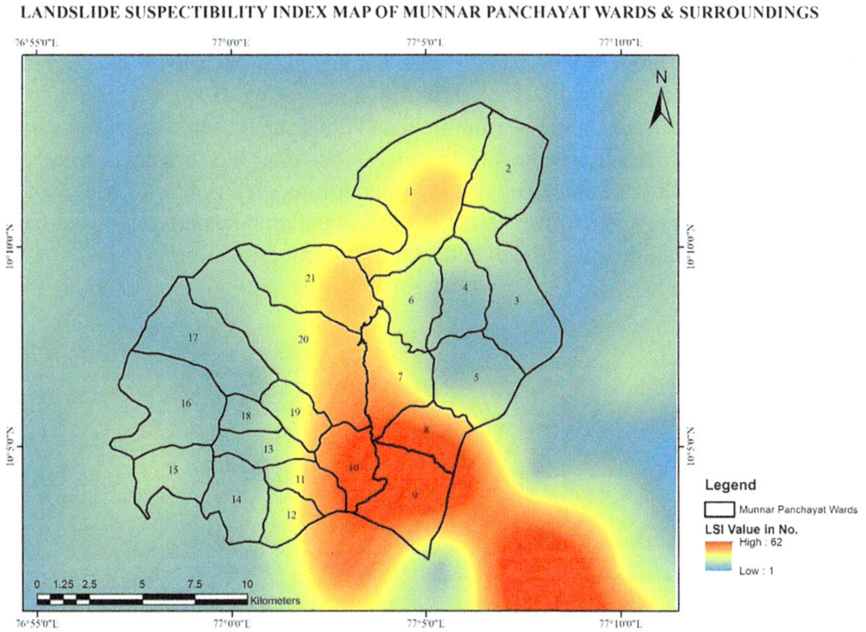

Fig. 21 Landslide susceptibility index map of Munnar Panchayat and surroundings (By the number of landslide scar)

6 are in moderately to high zone. Eastern part, western part, and north-eastern parts are in low susceptibility zones.

10 Buildings are in Vulnerable Zones

This map shows that some buildings are at risk and are vulnerable. On the basis of LSV map analysis, the buildings are classified into three classes highly vulnerable, moderately vulnerable, and less vulnerable. In this type of vulnerability analysis, we have considered whether it is near drainage channels or road side or near the landslide scars. This map shows the probability of risk of those houses near the vulnerable zones. The houses near or in the vulnerability value 221–243 class are in high risk. If these houses are near the road or channels, they are very much vulnerable. In Fig. 22, houses are near the highway and near the drainage channels and according to the LSV analysis these areas are coming into high LSV values and all houses are in great risk. In the central part of Munnar, some houses near the past landslide scars and near road, river, or intended land use like tea plantation are highly vulnerable but with increasing the distance from the roads the vulnerability decreases.

11 Validation Process

The ability of the model to anticipate values at new locations should be examined before selecting an interpolation model for decision-making. However, how can you determine whether your model is accurately and consistently forecasting the values if you don't know the true values at the places between the measured points? It appears that knowing the values of sites you haven't sampled is necessary to answer this question. Nonetheless, cross validation is a popular and extensively used technique for evaluating the quality and dependability of interpolation.

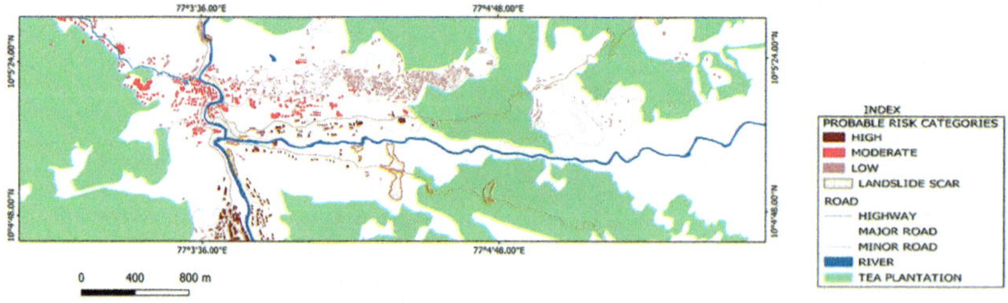

Fig. 22 Probable risk category

11.1 Cross Validation

As a leave-one-out resampling technique, cross validation first estimates the parameters of an interpolation model (e.g. the power value of inverse distance weighting or the semivariogram of kriging) using all input points. It then removes a single input point and uses the remaining points to predict the value at the location of the hidden point, and the predicted value is compared to the measured value (Fig. 23). The hidden point is then added back to the dataset, and a different point is hidden and predicted. This process repeats through all input points.

The cross-validation procedure for a single point is depicted in the diagram below. The value of the red point is concealed after the interpolation model has been estimated from all blue points, and the value of the hidden point is predicted using the remaining points. The measured value is then contrasted with the predicted. For each of the 10 spots, this procedure is repeated.

The value is inferred from the remaining points while a red point is hidden. For every point, this procedure is repeated. Because it mimics forecasting values at new, unmeasured locations—where the values are not unmeasured, merely hidden—cross validation is useful for evaluating interpolation models (Fig. 24). This allows the predicted values to be verified against their known values. The model should be able to predict the values at new, unmeasured sites with accuracy if it can predict the values of the concealed points. However, you should anticipate significant errors when making predictions for new locations if the cross-validation errors are quite high (Fig. 25).

When creating an interpolation model, cross validation is automatically carried out, and the results are displayed on the Geostatistical Wizard's final page. The Cross Validation tool can also be used to calculate cross validation on an existing geostatistical layer. If a geostatistical layer is present on a map, you may check the cross-validation statistics by clicking the cross-validation button in the Geostatistical layer's Data ribbon menu or by right-clicking the layer and selecting cross validation (Fig. 26).

The drawback of cross validation is that it first estimates the interpolation model parameters using all input points before successively hiding each point. They weren't completely buried in the process because every point helped determine the

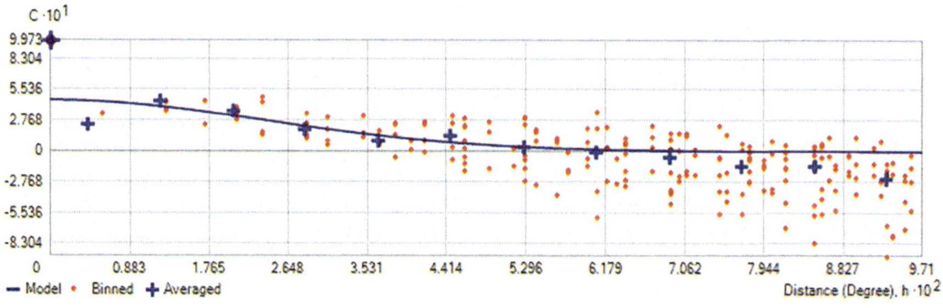

Fig. 23 Covariance (Observed landslides and calculated vulnerability)

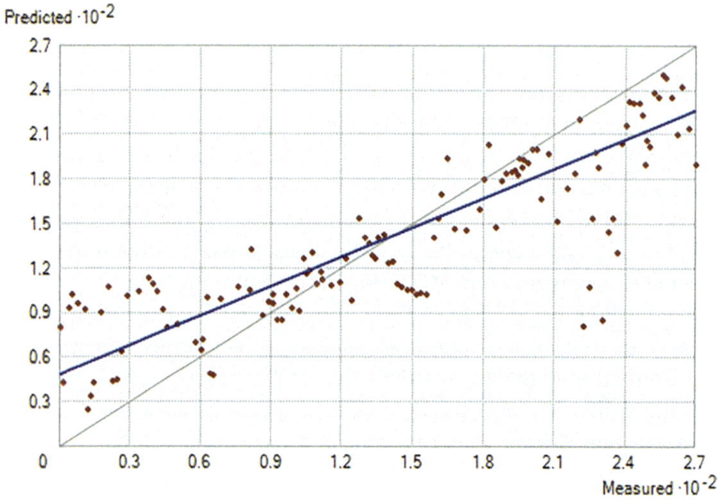

Fig. 24 Predicted and measured value between landslide location and prediction

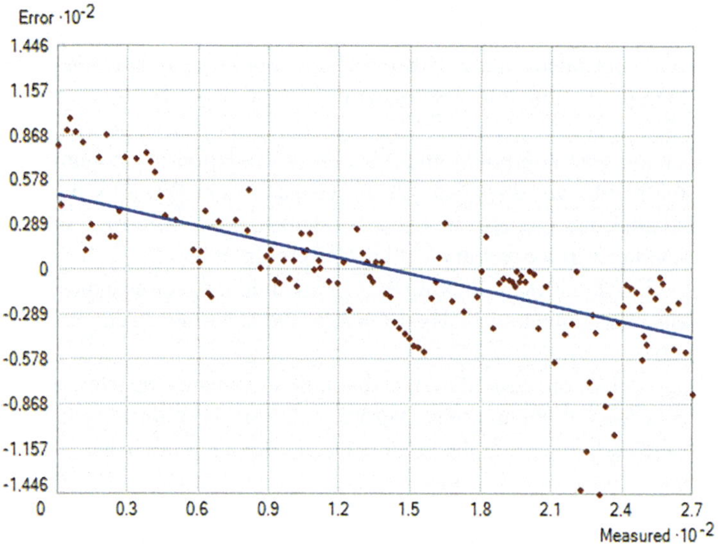

Fig. 25 Measured and standard error

interpolation settings. Even a single point can have a substantial impact on the estimations of interpolation parameters, whereas individual points often have little effect on the estimated values of the interpolation parameters. This is especially true for small datasets and datasets with outliers. Validation can be used to completely conceal point values and prevent data from being used twice.

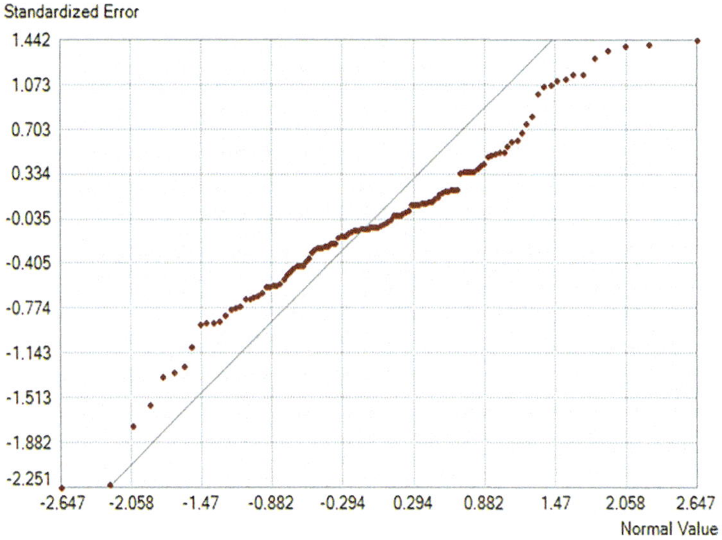

Fig. 26 Normal QQ plot

11.2 Validation

Similar to cross validation, validation starts by eliminating the test dataset, which is a complete subset of the input points. The remaining points, referred known as the training dataset, are subsequently used to estimate the interpolation model's parameters. The validation errors are then computed for every test point after the interpolation model has predicted to every location in the test dataset. Validation is the most rigorous method to determine how accurately and reliably the interpolation model will forecast to new locations with unknown values because the test dataset was not used in any way to estimate the interpolation parameters or make predictions. The main drawback of validation is that you can't utilise all of your data to create the interpolation model, which means that the parameter estimations might not be as precise and accurate as they would be if you had used all of the data. Unless the data is oversampled, cross validation is typically preferable over validation due to the need to minimise the size of your dataset.

The Subset Features tool can be used to generate the training and test datasets. The GA Layer To Points tool can be used for validation once an interpolation model (geostatistical layer) has been constructed on the training dataset. Predict the test dataset, validate on the field used for interpolation, and supply the geostatistical layer generated from the training dataset. The output feature class contains the validation errors and additional validation statistics. All principles are analogous to validation, but only cross validation will be covered in the remaining portion of this topic.

12 Conclusion

Around the world, landslides claim many lives and destroy a great deal of property in hilly regions each year. In order to identify places that are vulnerable to disasters, a landslip susceptibility zonation map is an unavoidable necessity. The data can be utilised by the local authorities to plan for disasters and put mitigation strategies into action. For the zonation of landslip hazards and susceptibilities, several approaches have been proposed. In the current study, the use of statistics-based modelling for landslide susceptibility mapping in a region of the Western Ghats that experiences frequent landslides is demonstrated.

The capabilities of GIS and remote sensing have been utilised during the integration and database preparation phases. By considering and compiling different landslip causative elements, a modified approach has been applied in this study to create an LSZ map of the Munnar Panchayat ward region. The primary goal of landslide susceptibility assessment and prediction, which addresses the spatial aspect of hazard, is to provide a response to the following query: Where can landslides occur in the future? For the local administrative authorities and decision-makers in charge of civil protection, infrastructure development, environmental management, and landslip mitigation methods, research on the spatial probability of landslip occurrence is essential. It is hoped that the zonation of landslide susceptible areas may help in effective and efficient management of landslide-related hazard.

References

1. Abraham, M. T., & Kuriakose, S. L. (2021). Multi-hazard vulnerability assessment framework for Western Ghats: Integration of landslide susceptibility with climate and anthropogenic factors. *Natural Hazards, 109*(3), 2567–2590.
2. Acharya, G., De Smedt, F., & Long, N. T. (2006). Assessing landslide hazard in GIS: A case study from Rasuwa Nepal. *Bulletin of Engineering Geology and the Environment, 65*, 99–107.
3. Ahmed, B. (2015). Landslide susceptibility mapping using multi-criteria evaluation techniques in Chittagong Metropolitan Area, Bangladesh. *Landslides, 12*(6), 1077–1095.
4. Aleotti, P., & Chowdhury, R. (1999). Landslide hazard assessment: Summary review and new perspectives. *Bulletin of Engineering Geology and the Environment, 58*, 21–44. https://doi.org/10.1007/s100640050066
5. Anbalagan, R., Kumar, R., Parida, S., & Lakshmanan, K. (2014). GIS based post earthquake landslide hazard zonation mapping of Lachung Basin, Sikkim. *International Journal on Emerging Technologies, Advanced Engineering, 4*(1), 431–441.
6. Bhargavi, G., & Arunnehru, J. (2025). *Landslide vulnerability assessment in Devikulam Taluk, Idukki District, Kerala using GIS and machine learning algorithms.*
7. Bhargavi, G., & Arunnehru, J. (2022). Identification of landslide vulnerability zones and triggering factors using deep neural networks—An experimental analysis. *Communications in Computer and Information Science*. https://doi.org/10.1007/978-3-031-12638-3_11
8. Bhattacharyya, P., DasGupta, S., Das, S., & Paul, S. (2020). *Landslide susceptibility analysis: A case study of Nainital municipal area.*

9. Cevik, E., & Topal, T. (2003). GIS-based landslide susceptibility mapping for a problematic segment of the natural gas pipeline, Hendek (Turkey). *Environmental Geology, 44*, 949–962. 12th Esri India User Conference 2011.
10. Chauhan, S., Sharma, M., & Arora, M. K. (2010). Landslide susceptibility zonation of the Chamoli region, Garhwal Himalayas, using logistic regression model. *Landslides, 7*, 411–423.
11. Chung, C. F., & Fabbri, A. G. (1999). Probabilistic prediction models for landslide hazard mapping. *Photogrammetric Engineering and Remote Sensing, 65*(12), 1389–1399.

Characterization and Pollution Risks of Surface Water Hydrochemistry in Mahanadi River Basin, Odisha (India): An Integrated EWQI, PCA, and WAPAS-TOPSIS Framework Approach

Abhijeet Das

1 Introduction

Rivers are utilized for industrial and agricultural processes, and they are a significant supply of drinking water. Water is requisite not only for humans but also for countless natural organisms, making it one of the most widely utilized environmental resources [1]. Surface water, a naturally occurring vital resource exhibits as a purer form of water as it is always clear, colourless, and odourless and is, however, currently overused in many regions of the world to satisfy the rising demand for household, industrial, agricultural, and drinking use [2, 3]. Additionally, surface water is a critical resource for maintaining global water security and contributes significantly to tackling the growing problems brought on by climate change and drought. Globally, climate change has resulted in a depletion of water supplies, which is exacerbated by the increasing demand for water due to urbanization [4]. Among the various water sources, surface water is particularly vulnerable and non-renewable and is extensively used for agriculture, potable water production, and industrial activities. The sustainable growth of humankind may be in danger because of the unbalanced supply of water and the increasing demand for it. Recent human activity has greatly increased the amount of nitrogen cycling between the atmosphere, soil, water, and living things [5, 6]. Particularly in arid and semi-arid areas, ensuring its sustainable usage is crucial to meeting the growing demand for irrigation and drinking water. Water pollution from both organic and inorganic chemicals is a major worry for the public. However, urbanization and population increase have resulted in a lack of freshwater resources worldwide in recent years, causing a

A. Das (✉)
Department of Civil Engineering, C.V. Raman Global University (CGU), Bhubaneswar, Odisha, India

S. Pathak et al. (eds.), *Geo-Data Revolution*, Advances in Geographic Information Science, https://doi.org/10.1007/978-3-031-94078-1_13

deterioration in both the quantity and quality of water supplies [7]. However, in many emerging economies, surface water is increasingly degraded, posing serious risks to human health. Human and ecological activities have been identified as major contributors to surface water contamination. Unfortunately, excessive use without a balanced recharge and unrestrained use of fertilizers and agrochemicals that seep into the aquifer system are causing surface water degradation in India to increase quickly [8, 9]. In Odisha, the Mahanadi River is an essential resource for agricultural operations and drinking water supply. However, the unsustainable exploitation of aquifers and inadequate management practices have led to a global decline in piezometric levels, threatening long-term water availability [10]. In fact, this also leads to worldwide concern due to a high supply of synthetic fertilizers, aquaculture wastes, sewage disposal, and animal wastes. However, increasing water demand, recurrent droughts, and overextraction have resulted in a concerning annual deficit of 3.3 million m^3, exacerbating water depletion. Therefore, water quality should be routinely inspected, especially in areas where sewage contamination is more likely and water treatment systems are less advanced or efficient. Therefore, preventing and managing river water pollution are crucial [11]. Addressing this issue requires a scientifically robust and data-driven evaluation of groundwater potential zones to guide strategic decision-making and ensure sustainable exploitation. To address these challenges, the development of advanced and precise methodologies for groundwater potential assessment is important for identifying high-yield zones and supporting sustainable water resource management [12, 13].

Assessing its quality traditionally relies on labour-intensive sampling and laboratory analysis, often hindered by high costs and logistical constraints, such as reagent shortages during the COVID-19 pandemic. Several scientists have developed water quality indices (WQIs) to classify and predict water quality [14]. Therefore, the water quality index (WQI), a mathematical tool that significantly simplifies such data sets, provides a single categorization value that characterizes the extent of contamination or the water quality state of water bodies. Hence, by transforming water quality into a single score for comprehensive assessment, WQI aids the public and policymakers in understanding it [15, 16]. They are often used to assess the availability of safe drinking water and its suitability for domestic use, with calculations based on the World Health Organization's (WHO) water quality standards and national regulations. However, calculations of WQI metrics have undergone several developments and enhancements to better and more precisely portray water quality [17]. As a result, these water quality studies provide valuable insights into the contamination status of water bodies, based on the concentration of physical, biological, and geochemical parameters.

Upon reviewing the literature, it is clear that many studies have focused on comparing conventional WQIs with country-specific indices to assess water quality in various regions. These WQI models are often tailored to the specific characteristics of the study area. Commonly used methods include the weighted arithmetic water quality index [18, 19]. This methodological procedure is typically labour-intensive, incurring higher costs for equipment, chemical solutions, and human resources. Various academics have demonstrated that experts have varying preferences, which leads to varied ambiguities in the outcomes of water quality evaluations [20]. However, these methods often lack innovation and may involve complex calculations that lead to computational errors, time consumption, and issues with uncertainty and sensitivity. The specific end use of the index, such as drinking, entertainment, or irrigation, determines the selection and weighting elements of the characteristics. Meanwhile, combining additive and geometric techniques can help solve issues like ambiguity and eclipses [21, 22]. To mitigate these challenges, it is suggested that only the most significant water quality factors be included in the computations. For this, Entropy (E)-based weights have become a practical method that uses information entropy to give weights to water quality metrics, calculates the index weight based on the degree of variance of each index value, and prevents deviations brought on by human causes [23, 24]. As a result, this method is an improvement over conventional WQIs, which frequently draw from the Delphi method, the expert survey method, and the analytical hierarchy process. These techniques assign weights to variables based on both subjective assessments and professional opinions [25, 26]. Prior to computing the normalized choice matrix, chance of the attribute occurring entropy, average entropy represented by each explanation, and entropy weight, the method starts with establishing objectives such as the decision matrix [27].

Conventional entropy weights are used to assess each response's significance without taking the decision maker's choice into account. Yet, they have certain shortcomings, such as instability traits caused by differences in sample size and value and an inability to account for the conflict between the indexes [28, 29]. As complex as the cumulative effects of the many variables that define water quality are the challenges in classifying the crucial measures used to quantitatively evaluate the state of water resources. This method disregards the spatial and temporal oscillations of surface water components as well as the randomness, complexity, and nonlinearity of environmental issues [30].

To compensate for the deficiencies of the entropy method, multi-criteria decision-making (MCDM) analysis was considered in this study. It is quite essential that a small change in weighting will affect the overall interpretation of water quality [27, 31]. Hence, this strategy is a crucial method for establishing a set of appropriate weights for the variables affecting water quality. In past decades, several authors have applied the MCDM model such as Weighted Aggregated Sum Product Assessment-Technique for Order Preference by Similarity to Ideal Solution (WAPAS-TOPSIS) method, to various purposes in water resource management, evaluation of water quality as well as in other domains, to address issues related to technology and information technology management, security and threat management, power and long-term viability, and the environment [32]. Consequently, entropy is employed to determine the weight of the criteria, whereas WAPAS-TOPSIS ranks the options and selects the optimal one. By following the computation theory, this approach indicates potentials to work and create a combined or integrated approach, which at the same time logically receives positive feedback and in other side is within a domain of as MCDM format [33]. It exhibits as a class of correlation-based methods based on analytical testing of the decision matrix to determine the information contained in the criteria by which criteria weights are evaluated [34]. This makes it necessary for this method to prioritize options in the event of an emergency and provide an overall assessment of the sites by considering both the physical and chemical criteria [20, 35]. The imbalance between the evaluation indicators, which is based on the correlation between indicators, and contrast intensity, which includes the value variance between the various evaluation schemes of the same index and is expressed in the form of standard deviation, are therefore the two main concepts upon which the objective weight of the indicators is based [36].

For spatial representation, understanding and addressing these intricacies are crucial for developing sustainable and efficient water management strategies that consider the multifaceted nature of global water challenges, including insights from inter-basin water transfer projects that highlight broader management perspectives [37]. Therefore, many themed spatial maps that are used to examine the variance in surface water quality are created using Geographic Information System (GIS) software. A researcher, namely, Ibrahim et al. [38] applied a GIS to digitize the maps based on several chemical criteria, then merged all of the maps to identify the drinking-safe water zones. Through spatial analysis, modification, and visualization, remote sensing (RS) in conjunction with GIS is a versatile and intuitive application that supports surface water management planning and decision-making [39]. In the present scenario, inverse distance weighted (IDW) interpolation techniques were used to derive spatial variation maps. Ghosh and Jha [40] conducted an investigation that implemented a restricted irrigation WQI in conjunction with GIS to determine whether surface water is suitable for irrigation in the West Pampa Plain, located in Argentina. The IDW method was employed because a linear weighting of the disparity between the surrounding points, which is generated from variogram analysis, is capable of predicting the value at the unsampled point. Taghavi et al. [41] assessed WQ in Central Anatolia, Turkey, using an analytic hierarchy (AH) and IDW. Again, in the Sivas Province of Turkey, Mbissik et al. [42] evaluated the irrigation water using GIS and IDW and found that the irrigation-suitable zone contained the bulk of the water samples. This method provides two categories of interpolation strategies, such as deterministic and geostatistical, by allocating weight to the unknown point during the IDW extraction process based on the surrounding known points. Researchers have been calculating vulnerable zones by weighing several thematic layers using decision-making processes for numerous decades [43].

Integrating entropy and WASPAS-TOPSIS weights with GIS is a practical and affordable approach to managing spatial data [44]. Remote sensing technologies, such as multispectral imaging and satellite-based mapping, in conjunction with MCDM, enhance monitoring capabilities in areas where traditional sampling is challenging, capturing land use changes and water stress indicators to infer surface water quality over large, inaccessible regions [13]. Thus, GIS has provided a sophisticated workspace for managing, integrating, visualizing, and analysing digital spatial data from several sources [45].

Apart from WQI, multivariate statistical methods like factor analysis (FA) and principal component analysis (PCA) are commonly employed to evaluate surface and surface water quality [46]. Studies like Gao et al. [47] have applied hierarchical cluster analysis (HCA) and PCA, to assess and evaluate suitability of groundwater for drinking in Vietnam. Similarly, Alnahit et al. [48] used PCA and FA to find latent factors that explained 65.39%, 73.18%, and 77.61% of the total variance in terms of pollution levels after grouping 13 sampling locations into three clusters depending on water quality. By assisting in the identification of particular pollutants and their origins, these techniques offer insight into the environmental processes influencing the quality of water. For instance, Research has employed PCA and FA to evaluate groundwater contamination in regions like Dinajpur, Bangladesh, and Taiwan that are susceptible to health problems like black foot disease [49]. Additionally, Jayasiri et al. [50] categorized 12 sampling locations into three groups using hierarchical cluster analysis (HCA) based on 22 characteristics in order to evaluate the contamination levels in the Jinshui River, China. Additionally, they employed FA, which identified five factors that accounted for 90.01% of the total variance. Therefore, multivariate analyses provide insightful information about hydrochemical processes and the sources of contamination, and aid in determining the key elements affecting water quality, including natural geological processes, agricultural runoff, and industrial effluents [51].

This study highlights the benefits of a new integrated approach, which fills a research gap because the subsequent methodology concentrated mostly on a single method of solving problems in any setting.

The key contributions of this paper are as follows: The principal objective of this study is to propose a water quality index (WQI) utilizing entropy (E) model that utilizes the minimal number of attributes. This model adheres to the WHO guidelines, which were specifically developed for the Indian state of Odisha. The combined strategy of entropy with WAPAS-TOPSIS method is typically founded on a combination of the exponentially weighted product technique and weighted sum technique. This study comprised both PCA and spatial variation maps of water quality measurements, and at the end, maps were made and compared. To date, no model has been developed based on these state-specific guidelines, making this research unique, in the Mahanadi River Basin (MRB), Odisha.

The proposed model computes a single WQI value using 20 key water quality parameters. Based on the entropy WQI value, the water quality is categorized as either "Satisfactory" or "Unsatisfactory," following the WHO regulations. This approach not only simplifies the WQI computation but also reduces the time required to assess and classify water quality. The use of PCA and MCDM theory was utilized to identify the drinking water quality susceptible zones and examine how the physicochemical parameters affecting the water quality vary annually and geographically.

Ultimately, the model aims to serve as a basis for developing a water quality index system applicable across India. The findings of this study will help in evaluating the current state of drinking water quality in India and its suitability for human consumption.

2 Background of the Study Area

Mahanadi River System (MRS), most often, known as the largest river in the State of Odisha, it spans an area of over 141,600 square kilometres and is about 851 km long overall. Out of 851 km, 357 km falls in Chhattisgarh and further, 494 km in Odisha [52]. This basin is divided into three smaller basins: the Upper Basin, comprising up about 21.34% of the total, the Middle Basin, which takes up about 37.16%, and the Lower Mahanadi, which accounts up the remaining 41%. Generally, in the last, the lower basin extends a geographic region of around 57,960 km^2. The topography of this basin is characterized by vast, flat desert plains interspersed with low ridges and isolated dunes, with elevations ranging from approximately 200–400 m above sea level, influencing groundwater and surface water flow and recharge patterns in this hyper-arid environment [53]. This river indicates that there are 1.85 million hectares of net land under irrigation, with 686 m^3 of irrigation outflow per person. It also implies that central Odisha, which is inside the basin, is more thickly forested and has smaller-scale rainfed farming systems. The research area is set up in Odisha state, India. It lies within latitudes 19°20′N to 23°35′N and longitudes 80°30′E to 86°50′E. The research area is located in a tropical location with two different seasons: the dry season, which lasts from December to March and has an average annual temperature of 24–27 °C, and the wet season, which lasts from June to November and has an annual rainfall range of roughly 1200–1400 mm [54].

In this basin, the irrigation intensity was found to be 112%, total live storage capacity as 14.244 BCM and the estimated ground water share in irrigation reaching up to 34%. This amounts to 28.4% of the 75% reliable usable water and 21.32% of the mean annual flow. Thus, due to extensive irrigation infrastructure provided by big and medium-sized projects, the basin's primary land uses are agriculture and forestry. However, foods such as cereals consumed around 76% of the irrigated area, while the irrigation efficiency is pegged at 47% [55]. The drainage network in the study area is sparse due to the hyper-arid climate, featuring intermittent wadis that rarely flow except during infrequent rain events; however, subsurface drainage, through traditional Foggaras and natural aquifer pathways, plays a critical role in channelling water across the region [56]. Therefore, this watershed, which is specifically referred to as the third largest river in the Indian Peninsula, supplies residential water to large cities and has a long history of assisting with agricultural irrigation and fisheries in the Indian states of Odisha and Chhattisgarh.

The study's coverage area is a rapidly expanding suburban district with a number of commercial and industrial establishments. The amount of waste produced per person has increased as a result of these activities. Regulating waste management plans and strategies is still a problem in the research area. A possible threat to a sustained, high-quality water supply is hinted at by this circumstance. The demand for drinkable water has increased due to socioeconomic activity and population growth. Surface and subsurface waterways are the main water sources in this region [57]. The combination of traditional knowledge and modern scientific understanding adds a unique dimension to the study area, making this watershed a compelling region for comprehensive research on arid environments, aquifer systems, and sustainable water resource management strategies. The detailed map of the sampling locations is presented in Fig. 1.

3 Sample Collection and Measurement

The stations' locations were indicated in the current work, with a handheld Global Positioning System (GPS). For the purpose to measure 20 physicochemical characteristics, 19 surface water samples were taken from the State Pollution Control Board (SPCB), Odisha, throughout the pre-monsoon season between 2016 and 2024. This organized sampling approach involved selecting diverse sources such as industrial sites, domestic sewage dump yards, and point and non-point sources, based on their spatial distribution, accessibility, and relevance to agricultural and domestic use [58]. This strategy ensured comprehensive coverage of the region's hydrogeological variability without relying on random selection or a strict statistical systematic method. Samples were gathered using plastic polyethylene bottles, precleaned with 10% HCl for 24 h and rinsed with double deionized water. Two sample sets were prepared in the field: one non-acidified set for physical measurements and another acidified with 2–3 drops of ultrapure HNO_3 for chemical analysis [21, 22].

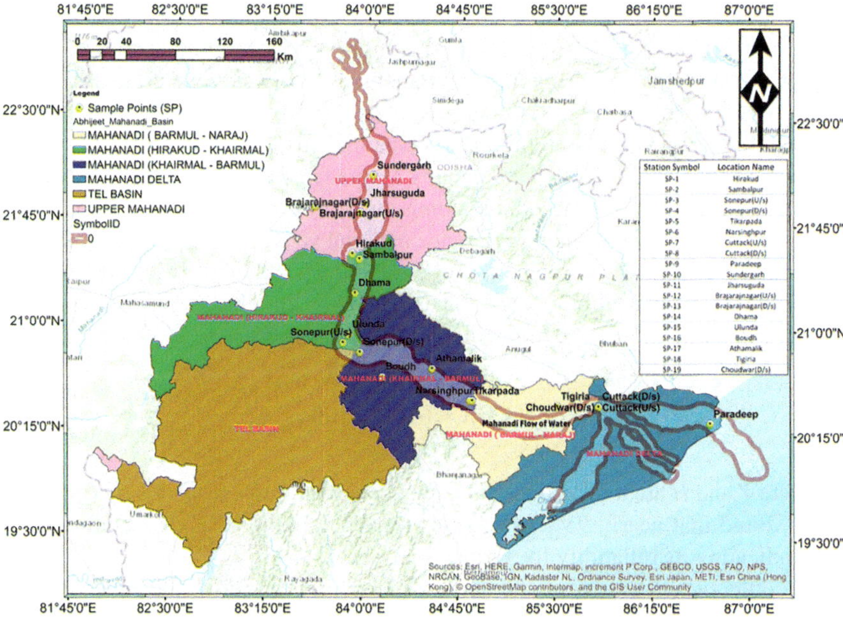

Fig. 1 Detailed overview of the study area

In the present work, 20 different water quality indicators such as pH, electrical conductivity (EC), NH_3-N (ammoniacal nitrogen), TKN (total Kjeldahl nitrogen), BOD (biochemical oxygen demand), free ammonia (Free NH_3), DO (dissolved oxygen), TC (coliform/total coliform), TDS (total dissolved solids), TH (total hardness), alkalinity, chloride (Cl^-), SAR (sodium adsorption ratio), Sulphate (SO_4^{2-}), Iron (Fe^{2+}), Fluoride (F^-), Boron (B^+), TSS (total suspended solids), COD (chemical oxygen demand), and TH (total hardness) have been used to measure and analyse the suitability of the Mahanadi River water quality for drinking. Water quality standards were kept in accordance with the American Public Health Association's [59] standard operating procedures since it identifies every element in a single category for all kinds of water and examines a broad range of water quality.

Anions, including nitrate (NO_3^-) and sulphate (SO_4^{2-}), were quantified with a Hach DR-5000 UV-VIS spectrophotometer employing a standardized turbidimetric method at 410 nm and 420 nm, respectively. BOD is determined by measuring oxygen consumption over 3 days at 27 °C. On-site measurements of pH, total dissolved solids (TDS), TSS, and electrical conductivity (EC) were conducted using a Hanna Instruments HI98130 portable water quality checker, with sampling locations georeferenced using a Garmin eTrex 10 handheld GPS device. Indicators like TKN, NH_4-N, and free-NH_3 were determined through Kjeldahl method, while DO is estimated using Winkler's approach. In the laboratory, alkalinity (HCO_3^-) was determined by titrimetric, and chloride (Cl^-) concentrations were measured using the Mohr method. Bacteriological indicator such as coliform or total coliform is obtained using multiple-tube fermentation technique, and is expressed in Most Probable Number (MPN)/100 mL. Cations (Mg^{2+} and Ca^{2+} in TH, Na^+ in SAR, F^-, B^+, and Fe^{2+}) were analysed under controlled conditions using a PerkinElmer Analyst 800 Graphite Furnace Atomic Absorption Spectrophotometer (GFAAS). The COD was measured using a titration method. Remaining water quality agents were examined, following the regulations set by APHA.

Prior to data analysis, a preliminary evaluation was carried out to guarantee the accuracy of the water quality data. Quality assurance and quality control (QA/QC) protocols were strictly followed, including instrument calibration, method blanks, duplicate analyses, and standard reference material verification to ensure data accuracy and precision. The accuracy of the analytical technique was measured by employing electric neutrality equation to estimate the charge ion balance error (CBE), expressed in Eq. (1), following Hasan and Rai [60].

$$\%\text{CBE} = \frac{Y - H}{Y + H} \times 100 \tag{1}$$

where Y and H are total cations and anion concentrations, respectively (in mg/L). It was found that nearly 93% of samples were below the CBE threshold value of 10% indicating its reliability for further analysis.

The outcomes demonstrate the data's dependability. Additionally, every sample in this study was tested multiple times to guarantee the correctness and dependability of the test results. In particular, this study computed the mean and standard deviation for each sample by running at least three parallel tests on it. Only when the standard deviation was within the acceptable range, it was considered to be reliable [61].

4 Methodology

The present study has chosen to analyse irrigation dangers and assess river water quality for drinking using a methodical approach. A summary of the suggested technique for evaluating and studying the WQI is illustrated in Fig. 2. The approach combines field and laboratory data collection with computational tools such as principal component analysis (PCA), and the entropy water quality index (EWQI), to offer a thorough assessment of the sources of contamination and appropriateness. Unlike traditional methods relying solely on sampling or basic statistics, this strategy offers a holistic perspective by synthesizing diverse data types, enhancing accuracy and reproducibility [62–64]. Moreover, decision-making model, namely WAPAS-TOPSIS, has been employed to disclose the spatiotemporal water quality for irrigation and consumption. It is particularly suited to resource-scarce arid regions, outperforming costlier, less integrative approaches by enabling rapid identification of key pollutants and supporting scalable water management decisions, with state-of-the-art computational decision support systems further enhancing efficiency in a changing climate.

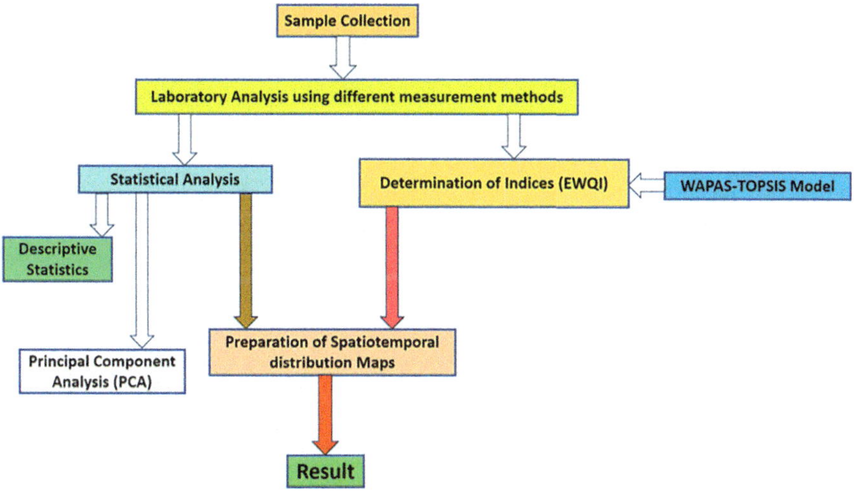

Fig. 2 Methodological flow chart of the current study

4.1 Water Quality Index Computing

The entropy-weighted water quality index (EWQI) method evaluates the temporal and spatial shifts of a water source's water quality by developing a simple WQI calculation process that uses less effort and is more accurate for assessing surface and subsurface water quality [27]. This method reduces oversimplification, enhances interpretability for policymakers, and outperforms costlier, less integrative techniques in arid regions with limited monitoring infrastructure. The EWQI can be thought of as simplified depictions of a complex reality or water quality models, where variables are selected and weighting and aggregation techniques are established. In contrast to the conventional WQI, the EWQI determines the weights of each water quality measure by using information entropy. By reducing the influence of human variables on weight distribution, this method improves the evaluation results' objectivity [44]. Therefore, by creating a water quality index, a trustworthy evaluation of the dataset of numerous variables was carried out at different sample locations. Despite the fact that this approach was used in surface water research, most studies on the different conditions concluded that it was successful when used in surface water studies [25]. As a result, when dealing with the stochastic component that causes uncertainty or unpredictability in a temporal study of time series data, this theory provides a concise definition of the system domain. Moreover, by taking into account the correlations between characteristics, the EWQI method provides a more thorough assessment of water quality by utilizing the information entropy approach [19]. The normalized choice matrix, the likelihood of the attribute occurring entropy, the average entropy contained by each explanation, and pronounced exceedances are computed after the objectives and decision matrix have been determined. Priority was given to those that have a significant impact on variations in surface water quality. Compared to the conventional single-factor assessment method, the EWQI provides a more thorough representation of the state of the water quality by taking into account and integrating the contributions of several water quality indices [65]. The approach was chosen because it allowed for a thorough evaluation of the cumulative parameter's concentration in aquifers impacted by a variety of natural and man-made sources by normalizing concentrations of potentially hazardous components against regulatory criteria. In the context of time and precision, entropy is frequently used to characterize the quantity of information or unpredictability in a series of data or events. Entropy tends to increase with time and evolution in a system. This suggests that the system becomes more unpredictable or ambiguous with time. Creating an initial water quality matrix, normalizing data, determining weight using the entropy weight technique, establishing quantitative grading standards, and computing and categorizing the water quality index are some of the phases in the process [29]. As a result, EWQI provides a more comprehensive perspective on water quality, lessens the impact of humans, and improves the accuracy and dependability of the assessment results. The suggested index's quality rating system, which was proposed by Kumar and Singh [24] designates five distinct types. The EWQI method was used in this study, consisting of five steps,

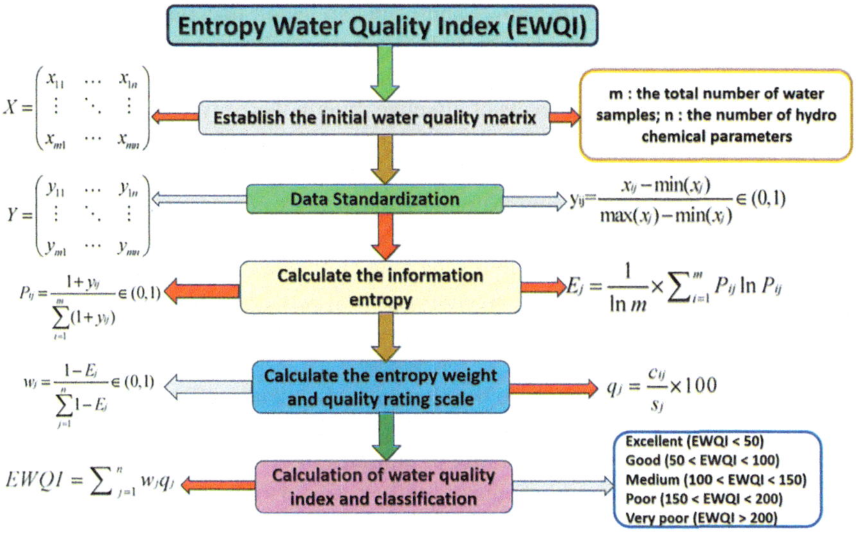

Fig. 3 A methodological workflow followed in this entropy study

which are as follows in Fig. 3. Conclusively, the method improves evaluation accuracy by analysing many parameters in different dimensions to provide an efficient assessment of water quality.

4.2 *Surface Water Quality Evaluation Utilizing WAPAS-TOPSIS Approach*

Multi-criteria evaluation is an essential step in the modelling of prospective zones for surface water. When estimating an EWQI, issues like rigidity, uncertainty, and eclipsing will always arise; however, these can be resolved using this method [35]. However, it is important to note that multi-criteria decision-making (MCDM) approaches that use Weighted Aggregated Sum Product Assessment-Technique for Order Preference by Similarity to Ideal Solution (WAPAS-TOPSIS/W-TOP) to evaluate the weight of each parameter separately can easily simplify the strengths and weaknesses of this process in establishing WQI for water quality assessment. Standardizing criterion weights, estimating ranks for each criterion, and aggregating the criteria are the three main phases in creating the surface water potential model [27]. An integrated, straightforward additive weighting and exponentially weighted product model serve as the foundation for the recommended methodology. As a result, the method suggests a list of compromise solutions. In the first phase, an evaluation was carried out to determine all the elements that influence the possibility of water recharge while accounting for various scales. Given the subjectivity and ambiguity of the stages involved in creating an EWQI, it is also reasonable to believe that this novel approach can lessen uncertainty in processes like ranking and parameter selection [34]. Therefore, a linear scaling analysis was performed to normalize the weight of each criterion, selected for its simplicity in the standardization process. All levels of effective criteria were given scores between 0 and 1, where 0 denotes no influence and 1 denotes the greatest influence on surface water potential. Therefore, the effectiveness of the methods suggested by this methodology is determined by the philosophical framework of process evaluations and the common language used to recognize and address complicated water challenges. Additionally, it makes it simple to decide how to handle the uncertainties that frequently characterize water management issues [66]. Criterion ranks were calculated according to how much each component contributed to the rise in surface water potential.

Field observations that clarified the elements influencing surface water potential were used to inform the selection of factors and their ranks. Thus, if the method being provided is predicated on a mix of attitudes of compromise, it is entitled to W-TOP. This makes it possible to query for the necessary data. Following the establishment of all necessary theme layers, a search can be made to find all appropriate sites [32]. Analytical hierarchy process (AHP) analysis allows for small judgmental errors in forecasting the W-TOP score, offering a strong theoretical foundation for multi-criteria evaluation using a normalized primary Eigenvector matrix. Plus, this technique's complexity is dependent on the Integrated Entropy WQI analysis, which establishes the criterion's weight. The MCDM technique will be used to produce a ranking preference order of the choices and discover the best alternative.

This technique's versatility in water quality monitoring was proved by combining the use of rough set theory, entropy weight, and the W-TOP approach [15]. The complete methodological framework is summarized in Fig. 4. Hence, the proposed

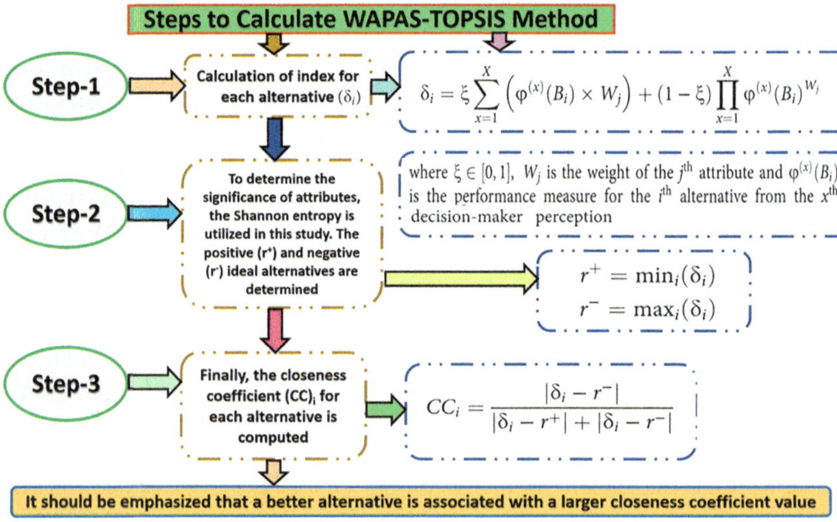

Fig. 4 Criteria and process for selecting the optimal procedure based on WAPAS-TOPSIS model

model will serve as a useful guide for future research in related fields and for educating stakeholders and policymakers about sustainable water management techniques in the study area.

4.3 Spatial Analysis

GIS spatial analyst techniques with Arc GIS 10.3 were used to perform the spatial analysis of several physicochemical characteristics. By using surface water proxy data, the research sought to use remote sensing (RS) and GIS to determine surface water potential zones. Using this method, a thorough assessment of surface water is essential for effectively maintaining and using this priceless natural resource. The inverse distance weighted (IDW) interpolation techniques have been responsible for producing the maps showing each physicochemical parameter's spatial distribution [48]. The method enhances the processing of data and the interpretation of the properties influencing surface water potential. Consequently, GIS and RS are additional methods for extracting, modifying, and assessing geographical and temporal information from satellite imagery [42]. For the Mahanadi River in Odisha, a spatial distribution map of the several water quality indicators has been created. Based on WHO drinking water standards, the technique in Fig. 3 comprised mapping physicochemical parameters and creating a drinking water quality index map for the studied time.

4.4 Principal Component Analysis (PCA)

Principal component analysis (PCA) was applied to reduce the dimensionality of the multivariate surface water dataset from the studied region, enabling the identification of orthogonal principal components (PCs) that represent the key axes of hydrochemical variability [67]. In order to minimize the dimensionality of the data, uncover latent patterns, and clarify the connections between items and their contributions to the total variance, this multivariate technique was selected. As a second-order statistical approach, PCA is utilized to mitigate multicollinearity and optimize variance explanation, as described by Hagage et al. [12]. Since the PCA is sensitive to scale variations, the analysis started by normalizing to make sure each component contributed equally. Nowadays, PCA has grown in popularity because it consistently identifies sensitive zones in a variety of scenarios, which makes it helpful for realistic assessments and forecasts. This robustness led to the prioritization of water quality in the study [8]. The methodological approach relies on the eigen decomposition of the sample covariance matrix, with the variance-covariance matrix X, expressed through Eqs. (2) and (3).

$$S = \frac{X^T X}{(n-1)} \tag{2}$$

$$S_{ij} = \frac{1}{(n-1)} \sum_{k=1}^{n} X_{kj} \cdot X_{ki} \tag{3}$$

where S turns into the correlation matrix sample, X^T culminates transpose of the standardized X variables, and $(S)_{ij}$ is the value of every covariance element. The PCA maximizes the variance of the principal component (PC), which is primarily associated with the correlation matrix or sample covariance eigenvalues. The correlation r between the X_{ij} of the PC is determined by dividing the variance of Z_j of the PCs by the covariance of the 'X' matrix, which can be obtained using Eq. (4).

$$r(x_i;z_j) = \frac{\text{cov}(x_i;z_j)}{\sqrt{\text{Var}(z_j)}} \tag{4}$$

where r is the correlation coefficient; Cov(x_i, z_j) is the covariance of X variables in the ith row and the jth row of the PCs; and Var(z_j) is the variance of jth PCs. According to PCA theory, a correlation is considered strong if its value is >0.90, good if it is between 0.75 and 0.90, and moderate if it is between 0.60 and 0.75. To identify the primary components that best explain the data's variance, PCA was then carried out independently. The first PC is oriented along the direction of maximum variance, with subsequent PCs orthogonal and ordered by decreasing eigenvalues. To evaluate the significance of each PC and ascertain the number of components required to capture the majority of the variability, the explained variance ratio and cumulative variance were computed [68]. To mitigate potential biases from multicollinearity, a pre-processing step implemented the variance inflation factor (VIF). Thus, variance inflation factor is determined to explore the level of multi-linear relationships that always exist in between variables. The method works by first measuring the relative distances between options and then assigning loadings to each conflicting criterion [5]. Each explanatory variable is regressed against the other remaining explanatory variables and response variable. Hence, the VIF is calculated as mentioned in Eq. (5):

$$\text{VIF} = R / (1 - R^2) \tag{5}$$

where R^2 is the regression model coefficient of determination. A standard threshold for identifying severe multicollinearity is a VIF >10. Sample variance and the dependency characteristics of the variables in the system are inflated by the variance inflation factor. The PCA approach is especially helpful for comprehending the predominant geochemical processes and seasonal fluctuations in surface water composition as a supplement to the water quality indexing phenomena [3]. Statistical analysis was performed using Python version, 2016.

5 Results and Discussion

5.1 Drinking Water Quality Mapping

The consequences arising from the study of water quality in India's water sources are examined in great detail. In-depth evaluations of the input parameters, statistical analysis, and correlation analysis were carried out in our study, which produced important new understandings of the main contaminants impacting the region's water quality. The article discusses the differences in WQIs between sampling resources, focusing especially on the health dangers brought on by microbial contamination [40]. Moreover, multivariate statistical analyses and decision-making techniques clarify the fundamental causes and mechanisms causing variations in water quality, providing guidance for public health protection and sustainable water resource management plans. By analysing the concentration of key contaminants using descriptive statistics and PCA analysis, this study establishes the groundwork for future studies and policymaking targeted at enhancing water quality throughout the region and offers insightful information on the current condition of river water quality. Surface water sample values were compared to drinking water quality requirements set by the World Health Organization (2017) for drinking and public health purposes (Fig. 5a–t).

The temperature is an important water quality parameter that can affect the water taste and odour, facilitate the growth of microorganisms and the corrosive ability of the water. One of the key characteristics of water is temperature, which has a significant impact on the water chemistry and biological processes of the aquatic species. Water temperature influences consumer perception, taste, and preference [69]. Elevated water temperatures accelerate chemical reactions, promote microbial development, decrease gas solubility, and intensify smells and tastes. Temperature values across the sites varied, with the detected ranging from 26.22 °C to 28.20 °C. The maximum temperature was recorded at SP-(9), and the minimum was at SP-(3). These fluctuations may be due to natural factors like weather and geography or anthropogenic influences such as wastewater discharge.

One of the most critical elements serving as a pollution index is pH. Thus, the pH indicator helps to identify whether the surface water is placed under alkaline or acidic group. Changes in pH can significantly alter the behaviour and toxicity of various pollutants, making the measurement of surface water pH essential. The pH of water controls many of the geochemical reactions, solubility of chemical compounds, disassociation of toxic heavy metals from their bound matrix, and their release into surface water [70]. In addition, the high pH of the river water may be caused by the photosynthesis process used by the algae and aquatic plants in the water. Analysis of the research period revealed that the value varied as 7.74–7.91 mg/L (Fig. 5a), indicating a somewhat alkaline character. These values are below the regulatory limit of 8.5 set for drinking by WHO [71].

Low DO levels (<3 mg/L) could suggest areas of eutrophication or excessive organic pollution, where oxygen is consumed rapidly by microorganisms decomposing organic material. Low DO levels can be harmful to aquatic life, leading to conditions like hypoxia or anoxia, which can result in fish kills and other adverse effects on the aquatic ecosystem [72]. Conversely, higher DO levels (up to 19.60 mg/L) might indicate areas with good water quality, possibly due to high aquatic plant growth or low organic load. Monitoring and managing organic pollution sources are crucial to maintaining balanced oxygen levels and supporting biodiversity in these water sources. In the river dataset, the computed DO value spanned between 7.25 and 7.81 mg/L (Fig. 5b). This indicates that the optimal value is within

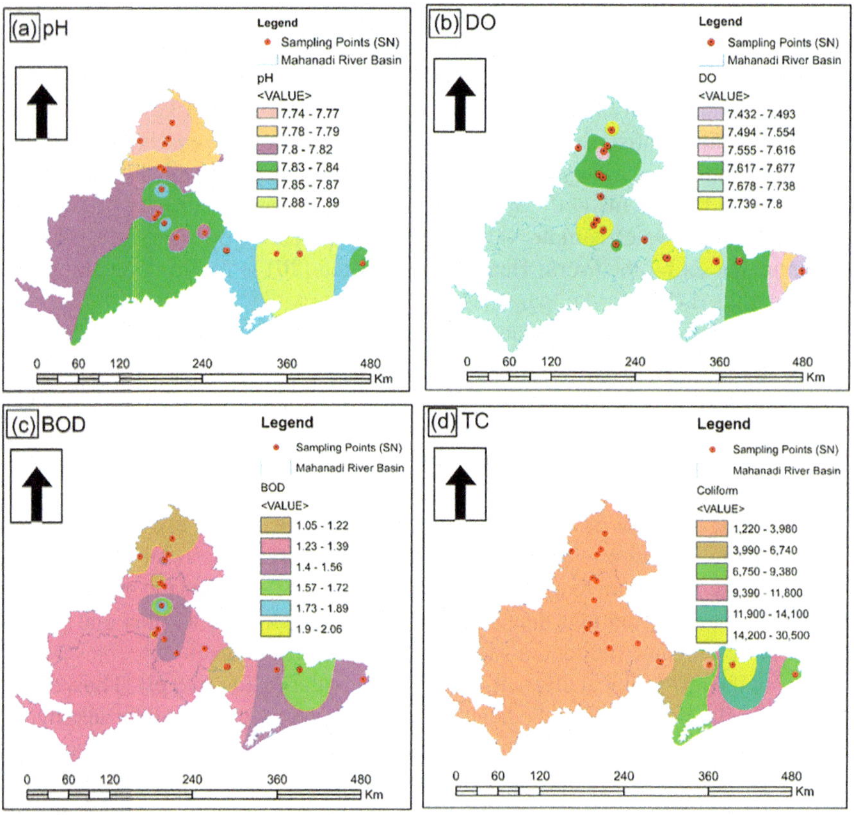

Fig. 5 Spatial representation of water sampling sites of parameters, (**a**) pH, (**b**) DO, (**c**) BOD, (**d**) coliform, (**e**) TSS, (**f**) alkalinity, (**g**) COD, (**h**) NH_3-N, (**i**) free NH_3, (**j**) TKN, (**k**) EC, (**l**) SAR, (**m**) B^+, (**n**) TDS, (**o**) TH, (**p**) Cl, (**q**) SO_4^{2-}, (**r**) F, (**s**) NO_3^-, and (**t**) Fe^2

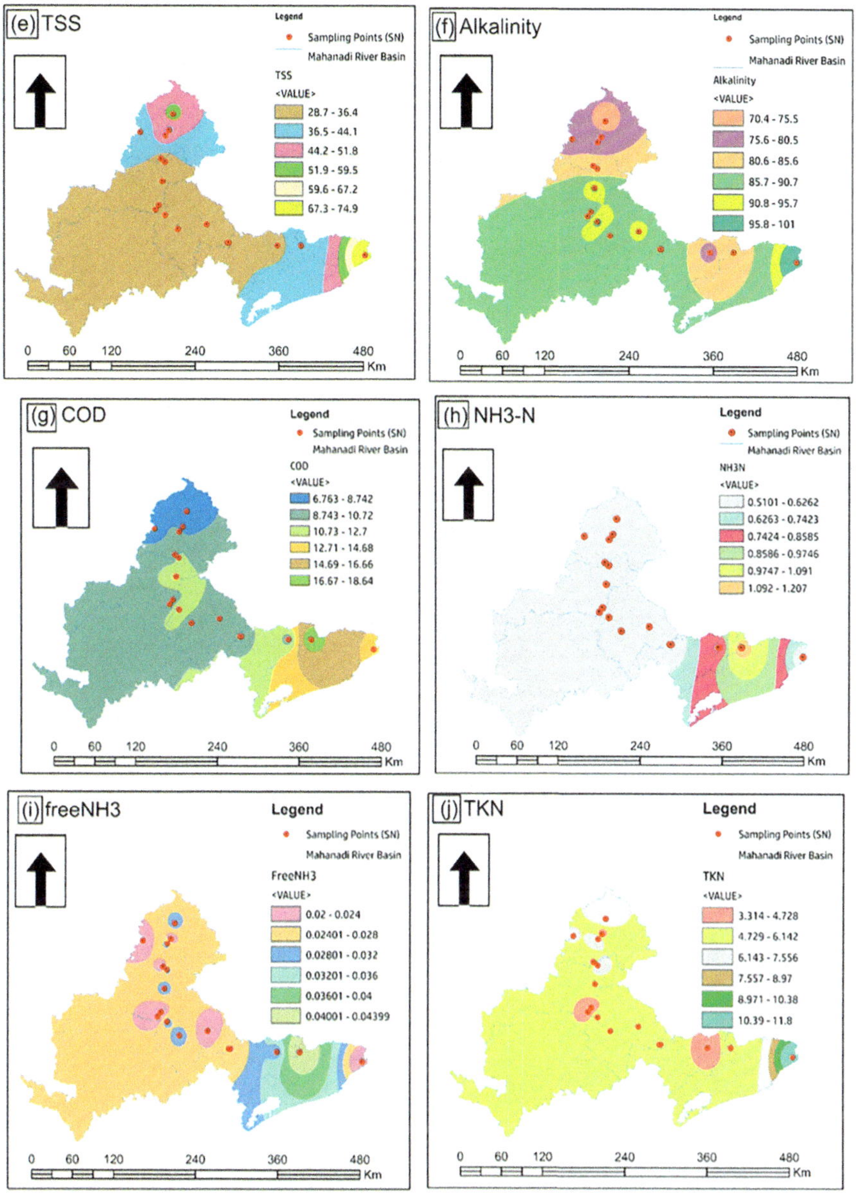

Fig. 5 (continued)

the typical range for freshwater systems (>6), suggesting that, on average, the river water maintains a moderate level of oxygen content suitable for aquatic life. The observed variation in data highlights the variability in oxygen availability across different water bodies.

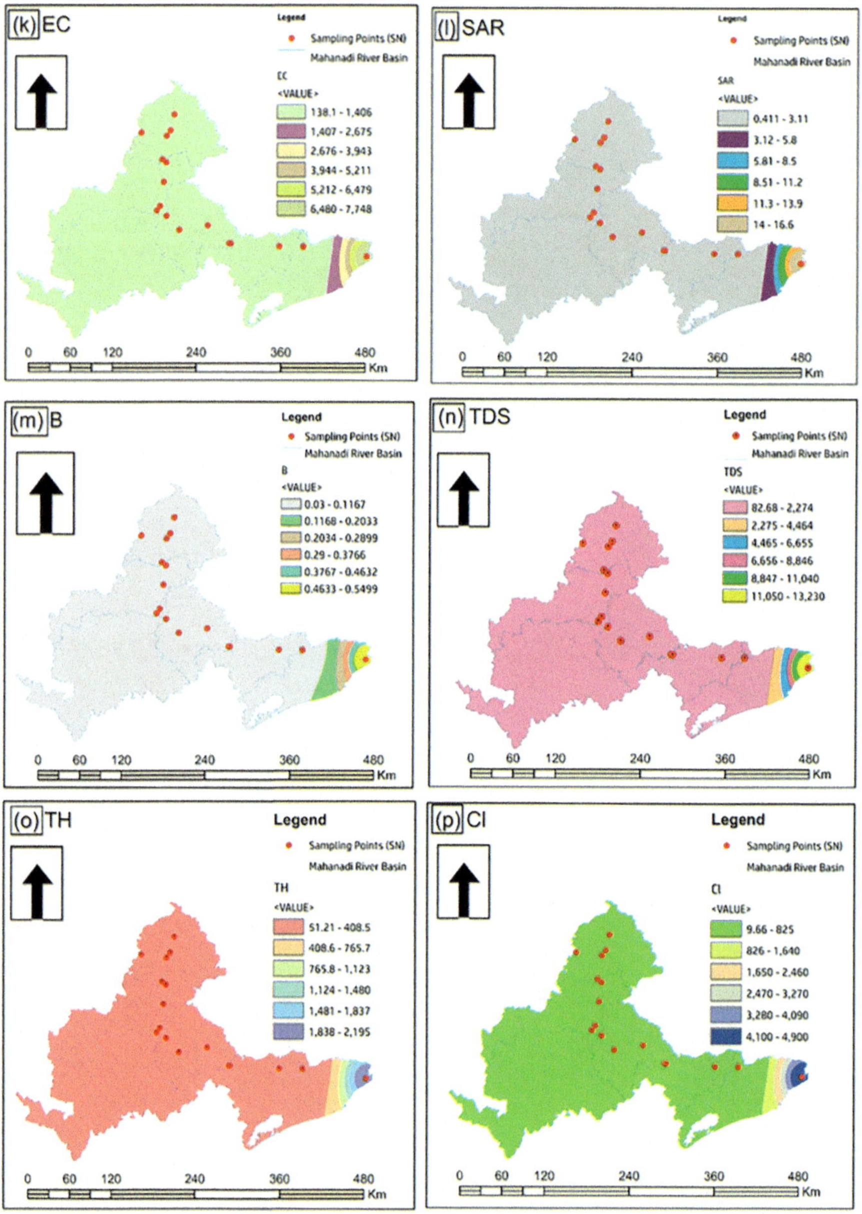

Fig. 5 (continued)

To evaluate the contamination of surface and ground water when it has happened as a result of the dumping of household and industrial effluents, water quality measurement using BOD is necessary [73]. A high BOD suggests that there is a large amount of organic material in the water, potentially from pollution sources such as

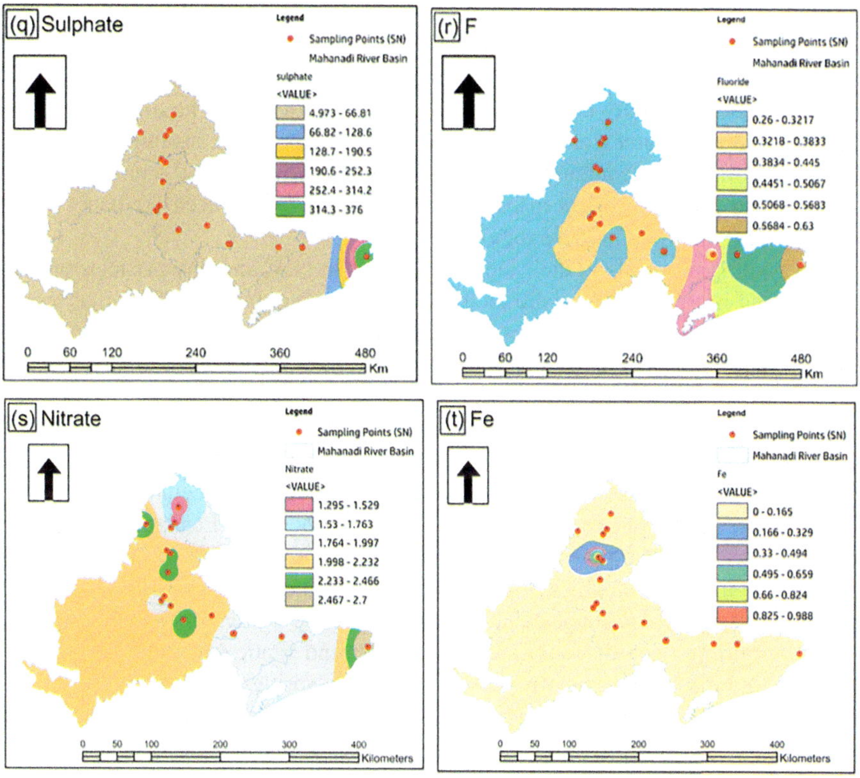

Fig. 5 (continued)

agricultural runoff or untreated sewage [74]. Sometimes, elevated BOD value suggests significant spatial variation in water quality, likely due to local pollution sources, land use patterns, and the overall health of the lake. Maintaining stream contamination levels, maintaining ecosystems, and regulating surface water quality all depend on accurate BOD assessments. The mean BOD concentration is 1.36 mg/L, which is relatively low, suggesting a minimal level of organic contamination in the sampled water bodies. The BOD values ranged (Fig. 5c) from 1.09 mg/L to as high as 2.38 mg/L, with considerable variability across the samples. This indicates that all examined locations are relatively clean (< 5 mg/L) and offer no show signs of pollution from organic matter.

Coliform serves as an indicator of faecal contamination as a reflection of the coliform bacteria present in the water. The total quantity of coliform bacteria, including those from non-faecal and faecal sources, is known as total coliform (TC). The present work reports a calculated value which ranged (from 1219 MPN/100 mL to 42,530 MPN/100 mL), that further supports the variability in contamination levels across the study sites. The mean concentration (Fig. 5d) of TC is 5151 MPN/100 ml (Most Probable Number per 100 millilitres), which is alarmingly high

(> 5000 as per WHO), indicating that a significant portion of the river water samples are contaminated with faecal matter. Sites SP-(8), (9), and (19) acquired readings that crossed the threshold limits, indicating a high degree of variability in contamination levels, potentially due to human or animal waste [75]. The findings suggest that three water sources are not suitable for direct human consumption without proper treatment. The contamination likely stems from agricultural runoff, untreated sewage, and improper waste disposal, which highlights the need for better waste management practices in these regions [76].

All floating materials in the water, including silt, clay, and tiny particles of both organic and inorganic debris, are included in TSS. The obtained TSS value scored as 28.62–74.88 mg/L (Fig. 5e), in the recent analysis. It is noticed that throughout the research region, the water quality is excellent. This phenomenon might be attributed to precipitation during the monsoon season, which facilitates surface runoff. Thus, the findings demonstrated that the river water's measured TSS concentrations were complying beautifully with WHO guidelines (<100 mg/L). Consequently, the combined discharge of agricultural and industrial effluents accumulates in river water bodies at few places, i.e. site (8) and (9), respectively. Leakage of untreated domestic wastes, open defecation, livestock wastes, septic tank, and toilet wastes leakage are important sources of microorganisms that contaminate surface water at site (9). This result also suggest that the need to regularly monitor water quality and strict controlling open defection, releasing toilet and septic wastes into rivers and streams, and discharging nearby water points likes springs and dug wells [77]. In another study conducted elsewhere, Das [78] have reported TSS ranged as 50–76 mg/L of Omo Kuraz drinking water which is slightly less than the present findings, may be due to difference in study locations.

Higher alkalinity in water improves its ability to neutralize acids and vice versa. The taste of water gets worse when its value is greater than 200 mg/L. The water at SP-(9) was found to be relatively more alkaline than at other stations, which may be due to the higher concentration of salts. Furthermore, at all locations, the good water quality and its homogeneity throughout the River Basin might be due to high flow rate, which aids in the dispersion and dilution of contaminants [79]. In this work, the reported alkalinity value ranged as 70.39–100.88 mg/L (Fig. 5f), signifying that the value fits well with 200 mg/L, as instructed by WHO norms. However, the findings show that the river can counteract acidic contamination from waste water or rainfall. Studies reveal that the underlying geology has a significant impact on the solute chemistry of Indian Rivers, for example, a river such as Mahanadi, which has dominant silicate mineral concentrations, has a distinct ionic composition with lower alkalinity concentrations [80].

COD is chemical oxygen demand for compounds discharged into water during oxidation processes. Thus, it is an essential indicator of organic pollution from sources such as partially or completely untreated industrial and residential wastewater in metropolitan areas [32]. The value acquired in this effort shows a range between 6.75 and 21.87 mg/L (Fig. 5g). For drinking water, the WHO [71] recommends a maximum allowable value of 30 mg/L. The lowest value was recorded at SP-(10) which might have contributed to reduced concentration by dilution effect, whereas the highest value was recorded at SP-(19), which indicates sampling point also lies in the area with dense human settlement. Based on the observation, because of the study's overall low COD, the water sample may contain less organic matter for bacteria to oxidize.

Ammonia (NH_3) is a by-product of organic matter decomposition and is often associated with wastewater discharge, agricultural runoff, or natural mineralization. The study region found that sewage disposal, chemical fertilizers, and the discharge of resident trash in open areas were the primary contributors of NH_3 contamination in surface water. The elevated levels of ammonia suggest significant anthropogenic influence, likely due to fertilizers, sewage, or animal waste [62–64]. High ammonia levels indicate potential nutrient pollution and organic matter contamination. Notably, the concentrations of Free-NH_3 and NH_3-N were found to be the most significant contaminants of surface water when considering both drinking and agricultural use. For human consumption, the desirable value is considered as 2 mg/L. The calculated NH_3-N score falls within the spectrum of 0.51–1.92 mg/L and the value of Free-NH_3 in the surface water samples recorded between 0.021 and 0.059 mg/L (Fig. 5h, i), respectively. The WHO criteria are nicely met by all tested observations. To restrict the contamination at SP-(9), management interventions such as controlling agricultural runoff, improving wastewater treatment, and restoring riparian vegetation are needed.

Increased TKN values were associated with areas that were irrigated by partially untreated wastewater, untreated wastewater, and solid waste disposal sites that had a higher nitrate value in the study area. The highest TKN concentration was recorded at SP-9 (11.80), followed by SP-2 (8.40 mg/L) and Sp-8 (6.05 mg/L). These values are considerably higher than those at the other sites. Sites SP-(1, 7, 12, 16, 18) showed much lower TKN concentrations (< 5 mg/L), indicating relatively better water quality. Notably, station SP-(9) had the highest variability, reflecting significant fluctuations in TKN concentrations (Fig. 5j) over time. SP-(2) and (8) also exhibited considerable variability, suggesting dynamic environmental or pollution factors at these locations [81]. However, TKN concentrations across 5 sites were within the recommended standards (WHO), indicating that the overall water quality is acceptable. The elevated levels at 14 sites are indicative of nutrient pollution, likely stemming from agricultural runoff, sewage, or mineral deposits. These sources often carry nitrates, which can be converted to nitrite under certain conditions [82].

As a measure of salinity or ions, electrical conductivity (EC) provides an estimate of the presence of specific ions indicative of a high concentration of dissolved solids. Because of its ability to accurately forecast the total quantity of mineral salts dissolved in water, EC is commonly used to evaluate salinity concerns related to agricultural irrigation [83]. The importance of EC lies in its ability to measure the concentration of dissolved salts, which can originate from both natural and anthropogenic sources. Prolonged use of high EC water for irrigation may necessitate special management practices, such as leaching salts or selecting salt-tolerant crops, to prevent long-term soil degradation. The study's EC values varied as 138.1–7779.30 mg/L (Fig. 5k). The findings demonstrated that water samples at every location fell within WHO's [71] recommended limits of 2250 mg/l and could be used as drinking water without additional treatment, with the exception of the SP-(9) site. Weathering of rocks and soil in the catchment area can release ions and dissolved solids into the water, increasing the EC and TDS values. The sampling site (9) is situated in downstream region of the basin, which can be the reason for its higher values of EC due to the increased concentration of dissolved solids resulting from evaporation and low precipitation.

The SAR is used to assess sodium hazard levels in irrigation water. It is also used to measure salinity hazard with the help of electrical conductivity. The irrigation water's bicarbonate and carbonate ions raise the permeability hazard as measured by SAR. Reduced osmotic pressure in drinking water with a high SAR value may limit plants' ability to absorb nutrients from the soil [62–64]. This disorder, known as sodicity, is brought on by an excessive buildup of Na^+ in soils. The SAR in irrigation water relies on the sodium (Na^+), calcium (Ca^{2+}), and magnesium (Mg^{2+}) concentrations and their ionic exchange. The calculated value of SAR has been assessed and varied as 0.412–16.589 (Fig. 5l). The results shows that the value of SAR was found <10 at 18 places, which indicates excellent water quality suitable for irrigation. The influx of rainwater and anthropogenic inputs from increased agricultural activities during the sampling season tends to increase the dissolved salt content in surface water at SP-(9), leading to a decrease in water quality. The cation exchange mechanism might be connected to higher values.

Numerous sources, such as sewage, manure, and other synthetic fertilizers used in agricultural fields, can introduce metals like B^+ into water systems. Higher boron values are primarily due to concentration effects from lower water levels, reduced dilution from less rainfall, and greater evaporation concentrating alkaline compounds. In addition, more B^+ concentration causes river water's oxygen content to drop, which affects the nutrient cycle activities and the consistency of the substance [84]. The observed B^+ values of the river were estimated as 0.03 to 0.55 mg/L (Fig. 5m). The observed low level of boron in the river system might also be due to its highly turbulent nature, photosynthesis, and decrease in temperature. Relatively, the highest concentration of B detected at sample site (SP-9) may be attributed to anthropogenic inputs of nitrate containing wastes from the surrounding residential and municipal wastes, sewerage, cattle and dairy wastes, leakage from toilet and septic tanks. However, the existing values in this investigation complied at-par with WHO recommendation (2 mg/L). A study reported by Kassegne and Leta [85] that demonstrated boron concentrations ranging from 0.01 to 0.67 mg/L, which are slightly higher than the present findings, may be due to varied sources of human-induced and abundant fertilizer inputs at different places.

Basically, TDS measures the concentration of all dissolved substances in groundwater, these include salts; organic compounds; dissolved calcium, magnesium, sodium, potassium, carbonate, bicarbonate, chloride, and sulphate. The computed value ranged as 82 to 13,230 mg/L (Fig. 5n). Based on the estimated score, it is noted that the TDS value is higher at locations which includes SP-(9) & (19). The lowest TDS value was recorded at many sites, possibly due to low inputs from surrounding farmlands and dissolution of minerals from underlying rocks [62–64].

From this result analysis, TDS values fits well, accompanied with the WHO threshold of 100 mg/L. This pattern lowers the river's TDS concentration and supports the diluting effect. Highest TDS value detected at sample site (SP-9) which located at the centre of the town where anthropogenic inputs of dissolved substances infiltrate from decomposed plant materials, domestic and municipal wastes, as well as, natural inputs from dissolution of geologic materials. Sometimes, elevated levels at SP-(8) are transported on by the release of chemical waste, rainfall from farms, and household sewage into rivers.

During the sampling timeframe, numerous dissolved polyvalent metallic ions, primarily Ca^{2+} and Mg^{2+}, are the source of water hardness or TH. Thus, calcium (Ca^{+2}) and magnesium (Mg^{+2}) are normally found in surface water in varied quantities; however, an elevated concentration of metals usually occurred due to human activity. High magnesium concentration in irrigation water impacts soil structure and reduces soil infiltration rates which affect soil quality and decrease crop production. Furthermore, magnesium may also be derived from surface water containing an elevated concentration of limestones that may infiltrate into surface water. Therefore, at higher TH levels in the soil, low potential yield and nutritional advantages may result from it. According to WHO [71], the maximum permissible level of TH in drinking water is 100 mg / L. The calculated total hardness (TH) concentration reports as 51.18 to 2194.90 (Fig. 5o). The lowest concentration occurs at 17 samples, which is found in the peri-urban area and free from urban wastes. Again, the highest TH concentrations detected at sample site (SP-9) may be due to the presence of Ca^{+2} bearing rocks from which Ca^{+2} is leached by water, increasing its concentration in surface water. In recent study, Dubey et al. [86] have reported TH concentrations ranged from 50 to 120 mg/L, which is higher than the present findings that may be due to varied locations of the study and sources of input.

Typically, the most common sources of Cl^- in drinking water include saline intrusion, fertilizers, leachate, sewage and industrial effluents, and natural sources. Higher concentration of Cl^- is highly toxic to aquatic life and to human health; hence, long-term exposure and intake of Cl^- may cause various respiratory problems likes respiratory problems and damages human lung [87]. Excessive occurrence of Cl^- in water can lead to environmental concern and cause eutrophication of lakes and rivers which in turn cause the depletion of dissolved oxygen, leading to create a toxic environment, kill aquatic life with oxygen starvation, and generate bad or unpleasant smell [88]. The present work indicates that the concentration of Cl^- in all sample sites ranging from 9.64 to 4904.88 mg/L (Fig. 5p). The allowable concentration of Cl^- in drinking water is 250 mg/L. Relatively, highest concentration of Cl^- detected in the sample site (SP-9) which is located at centre of town. High concentration of Cl^- attributed to nitrogen containing wastes from nearby central hospital and where anthropogenic inputs of nitrogen like domestic organic wastes, septic and toilet wastes, commercial and municipal wastes are widespread. It can also be generated from degradation of naturally occurring organic matter, livestock waste, sewage, food, and animal feeding wastes.

Sulphate (SO_4^{2-}) in water is usually originated from earth crust (geologic materials) through the processes of weathering of sulphide bearing metamorphic rocks and anthropogenic sources like domestic detergents, sewage, and groundwater extraction. An elevated level of sulphate can be toxic to plants and aquatic life. According to WHO Guidelines, on the basis of aesthetic objective (AO), the value of sulphate is considered as 200 mg/L. Sulphate concentration in the surface water study falls within the benchmark of 4.97–376.07 mg/L (Fig. 5q). The highest concentration of SO_4^{2-} occurred at sample site (SP-9), possibly due to discharge of domestic wastes, detergents, sewerage, institutional and municipal wastes, dairy animal wastes, decomposed plant materials, decaying biomass, animal wastes, laundry wastewater, detergents, wastewater, and leakage from septic tanks that percolate into surface water. Furthermore, in literature (Krishnakumar et al. [89], it is stated that higher concentration of sulphate that exceeded the maximum acceptable concentration (MAC) of 200 mg/L in drinking water may cause unpleasant odours and affect the palatability of the water and influence redox and pH water quality. Additionally, literature showed that surface water can also be contaminated with SO_4^{2-} generated from natural source of sulphides and phosphates; these include natural weathering of rocks containing phosphate and dissolution of soluble inorganic materials.

Under natural condition, fluoride can occur in surface water in varied quantities. The weathering of basic rocks and the liquidation of fluoride-containing minerals are the primary reasons of high fluoride rates in India, which are typically linked to low calcium levels and high bicarbonate ions. The main natural reservoirs for F^- ions are OH^- ions that contain silicate rock-forming minerals including biotite, hornblende, and alkali amphiboles, according to published research [90]. The recommended a maximum acceptable level of fluoride in drinking water is 1.5 mg/L [71]. Thus, this indicator is mostly added by several geological processes to drinking water. As tested in this work, the estimated F^- concentration in the surface water exhibits a value between 0.25 and 1 mg/L (Fig. 5r). The measured concentrations F^- in all sample sites were below the maximum acceptable limit value; hence, the water quality not influenced with F^- and may not pose adverse health effects. Atumo Ante et al. [91] have described that permissible level of fluoride is essential for the formation of dental enamel. In another study, Sunkari et al. [92] have reported fluoride concentration value ranged from 0.03 to 1.34 mg/L for Chagni Town's groundwater which is in agreement with the present findings.

Nitrate (NO_3^-) concentrations in water are primarily influenced by the activity of nitrifying bacteria, which depend on the availability of dissolved oxygen. The primary culprit is the nitrate that enters the permeate water from a variety of agricultural and human-caused processes. While ground water nitrate levels typically change slowly, bacterial nitrate denitrification, phytoplankton absorption, and surface washing can all cause nitrate concentrations to rise quickly [93]. Additionally, it is also vital for aquatic ecosystems, but excessive levels can cause eutrophication, depleting oxygen, and endangering aquatic life. The reported nitrate value in the surface water study falls with the boundary between 1.28 and 2.68 mg/L (Fig. 5s). The maximum threshold as instructed by WHO norms is noted as 45 mg/L. Lower nitrate levels (< 10 mg/L) at all sampling locations suggest reduced organic pollution and minimal anthropogenic impact, contributing to better water quality at these locations. Subsequently, high nitrate levels at sites SP-(9), (19), and (8) likely stem from anthropogenic sources such as agricultural runoff, wastewater, or organic matter decomposition.

Iron (Fe^{2+}) is a vital trace element that comes from natural and industrial waste water and is necessary for optimal health. Elevated Fe^{2+} levels in water can lead to salt accumulation in agricultural soils, adversely affecting soil structure, fertility, and plant growth. Even while this indicator helps the blood carry oxygen, high concentrations of it can lead to bleeding disorders, exhaustion, weight loss, tenderness in the joints, stomach issues, vomiting, and impairment of DNA. This work reports iron concentration, which spanned a score as 0.59–2.609 mg/L (Fig. 5t). Relying on findings, 18 sites exhibited relatively lower EC values compared to other sites, suggesting possible natural dilution or lower input of pollutants in these areas. The highest values at SP-(9) may reflect the influence of wastewater discharge or other anthropogenic activities contributing to increased salinity. This could result from erosion and the unstable sides of water bodies. It is noticed that high standard deviations, at SP-(9), reveal significant fluctuations in turbidity, potentially linked to episodic events such as heavy rainfall or localized disturbances. The elevated iron levels observed are consistent with findings reported by Ubuoh et al. [94], who attributed high Fe^{2+} to soil particles and other materials entering water systems from unstable sides and erosion.

The relative abundance order of cations (Fe^{2+} > B^{+}) and anions ($Cl^{-} > SO_4^{2-} > NO_3^{-} > F^{-}$) implies that the primary supply of cations from chemical weathering processes is mineral phases that contain ferrites, calcium, and mainly carbonates. These findings are also consistent with previous findings from other studies on the same river and other rivers with similar basin characteristics [95]. According to physicochemical data, the concentrations of TKN and total coliform are higher than the WHO-recommended thresholds due to the numerous geogenic and anthropogenic sources of availability. Along the river stretch, the geospatial maps of several water quality criteria are plainly visible in Fig. 5a–t. Based on these findings, several water management strategies should be considered. To detect contamination early and mitigate risks to human and ecosystem health, regular monitoring of water quality parameters is essential across all water bodies. Efforts to reduce organic and microbial pollution, such as improving sewage treatment, controlling agricultural runoff, and promoting better waste disposal practices, are needed [96]. Raising awareness about the importance of maintaining clean water sources and implementing sanitation systems in rural and urban areas is crucial for sustainable water management [97].

5.2 Estimation of GIS-Based EWQI

EWQI was used to evaluate how each water quality parameter affected the overall quality of the water. The research area's surface water pollution was assessed using this index. As a result, when we do more study using traditional WQI methods, we are prompted to adopt the perspective that the problematic condition of any surface water body is not indicated just by analysis, monitoring, and index development. Therefore, the EWQI approach can be used to quickly identify the relative pollution levels of the sampling locations in relation to drinking water quality criteria. Observations obtained from WHO drinking water quality guideline, the EWQI values were represented in Table 1. The calculated EWQI values ranged from 16.3 to 310, signifying excellent to unsuitable water category, in the study area. The developed bar plot is displayed in Fig. 6a. The surface water was divided into five different categories using the EWQI categorization system. Approximately in Fig. 6b, 52.63% of the chosen sites ($n = 10$) indicates excellent water quality class, 10.53% stations ($n = 2$) reported to be medium, 31.58% of investigated sites ($n = 6$) exhibited poor conditions, and ultimately, 10.53% of the examined samples (1 location) indicated under extremely poor water conditions. Significant pollution was found in the results, with the central and southeast regions having the highest levels of contamination [98, 99]. It identifies hotspots that may be impacted by increased surface water extraction, decreased hydrological dilution, and possible human inputs such untreated industrial wastes or pesticide leaching. On the contrary, along the Hirakud division, surface water contamination was generally low, and the water quality was found improved in the upstream stretches surrounding the sampling locations. The result can be credited to more rainfall in the monsoon season, and water infiltrating the ground, diluting the surface water system. Comparing the EWQI results from the 10 upstream sites to those from other parts of Bangladesh, the research area's

Table 1 Summary of water quality classification based on entropy and WAPAS-TOPSIS

	EWQI		W-TOP	
Sample no (SP)	Value	Water type	Performance score (T_i)	Rank
1	42	Excellent	1.59	15
2	154	Poor	2.57	4
3	25	Excellent	1.71	14
4	32	Excellent	1.50	17
5	20	Excellent	1.93	12
6	21	Excellent	2.04	11
7	134	Medium	2.02	5
8	196	Poor	3.90	2
9	310	Extremely poor	5.48	1
10	160	Poor	2.28	6
11	171	Poor	2.19	8
12	121	Medium	1.74	13
13	165	Poor	2.26	7
14	20	Excellent	1.20	18
15	16.3	Excellent	1.55	16
16	18.4	Excellent	2.17	9
17	19.5	Excellent	2.11	10
18	17.9	Excellent	1.10	19
19	160	Poor	3.40	3

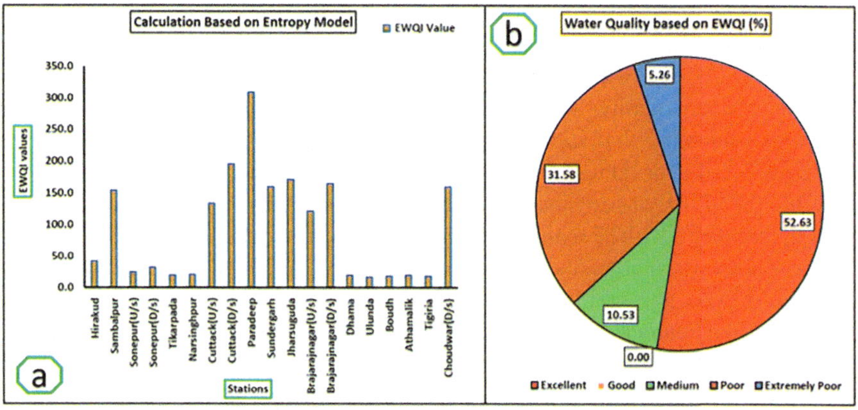

Fig. 6 EWQI categorization (**a**) Bar graph depicting values across 19 sampling sites, (**b**) water quality classification, expressed in %

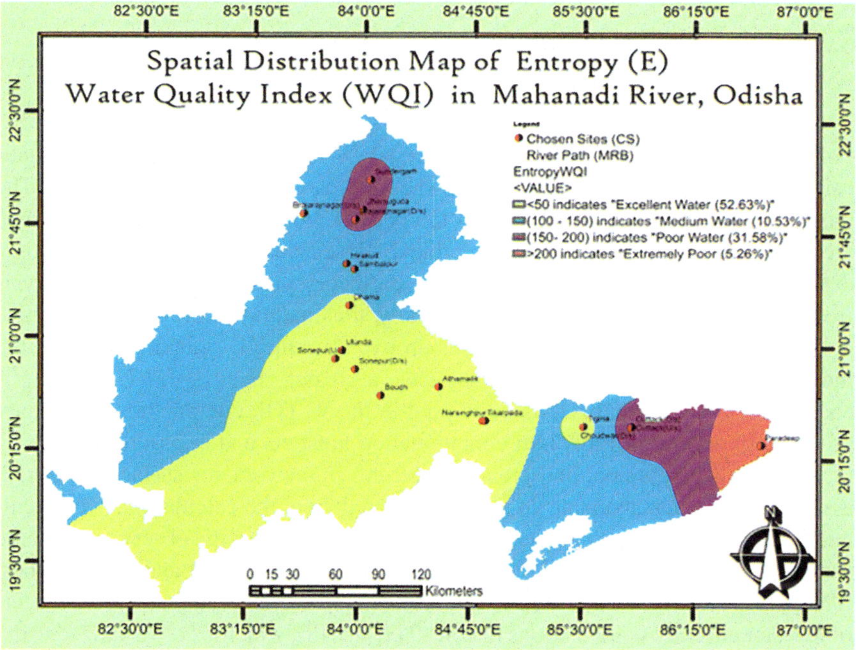

Fig. 7 Interpolated EWQI distribution map of surface water

surface water is generally safe and pollution-free [100]. In line with Sarker and Rahman [101], who observed that riverine dilution mitigates solute accumulation in nearby aquifers, the results demonstrate the Mahanadi River System's geochemical buffering. The spatial distribution of EWQI values can be observed in Fig. 7. Thus, the computed surface water quality score belongs to the excellent and good category in a major portion of the study area and is suitable for drinking as well as domestic uses except for these eight stations. Further, throughout the whole period, the locations—sites (2), (7), (8), (9), (10), (11), (13), and (19)—exhibited extremely poor/unsuitable water quality. These high EWQI values highlight localized pollution issues by indicating that surface water in the aforementioned study geographical locations is unfit for human consumption. Due to EC, SAR, TKN, and TC characteristics, the SP-(9) once again showed the highest EWQI value during the detection period, reaching 310. The majority of samples did not surpass an EWQI value of 50. The increased parameters found here are higher above the WHO-established acceptable limits. This is a vivid example of both organic and inorganic pollution from human-caused sources, including untreated municipal sewage discharge, industrial effluents, water treatment facilities, and household waste water [102]. Surface water quality at a few locations in the research area was judged unfit for use throughout the sample season based on the EWQI categorization put forth by Edet et al. [103], demonstrating seasonal fluctuation in pollution levels.

5.3 Drinking Water Quality Assessment by WAPAS-TOPSIS/W-TOP

Pollution evaluation MCDM approach, namely W-TOP mechanism, is systematically computed using MATLAB spread sheets to assess surface water quality. The ranking of the sampling locations and the performance score that was determined are displayed in Fig. 8 a and b and Table 1. Based on integrated weights, it is a dependable approach, and applying rough set theory yields a trustworthy analysis from various parameter weights. Additionally, this method determined the relative pollution level and assigned these sites an overall rank (T_i), ranking them as the most polluted places. However, it may have a greater effect on EWQI ranking if each parameter's normalized value is used in the W-TOP computation. The study provided important insights into contamination hotspots and exposure hazards by including this assessment and creating a strong framework for surface water risk assessment and management. In this investigated area, the calculated T_i value culminates between 1.10 and 5.48. According to the analysis of the observations in the studied area, the locations SP-(8) and site (19) renders 'second and third' highest EWQI, respectively, with scores of $T_i = 3.90$ and $T_i = 3.40$, designating them as terrible water classes. As for stations like SP-9 ($T_i = 5.48$), which were visualized using the IDW interpolation technique, they are therefore located in the heavily polluted location zone and have an overall rank of 1. This is because the concentrations of these parameters, such as EC, SAR, TDS, TH, Cl^-, SO_4^{2-}, TKN, and TC, demonstrating the largest concentration of any location and greater than their intended concentration. The trend of increasing TKN and TC or coliform is followed by a higher value in three locations. Poor quality stems from natural mineral dissolution and human activities like agricultural runoff and poor waste management.

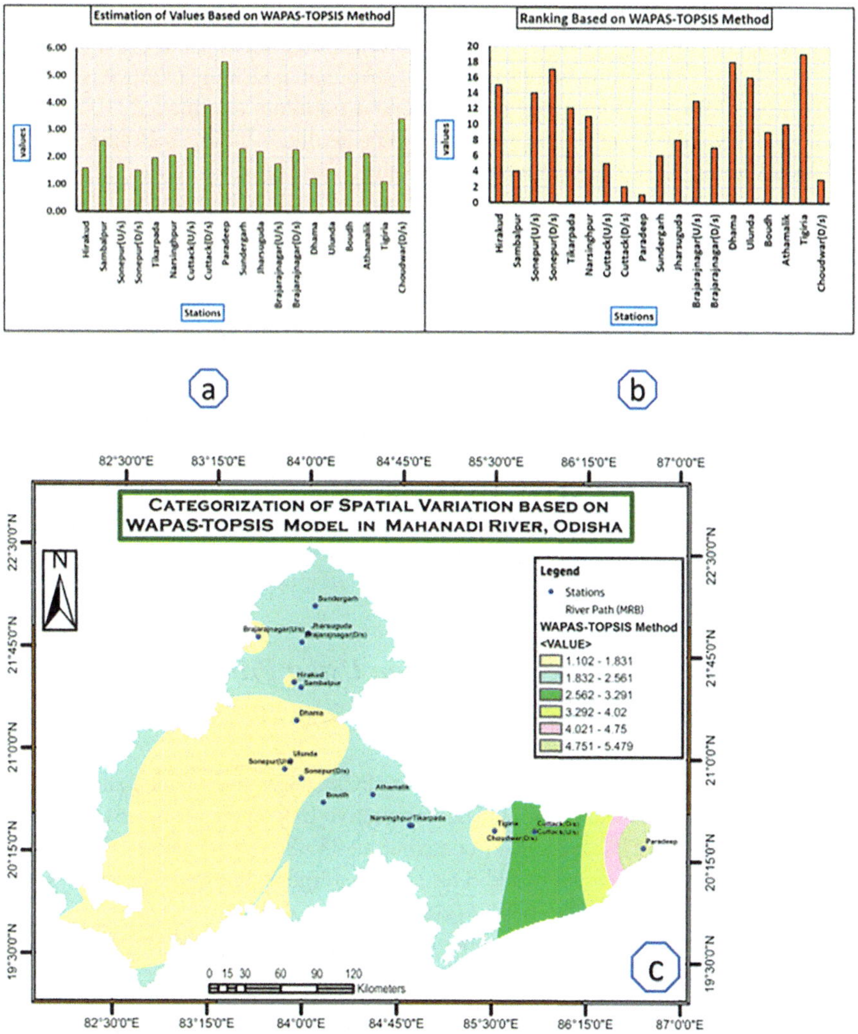

Fig. 8 Employing W-TOP approach showing (**a**) bar plot for different performance scores across all sites (T_i), (**b**) ranking classification, and (**c**) spatial variation map

Additionally, high values of TDS, SAR, and EC also accompany with it. These findings align with studies in arid regions [98, 99], where mineral dissolution and agricultural runoff elevate TDS, SAR, and NO_3^-, compromising suitability for drinking and irrigation. Finally, to identify regions with good and poor water quality, the retrieved findings from both methods were then contoured using geostatistical techniques. The spatial variability is shown in Fig. 8c. Overall, the study shows that "poor and very poor locations" are those that have significant levels of pollutants from sewage disposal, agricultural runoffs, textiles, and related industrial effluents.

These areas are unfit for human consumption without treatment. This assessment underscores catchment's severe contamination profile, necessitating tailored interventions like reverse osmosis for salinity and sustainable farming practices to reduce TKN inputs, distinct from regions with singular pollutant dominance [104]. These findings carry significant implications for public health, drinking, and agriculture in the tested zones, where surface water is a lifeline. The urgency of remediation aligns with recommendations from Naama aquifer studies [105], which advocate advanced treatment to mitigate mineral-driven salinity, while differing from Odisha's focus on recharge management due to the higher parameters loading. Addressing this requires detailed pollutant studies, advanced treatment, sustainable practices, and strict monitoring. Ongoing W-TOP evaluation, community involvement, and technologies like GIS and remote sensing are vital for sustainable management, highlighting the urgent need for comprehensive strategies to improve surface water quality in Odisha, further supported by explainable artificial intelligence approaches that optimize data-driven water quality assessments. Unlike 'Drink from Tap Mission, and BASUDHA' emphasis on piped drinking water and extraction regulation, Odisha requires a balanced approach addressing both natural and anthropogenic factors to safeguard water resources.

5.4 Hydro-Geostatistical Evaluation Using Principal Component Analysis (PCA)

The PCA yielded five principal components (PCs), with eigenvalues and variance contributions tabulated in Table 2. The cumulative explained variance for the first five PCs reaches 93.77%, with PC-1 contributing 35.88% (eigenvalue = 5.67), PC-2 as 25.23% (eigenvalue = 3.35), PC-3 as 16.20% (eigenvalue = 2.41), PC-4 9.67% (eigenvalue = 1.67), and PC-5 6.79% (eigenvalue = 1.45).

To objectively determine the number of retained PCs, a scree plot was generated (Fig. 9) plotting eigenvalues against PC indices. The plot exhibits a pronounced inflection at PC-2, where the eigenvalue decreases from 5.67 to 3.35, corresponding to a cumulative variance of 61.11%. This elbow, consistent with Gupta et al. [106] scree test criterion, indicates that PC-1 and PC-2 account for the dominant hydrochemical signals, justifying their prioritization. PCs > 5, with eigenvalues declining to 0.00235, contribute only <7% and were excluded from detailed interpretation due to their marginal explanatory power.

Hydrochemical patterns are elucidated through Table 2, which present PCA loadings and cumulative variance visualizations. PC-1, explaining 35.88% of the variance, is characterized by significant positive loadings: coliform (0.78), TKN (0.94), sulphate (0.82), EC (0.85), Cl^- (0.89), and moderate loadings of DO (0.52). These coefficients, derived from the factor matrix, suggest a strong association with salinity and mineralization processes. TKN has grown to be a significant global surface water pollution, mostly from home waste water and industrial and

Table 2 Extraction method for principal component analysis

Chemical variables	Coefficients of PC1	Coefficients of PC2	Coefficients of PC3	Coefficients of PC4	Coefficients of PC5
Coliform	0.78	0.33	0.65	0.09	0.06
TH	0.43	0.88	0.38	0.18	0.18
Alkalinity	0.23	0.55	0.81	0.15	0.24
COD	0.17	0.24	0.12	0.26	0.58
NH_3-N	0.05	0.71	0.05	0.41	0.15
Free NH_3	0.10	0.62	0.38	0.18	0.21
TKN	0.94	0.65	0.23	0.42	0.33
F_-	0.41	0.53	0.75	0.36	0.43
NO_3^-	0.22	0.47	0.16	0.66	0.23
Fe^{2+}	0.19	0.32	0.04	0.53	0.66
SO_4^{2-}	0.82	0.51	0.19	0.54	0.11
TSS	0.51	0.92	0.36	0.41	0.34
TDS	0.48	0.77	0.44	0.59	0.16
SAR	0.33	0.61	0.59	0.44	0.04
B^+	0.26	0.62	0.31	0.29	0.37
pH	0.18	0.73	0.15	0.17	0.12
EC	0.85	0.69	0.23	0.11	0.22
DO	0.52	0.59	0.032	0.55	0.36
BOD	0.41	0.46	0.28	0.61	0.05
Cl^-	0.89	0.33	0.11	0.27	0.42
Eigen values	5.67	3.55	2.41	1.67	1.45
% of variance	35.88	25.23	16.2	9.67	6.79
Total variance (%)	35.88	61.11	77.31	86.98	93.77

agricultural processes. Sulphate and EC are attributable to geogenic mechanisms, including silicate and evaporite mineral dissolution and rock-water interaction, prevalent in the studied aquifer's hyper-arid environment. Cl^- reflects the impact of human activities on the hydrochemistry of aquatic systems. Geological data indicate a lithological composition dominated by evaporite deposits and limestone, enhancing ionic leaching. Comparative studies by Nisi et al. [107] in arid aquifer systems corroborate this, reporting analogous ion enrichment patterns. The resultant coliform and TKN concentrations, as proxied by SO_4^{2-}, EC, and Cl^-, render the surface water suboptimal for potable use, necessitating advanced treatment protocols. The discovery of high loadings in water quality measures indicates that the research area's surface water is impacted by both home sewage discharge and agricultural operations.

Again, PC-2, accounting for 25.23% of the variance, exhibits notable loadings for TH (0.88), TSS (0.92), TDS (0.77), and pH (0.73), indicative of biogeochemical and anthropogenic influences. The elevated total hardness (TH) loading reflects carbonate equilibrium and ion exchange processes, while pH, TDS, and TSS suggest

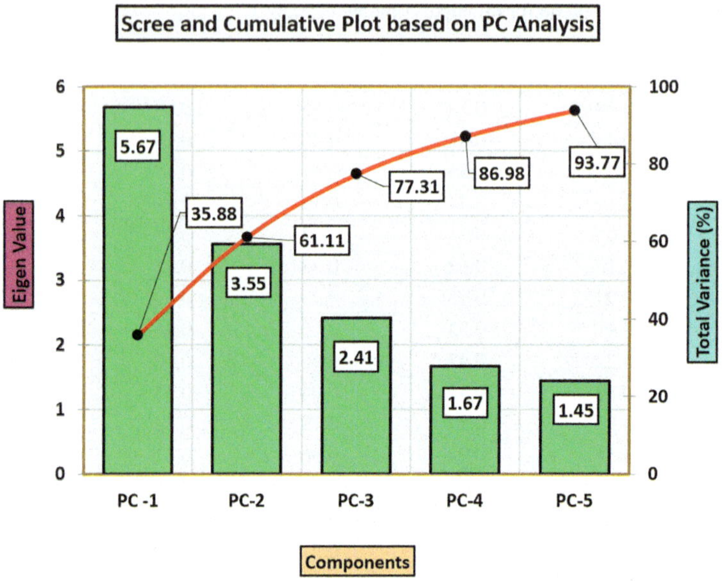

Fig. 9 Scree and total variance (%) plot of the component

fertilizer-derived inputs. The catchment's aquifer's high hydraulic conductivity and shallow water table (<5 m depth) facilitate rapid infiltration of agrochemicals, a hypothesis supported by spatial clustering of high PC-2 scores. Furthermore, TDS contributes heavy loading, predominantly arises because of agricultural activity. This aligns with [108], who documented nitrate mobilization from agricultural runoff, and [109], who modelled nutrient transport in permeable aquifers. For instance, surface water pollution may worsen as a result of fertilizer and pesticide use. These findings underscore the anthropogenic perturbation of surface water chemistry, particularly in agricultural zones.

Subsequently, remaining principal components such as PCs 3 to 5 provide additional insights: PC-3 (16.20%) is dominated by alkalinity (0.81), F^- (0.75), and SAR (0.59), suggesting organic matter decomposition; PC-4 (9.67%) is driven by NO_3^- (0.66), DO (0.55), BOD (0.61), and weak pH (0.17), indicating buffering capacity; and PC-5 (6.79%) shows moderate B^+ (0.37), Fe^{2+} (0.66), and COD effects (0.58). Furthermore, nitrate pollution of surface water is also largely caused by problems with sewage discharge and waste management in rural regions. The PCA thus delineates a dual-source variability model, with geogenic salinity (PC-1) and anthropogenic nutrient loading (PC-2 to PC-3) as primary drivers. Therefore, it is imperative to implement measures to reform agricultural practices in the study area. This necessitates rigorous monitoring and pre-utilization treatment, with implications for sustainable agricultural management.

Therefore, the investigation underscores the necessity of monitoring and treating surface water prior to utilization. Elevated salinity, EC, TKN, and TSS concentrations, as emphasized in PC-1 and PC-2, are significant issues. Geogenic sources mostly influence salinity, but agricultural activities lead to nitrate, coliform, and TKN contamination, indicating the necessity for sustainable farming methods to safeguard surface water quality. Therefore, one efficient way to lessen water pollution is to rationalize irrigation schedules and water amounts to lessen the leaching impacts of rainfall and irrigation water. Reducing fertilizer usage and optimizing application times and methods are crucial steps in minimizing contamination. For instance, the use of agricultural fertilizers and pesticides should be strictly regulated, and water purification equipment should be installed in high-risk areas to reduce the concentration of pollutants such as TKN and coliform in water.

In summary, the study's findings provide important information about how to manage surface water resources sustainably and safeguard the ecosystem in the area under study. Agricultural activities are the principal source of two characteristics, namely TKN and coliform, which were found as the primary pollutant in the analysed analysis. Therefore, local governments should put policies in place to limit the excessive use of chemical pesticides and fertilizers in agriculture, control the release of waste and wastewater from industry, and regularly evaluate the quality of surface water. Additionally, considering the results of the EWQI evaluation, the study area's high coliform level presents a comparatively high health risk, especially to youngsters. Children's continuous growth and development may be the cause of this, as their metabolisms are faster than those of adults, making them more susceptible to pollutants. Additionally, children are more vulnerable to health concerns because of their lower body weight and higher intake of pollutants per unit of body weight. It is essential to put preventative measures in place, such as providing safe drinking water in educational institutions and children's play areas, given that children are more sensitive to some toxins. For the prompt identification of health problems that may be brought on by TKN exposure, teenagers should have regular physical examinations with special attention to the blood and digestive systems. To encourage safe drinking practices, deter direct ingestion of untreated surface water, and raise knowledge of the risks associated with nitrate contamination, health education for teenagers and their parents should be improved.

6 Conclusion

Surface water is an important source of drinking water in the Mahanadi River Basin (MRB), Odisha, India. The Inverted Distance Weighted (IDW) interpolation method was applied through ArcGIS 10.5 to produce the spatial distribution maps for all parameters of water quality. This study addresses the growing issue of water contamination in Odisha, particularly at specific monitoring stations, and surface water sources, and explores a practical approach to assess water quality for potable use. For this purpose, 20 water quality (WQ) parameters collected during pre-monsoon season, from 19 water sources, for a timeframe taken as 2016–2024 (8 years). By focusing on the Indian states, the research developed an alternative, cost-effective method for classifying water quality using the Entropy-Based Water Quality Index (EWQI). Also, the study introduced the decision-making approach (WAPAS-TOPSIS (W-TOP)), in association with multivariate statistical analysis (MSA), which enables efficient prediction and classification of water quality conditions across various regions. The river water in the ongoing study contributes slightly alkaline in nature, and the DO is predominantly healthy, according to the physico-chemical results that were compared with the standard values. It was discovered that over 90% of the criteria under evaluation fell under the WHO criterion. The findings revealed that the majority of the water samples (57.89%) did not meet the acceptable water quality classification, with EWQI scores of ≥150, indicating poor or extremely poor water quality for human consumption. Notably, the locations, which are on the edge of an industrial area, are largely impacted by high concentrations of different elements in addition to the impacts of intense industry and human activity. However, a significant portion of the samples (31.58%) fell below this threshold (EWQI < 50), highlighting the need for further interventions in regions where water quality is suboptimal. MCDMs carried out the geographical assessment of the relative water quality based on physicochemical properties. The site, or SP-9 ($T_i = 5.48$), was determined to be the most polluted sampling location based on the extent of performance score or by the closeness coefficients, established W-TOP model. The next most contaminated sampling points were SP-8 (T_i = 3.90) and Site-19 (T_i = 3.40). High EWQI values were seen in this environment due to a number of markers, including TC, TKN, Cl^-, SO_4^{2-}, and Fe^{2+}. Seasonal fluctuations, agricultural practices, and other human-induced activities may be responsible for this top ranking. Therefore, the entropy WQI and MCDM model, which evaluates water quality based on 20 key parameters, proved to be a reliable tool for water quality assessment and offers a standardized framework that can be adapted to other regions with similar environmental and climatic conditions. On the contrary, PCA showed that the highest pollution was detected in the centre of the town as compared to peri-urban/ rural part/ of the town. According to PCA's findings, the primary sources for surface water contamination in the research area were rock-water interactions, combined mill effluent, pharmaceutical, textile, and agricultural nitrogenous fertilizer use. Finally, the study has validated the viability and accuracy of the EWQI, MCDM, and PCA techniques for interpreting surface water quality analysis data. In general, this is the first study in the river basin to present findings that point to significant discrepancies between the indices at different sampling sites.

This research underscores the need for continuous monitoring and adaptive measures to sustain water quality, informing policymakers, water resource managers, and local communities. Efforts to mitigate anthropogenic influences are crucial for maintaining water quality for various uses. Moreover, local governments should strictly regulate surface water contamination and put protective measures in place for children and adolescents, according to the health risk assessment. The comprehensive assessment can serve as a model for similar studies in other river basins, enhancing water management strategies in the context of global climate change impacts and anthropogenic interferences in the catchment. Although a monsoonal examination of the river basin was not carried out in this study, future research will compare the wet and dry seasons. The addition of hydrological modelling and GIS-based spatial mapping would further refine predictive capabilities and spatial resolution.

7 Limitations

However, the study's insights are tempered by limitations that warrant consideration. Potential sampling bias arises from uneven distribution across regions vast 65,600 km^2 area, possibly missing localized variations in quality. Analytical errors (±5% CBE) and the exclusion of organic contaminants may underestimate risks, particularly for health hazards beyond TKN, TDS, coliform, nitrate, and SAR. Seasonal fluctuations, not captured in this single-season snapshot, could also skew results, as seen in the investigated river basin, where multi-year monitoring revealed various water quality parameter's trend, which tied to rainfall variability, suggesting our study might overestimate chronic contamination. Compared to the Mahanadi plain, where isotopic analysis pinpointed TKN sources (e.g. manure vs. fertilizers), our reliance on physicochemical data limits source attribution precision. These methodological gaps contrast with best practices in arid-zone surface and ground water assessment, highlighting the need for future studies to adopt temporal sampling, broader pollutant screening such as isotopes, microbes, and GIS-based spatial analysis. Incorporating real-time hydrological modelling could further align this work with advanced regional studies, enhancing reliability and addressing these shortcomings for a comprehensive understanding of surface water dynamics in hyper-arid environments.

8 Recommendations

The following suggestions are made for the preservation, management, and sustainability of water resources as well as public health after it was established that the research area's drinking water quality is seriously threatened by pollution.

- The contaminated/polluted waters ought to receive proper treatment before being consumed by humans. It is advised to use techniques like ion exchange to treat waterways contaminated with lead.
- All the polluting parameters were found to be primarily impacted by human-caused factors, like inadequate garbage disposal. Thus, it is highly recommended that lifestyle patterns be changed in order to reduce the careless and uncontrolled release of trash from homes, businesses, and cities.

- The local inhabitants should be made aware of the risks that careless garbage disposal poses to the quality of water. They should also be informed about the ecological and health hazards that come with drinking contaminated water.
- For appropriate waste management and disposal, governments, pertinent authorities, and interested citizens should think about building landfills.
- Before being discharged into the environment, chemical wastes must be stabilized, solidified, and detoxified.

Acknowledgements The authors wish to acknowledge and thank the C.V. Raman Global University (CGU), Bhubaneswar, and Ranu-Prativa Research Agency, Bhubaneswar, for their support and cooperation. The author would like to extend their gratitude to the Editor of the Journal and the five anonymous reviewers for their valuable and constructive feedback on an earlier version of this manuscript.

Declaration The authors have read, understood, and compiled as applicable with the statement on ethical responsibilities of author, as found in the instruction for authors and are aware that with minor exceptions, no changes can be made to authorship once the paper is submitted.

Author Contributions Dr. Abhijeet Das: Conceptualized the whole article, developed the methodology, rendered support in formal analysis and investigation, wrote the original draft, executed the review and editing, and supervised the article.

Funding No funding is available.

Data Availability Statement All relevant data are included in the paper.

Conflict of Interest The authors declare there is no conflict.

References

1. Su, C., Liu, Y., Cheng, Z., Wang, W., & Zheng, Z. (2022). Hydrochemical process and controls on the hydrochemistry of river water in the Muling-Xingkai plain, Northeast China. *Frontiers in Environmental Science, 10*. https://doi.org/10.3389/fenvs.2022.1010367
2. Feng, W., Lu, H., Yao, T., Guan, Y., Xue, Y., & Yu, Q. (2022). Water environmental pressure assessment in agricultural systems in Central Asia based on an integrated excess nitrogen load model. *Science of the Total Environment, 803*, 149912. https://doi.org/10.1016/j.scitotenv.2021.149912
3. Hegab, M. H., El Sayed, S. M., Ahmed, N. M., Abdel-Aal, E. I., Kassem, D. A., Gaber, K. M., et al. (2025). Evaluating the spatial pattern of water quality of the Nile River, Egypt, through multivariate analysis of chemical and biological indicators. *Scientific Reports, 15*(1), 7626.
4. Rawat, K. S., Singh, S. K., & Gautam, S. K. (2018). Assessment of groundwater quality for irrigation use: A peninsular case study. *Applied Water Science, 8*, 233. https://doi.org/10.1007/s13201-018-0866-8
5. Chen, Y., Chen, H., Zhou, J., Liu, F., Wang, Z., Qian, C., & Sun, X. (2025). Correlation analysis between pH value of Yingna River with environmental factors and phytoplankton in China based on multivariable statistics analysis. *Algal Research*, 103991.

6. Taloor, A. K., Bala, A., & Mehta, P. (2023). Human health risk assessment and pollution index of groundwater in Jammu plains of India: A geospatial approach. *Chemosphere, 313*, 137329. https://doi.org/10.1016/j.chemosphere.2022.137329
7. Tiyasha, T. T. M., & Yaseen, Z. M. (2020). A survey on river water quality modelling using artificial intelligence models: 2000–2020. *Journal of Hydrology, 585*, 124670. https://doi.org/10.1016/j.jhydrol.2020.124670
8. Pang, H., Ben, Y., Cao, Y., Qu, S., & Hu, C. (2025). Time series-based machine learning for forecasting multivariate water quality in full-scale drinking water treatment with various reagent dosages. *Water Research, 268*, 122777.
9. Tiwari, A. K., Singh, A. K., Phartiyal, B., & Sharma, A. (2021). Hydrogeochemical characteristics of the Indus River water system. *Chemistry and Ecology, 37*, 780–808. https://doi.org/10.1080/02757540.2021.1999425
10. Liu, R., Li, Z., Liu, F., Dong, Y., Jiao, J., Sun, P., & El-Wardany, R. M. (2021). Microplastic pollution in Yellow River, China: Current status and research progress of biotoxicological effects. China Geology, 4, 585–592. https://doi.org/10.31035/cg2021081
11. Richards, L. A., Fox, B. G., Bowes, M. J., Khamis, K., Kumar, A., Kumari, R., Kumar, S., Hazra, M., Howard, B., Thorn, R. M. S., Read, D. S., Nel, H. A., Schneidewind, U., Armstrong, L. K., Nicholls, D. J. E., Magnone, D., Ghosh, A., Chakravorty, B., Joshi, H., Dutta, T. K., Hannah, D. M., Reynolds, D. M., Krause, S., Gooddy, D. C., & Polya, D. A. (2022). A systematic approach to understand hydrogeochemical dynamics in large river systems: Development and application to the River Ganges (Ganga) in India. Water Research, 211, 118054. https://doi.org/10.1016/j.watres.2022.118054
12. Hagage, M., Hewaidy, A. G. A., & Abdulaziz, A. M. (2025). Groundwater quality assessment for drinking, irrigation, aquaculture, and industrial uses in the waterlogged northeastern Nile Delta, Egypt: A multivariate statistical approach and water quality indices. Modeling Earth Systems and Environment, 11(1), 59.
13. Rai, S. K., Sahu, R., Dhar, S., & Kumar, A. (2023). Four decades of glacier and glacial lake dynamics in Kishtwar high altitude National Park, Chenab Basin, Jammu and Kashmir, India. Modeling Earth Systems and Environment, 10, 1171–1189. https://doi.org/10.1007/s40808-023-01836-w
14. Lkr, A., Singh, M. R., & Puro, N. (2020). Assessment of water quality status of Doyang River, Nagaland, India, using water quality index. Applied Water Science, 10, 46. https://doi.org/10.1007/s13201-019-1133-3
15. Ha, D. D., Nguyen, D. H., Nguyen, B. C., Nguyen, T. Q., & Le, G. T. N. (2025). Application of multivariate statistical techniques in the assessment of long-term surface water quality in Dong Thap Province, Vietnam. Environmental Monitoring and Assessment, 197(1), 1–19.
16. Khouni, I., Louhichi, G., & Ghrabi, A. (2021). Use of GIS based inverse distance weighted interpolation to assess surface water quality: Case of Wadi El Bey, Tunisia. Environmental Technology & Innovation, 24, 101892. https://doi.org/10.1016/j.eti.2021.101892
17. Dandge, K. P., & Patil, S. S. (2021). Spatial distribution of ground water quality index using remote sensing and GIS techniques. Applied Water Science, 12, 7. https://doi.org/10.1007/s13201-021-01546-7
18. Bedim Godoy, R. F., Crisiogiovanni, E. L., Trevisan, E., & Dias Radomski, F. A. (2021). Spatial and temporal variation of water quality in a watershed in center-West Paraná, Brazil. Water Supply, 21, 1718–1734. https://doi.org/10.2166/ws.2021.026
19. Gu, Y., Shi, X., Jin, P., & Wang, X. C. (2025). Assessing entropy changes in urban sewer systems during pollutant degradation. Water Science & Technology, 91(4), 424–437.
20. Giri, A., Bharti, V. K., Kalia, S., Kumar, K., & Khansu, M. (2022). Hydrochemical and quality assessment of irrigation water at the trans-Himalayan high-altitude regions of Leh, Ladakh, India. Applied Water Science, 12, 197. https://doi.org/10.1007/s13201-022-01716-1
21. Saha, A., Ramya, V. L., Jesna, P. K., Mol, S. S., Panikkar, P., Vijaykumar, M. E., Sarkar, U. K., & Das, B. K. (2021a). Evaluation of spatio-temporal changes in surface water quality and their suitability for designated uses, Mettur Reservoir, India. Natural Resources Research, 30, 1367–1394.

22. Saha, A., Ramya, V. L., Jesna, P. K., Mol, S. S., Panikkar, P., Vijaykumar, M. E., Sarkar, U. K., & Das, B. K. (2021b). Evaluation of spatio-tem-poral changes in surface water quality and their suitability for designated uses, Mettur Reservoir, India. Natural Resources Research, 30, 1367–1394.
23. Ahmad, S., & Ansari, Z. (2020). Characteristics of rock–water interaction in Gangotri proglacier meltwater streams at higher altitude catchment Garhwal Himalaya, Uttarakhand, India. Journal of Earth System Science, 129, 173. https://doi.org/10.1007/s12040-020-01426-9
24. Kumar, A., & Singh, A. (2024). Geospatial mapping and entropy-based analysis for groundwater evaluation with estimation of potential health risks due to nitrate and fluoride exposure. Environmental Science and Pollution Research, 31(59), 66953–66976.
25. Yuan, H., Han, S., Tian, Y., Zhang, Y., Bo, Z., & Liu, J. (2025). Water quality evaluation with entropy weight–grey correlation technique for fishing operation area in Tianjin, China. CLEAN–Soil, Air, Water, 53, e202300356.
26. Singh, V. B., & Kumar, P. (2022). Hydrogeochemical characteristics of meltwater draining from Himalayan glaciers: a critical review. Arabian Journal of Geosciences, 15(8), 680.
27. Li, P., Zhou, C., & Du, X. (2025). Evaluation and prediction of surface water pollution in China based on improved entropy-weighted TOPSIS algorithm and metabolism GM (1, 1) grey model. Journal of Industrial and Management Optimization, 21(2), 1600–1629.
28. Tiwari, A. K., Singh, A. K., Giri, S., & Mahato, M. K. (2023). Major ion chemistry and hydrochemical processes controlling water composition of Teesta River catchment, Sikkim Himalaya, India. International Journal of Environmental Analytical Chemistry, 103, 8597–8615. https://doi.org/10.1080/03067319.2021.1992402
29. Zhao, C., Li, X., Wang, Z., Hou, X., & Ma, J. (2024). Integration of self-organizing maps and entropy-weighted quality index methods to study hydrochemical processes and quality assessment in Qilian coal base of China. Environmental Earth Sciences, 83(3), 92.
30. Sharma, G., Lata, R., Thakur, N., Bajala, V., Kuniyal, J. C., & Kumar, K. (2021). Application of multivariate statistical analysis and water quality index for quality characterization of Parbati River, Northwestern Himalaya, India. Discover Water, 1, 5. https://doi.org/10.1007/s43832-021-00005-3
31. Mahammad, S., Islam, A., & Shit, P. K. (2023). Geospatial assessment of groundwater quality using entropy-based irrigation water quality index and heavy metal pollution indices. Environmental Science and Pollution Research, 30, 116498–116521. https://doi.org/10.1007/s11356-022-20665-5
32. Yang, L., Hao, Y., Wang, B., Li, X., & Gao, W. (2025). Evaluation of the water resources carrying capacity in Shaanxi Province based on DPSIRM–TOPSIS analysis. Ecological Indicators, 173, 113369.
33. Snousy, M. G., Wu, J., Su, F., Abdelhalim, A., & Ismail, E. (2022). Groundwater quality and its regulating geochemical processes in Assiut Province, Egypt. Exposure and Health, 14, 305–323. https://doi.org/10.1007/s12403-021-00445-1
34. Sahoo, S. K., Pamucar, D., & Goswami, S. S. (2025). A review of multi-criteria decision-making applications to solve energy management problems from 2010–2025: Current state and future research. Spectrum of Decision Making and Applications, 2(1), 219–241.
35. Han, F., Alkhawaji, R. N., & Shafieezadeh, M. M. (2025). Evaluating sustainable water management strategies using TOPSIS and fuzzy TOPSIS methods. Applied Water Science, 15(1), 1–13.
36. Kouser, B., Bala, A., Verma, O., Prashanth, M., Khosla, A., & Pir, R. A. (2022). Hydrochemistry for the assessment of groundwater quality in the Kathua region, Jammu and Kashmir, India. Applied Water Science, 12, 143. https://doi.org/10.1007/s13201-022-01673-9
37. Bu, J., Liu, W., Pan, Z., & Ling, K. (2020). Comparative study of hydrochemical classification based on different hierarchical cluster analysis methods. International Journal of Environmental Resear*ch and Public Health, 17*, 9515.

38. Ibrahim, H., Yaseen, Z. M., Scholz, M., Ali, M., Gad, M., Elsayed, S., et al. (2023). Evaluation and prediction of groundwater quality for irrigation using an integrated water quality indices, machine learning models and GIS approaches: A representative case study. Water, 15(4), 694.
39. Bhat, S. U., Bhat, A. A., Jehangir, A., Hamid, A., Sabha, I., & Qayoom, U. (2021). Water quality characterization of Marusudar River in Chenab Sub-Basin of North-Western Himalaya using multivariate statistical methods. Water, Air, & Soil Pollution, 232, 449. https://doi.org/10.1007/s11270-021-05394-8
40. Ghosh, S., & Jha, M. K. (2023). Hydrogeochemical characterization of groundwater and critical assessment of its quality in a coastal basin. Environment, Development and Sustainability, 27, 1–66.
41. Taghavi, M., Bakhshi, K., Zarei, A., Hoseinzadeh, E., & Gholizadeh, A. (2024). Soil pollution indices and health risk assessment of metal (loid) s in the agricultural soil of pistachio orchards. Scientific Reports, 14(1), 8971.
42. Mbissik, A., Elghali, A., Raji, O., Salifu, M., & Bodinier, J. L. (2024). Novel approach evaluating powdered raw and hydrothermally treated syenite as stonemeal for tropical environments: Potential effects on groundwater quality and agricultural suitability. Science of the Total Environment, 944, 173764.
43. Khanday, S. A., Bhat, S. U., Islam, S. T., & Sabha, I. (2021). Identifying lithogenic and anthropogenic factors responsible for spatio-seasonal patterns and quality evaluation of snow melt waters of the river Jhelum Basin in Kashmir Himalaya. Catena, 196, 104853. https://doi.org/10.1016/j.catena.2020.104853
44. Han, X., Tang, F., & Liu, A. L. (2024). Drinking water quality evaluation in supply systems in Wuhan, China: Application of entropy weight water quality index and multivariate statistical analysis. Environmental Science and Pollution Research, 31(1), 280–292.
45. Kaushik, H., Soheb, M., Biswal, K., Ramanathan, A. L., Kumar, O., & Patel, A. K. (2023). Understanding the hydro chemical functioning of glacierized catchments of the upper Indus Basin in Ladakh, Indian Himalayas. Environmental Science and Pollution Research, 30, 20631–20649. https://doi.org/10.1007/s11356-022-23477-9
46. Potapowicz, J., Szumińska, D., Szopińska, M., Bialik, R. J., Machowiak, K., Chmiel, S., & Polkowska, Z. (2020). Seashore sediment and water chemistry at the Admiralty Bay (King George Island, maritime Antarctica)—Geochemical analysis and correlations between the concentrations of chemical species. Marine Pollution Bulletin, 152, 110888. https://doi.org/10.1016/j.marpolbul.2020.110888
47. Gao, M., Qian, J., Li, X., Wang, Z., Hou, X., Gui, C., et al. (2024). Assessment of groundwater quality using entropy-weighted quality index (EWQI) and multivariate statistical approaches in Heilongdong Spring Basin, Northern China. Environmental Earth Sciences, 83(7), 196.
48. Alnahit, A. O., Mishra, A. K., & Khan, A. A. (2022). Stream water quality prediction using boosted regression tree and random forest models. Stochastic Environmental Research and Risk Assessment, 36(9), 2661–2680.
49. Rashid, A., Ayub, M., Bundschuh, J., Gao, X., Ullah, Z., Ali, L., et al. (2023). Geochemical control, water quality indexing, source distribution, and potential health risk of fluoride and arsenic in groundwater: Occurrence, sources apportionment, and positive matrix factorization model. Journal of Hazardous Materials, 460, 132443.
50. Jayasiri, M. M. J. G. C. N., Yadav, S., Dayawansa, N. D. K., Propper, C. R., Kumar, V., & Singleton, G. R. (2022). Spatio-temporal analysis of water quality for pesticides and other agricultural pollutants in Deduru Oya river basin of Sri Lanka. Journal of Cleaner Production, 330, 129897.
51. Mousavi, S. A., Sarshad Ghahfarokhi, M., & Soltani Koupaei, S. (2020). Negative impacts of nomadic livestock grazing on common rangelands' function in soil and water conservation. Ecological Indicators, 110, 105946. https://doi.org/10.1016/j.ecolind.2019.105946
52. Pattanayak, S. K., Rajesh, N., Dash, P. K., Das, S. K., & Ekka, K. (2025). Mahanadi River: Environmental challenges and needs for its sustainability. In Mahanadi River: The environmental challenges and *way forward* (pp. 1–16). Singapore.

53. Ray, S. L., Sahu, A. P., Paul, J. C., Das, D. M., Raul, S. K., & Kundu, S. K. (2024). Climate change impact on hydro-climatic fluxes in Kantamal catchment of the middle Mahanadi river basin, India. Journal of Agricultural Engineering, 61(6), 890–909.
54. Mishra, A. K., Mishra, S., Sahoo, M. R., & Sharma, S. (2025). Nature-based solutions for Mahanadi River basin protection and conservation. In Mahanadi River: The environmental challenges and way forward (pp. 113–123). Singapore.
55. Pal, D., Marttila, H., Ala-Aho, P., Lotsari, E., Ronkanen, A. K., Gonzales-Inca, C., et al. (2025). Blueprint conceptualization for a river basin's digital twin. Hydrology Research, 56(3), 197.
56. Shanmugam, H., & Lakshmanan, V. R. (2025). Development of stacking algorithm for bias-correcting the precipitation projections using a multi-model ensemble of CMIP6 GCMs in a semi-arid basin, India. Theoretical and Applied Climatology, 156(2), 1–28.
57. Subbarayan, S., Youssef, Y. M., Singh, L., Dąbrowska, D., Alarifi, N., Ramsankaran, R. A. A. J., et al. (2025). Soil and water assessment tool-based prediction of runoff under scenarios of land use/land cover and climate change across Indian agro-climatic zones: Implications for sustainable development goals. Water, 17(3), 458.
58. Regier, P. J., González-Pinzón, R., Van Horn, D. J., Reale, J. K., Nichols, J., & Khandewal, A. (2020). Water quality impacts of urban and non-urban arid-land runoff on the Rio Grande. Science of the Total Environment, 729, 138443. https://doi.org/10.1016/j.scitotenv.2020.138443
59. APHA. (2017). Standard method for the examination of water and waste-water (20th ed.). American Water Works Association, and Water Environmental Federation.
60. Hasan, M. S. U., & Rai, A. K. (2020). Groundwater quality assessment in the lower Ganga Basin using entropy information theory and GIS. Journal of Cleaner Production, 274, 123077. https://doi.org/10.1016/j.jclepro.2020.123077
61. Immerzeel, W. W., Lutz, A. F., Andrade, M., Bahl, A., Biemans, H., Bolch, T., Hyde, S., Brumby, S., Davies, B. J., Elmore, A. C., Emmer, A., Feng, M., Fernández, A., Haritashya, U., Kargel, J. S., Koppes, M., Kraaijenbrink, P. D. A., Kulkarni, A. V., Mayewski, P. A., Nepal, S., Pacheco, P., Painter, T. H., Pellicciotti, F., Rajaram, H., Rupper, S., Sinisalo, A., Shrestha, A. B., Viviroli, D., Wada, Y., Xiao, C., Yao, T., & Baillie, J. E. M. (2020). Importance and vulnerability of the world's water towers. Nature, 577, 364–369. https://doi.org/10.1038/s41586-019-1822-y
62. Das, A. (2025). Assessment of surface water quality modelling for drinking purposes in Brahmani River basin (BRB), Odisha, using geographical information system (GIS), IDOCRIW (I) and COPRAS (Ui) techniques. In Surface, sub-surface hydrology and management: Application of geospatial and geostatistical techniques (pp. 185–217). Cham.
63. Das, A. (2025c). Surface water potential zones delineation and spatiotemporal variation characteristics of water pollution and the cause of pollution formation in Brahmani River basin, Odisha. HydroResearch, 8, 99–112.
64. Das, A. (2025d). Surface water quality evaluation, apportionment of pollution sources and aptness testing for drinking using water quality indices and multivariate modelling in Baitarani River basin, Odisha. HydroResearch, 8, 244–264.
65. Poursaeid, M., Poursaeed, A. H., & Shabanlou, S. (2025). Water resources quality indicators monitoring by nonlinear programming and simulated annealing optimization with ensemble learning approaches. Water Resources Management, 39(3), 1073–1087.
66. Javanbakht, T. (2025). TOPSIS method for optimization of an apparatus for water and soil treatments (Vol. 12, p. H10).
67. Algamal, Z. Y., Shehab, Z. N., & Faisal, R. M. (2025). Spatiotemporal variation analysis of Tigris River water quality in Mosul, Iraq during 2020–2023 based on environmetric techniques. Environmental Earth Sciences, 84(2), 69.
68. Yogeshwaran, V., & Priya, A. (2025). Assessment of surface water quality parameters using multivariate analysis—A case study of Kurichi and big lakes in Coimbatore. Water Environmen*t Research, 97*(3), e70055.

69. Zhang, Q., & You, X. Y. (2024). Recent advances in surface water quality prediction using artificial intelligence models. Water Resources Management, 38(1), 235–250.
70. Mutlu, E., Tokatlı, C., Islam, A. R. M. T., Islam, M. S., & Muhammad, S. (2024). Water quality assessment of Şehriban stream (Kastamonu, Türkiye) from a multi-statistical perspective. International Journal of Environmental Analytical Chemistry, 104(19), 8229–8245.
71. WHO. (2017). Guidelines for drinking-water quality (4th ed.; Incorporating the 1st Addendum). WHO.
72. Nallakaruppan, M. K., Gangadevi, E., Shri, M. L., Balusamy, B., Bhattacharya, S., & Selvarajan, S. (2024). Reliable water quality prediction and parametric analysis using explainable AI models. Scientific Reports, 14(1), 7520.
73. Uddin, M. G., Rahman, A., Taghikhah, F. R., & Olbert, A. I. (2024). Data-driven evolution of water quality models: an indepth investigation of innovative outlier detection approaches-A case study of Irish Water Quality Index (IEWQI) model. Water Research, 255, 121499.
74. Hussain, A., Deshwal, A., Priyadarshi, M., Pathak, S., Sambandam, G., Chand, S., & Shukla, A. K. (2024). Landfill leachate analysis from selected landfill sites and its impact on groundwater quality, New Delhi, India. Environment, Development and Sustainability, 1–26.
75. Tong, N. X., Hoa, N. K., Tram, N. T. T., & Khang, L. T. P. (2025). Water quality index, heavy metals, and endocrine disruptors in the saigon river basin: Pollution assessment and correlation analysis. Environmental Quality Management, 34(3), e70063.
76. Gani, A., Hussain, A., Pathak, S., & Banerjee, A. (2024a). An empirical investigation on the elimination of heavy metals using bioremediation method for selected plant species. Physics and Chemistry of the Earth, 134, 103568.
77. Hammoumi, D., Al-Aizari, H. S., Alaraidh, I. A., Okla, M. K., Assal, M. E., Al-Aizari, A. R., et al. (2024). Seasonal variations and assessment of surface water quality using water quality index (WQI) and principal component analysis (PCA): A case study. Sustainability, 16(13), 5644.
78. Das, A. (2025). Applying the water quality indices, geographical information system, and advanced decision-making techniques to assess the suitability of surface water for drinking purposes in Brahmani River Basin (BRB), Odisha. Environmental Science and Pollution Research, 1–36.
79. Naz, I., Ahmad, I., Aslam, R. W., Quddoos, A., & Yaseen, A. (2023). Integrated assessment and geostatistical evaluation of groundwater quality through water quality indices. Water, 16(1), 63.
80. Shukla, A. K., Ojha, C. S. P., Mijic, A., Buytaert, W., Pathak, S., Garg, R. D., & Shukla, S. (2018). Population growth, land use and land cover transformations, and water quality nexus in the upper Ganga River basin. Hydrology and Earth System Sciences, 22(9), 4745–4770. https://doi.org/10.5194/hess-22-4745-2018
81. Mohseni, U., Pande, C. B., Pal, S. C., & Alshehri, F. (2024). Prediction of weighted arithmetic water quality index for urban water quality using ensemble machine learning model. Chemosphere, 352, 141393.
82. Shukla, A. K., Ojha, C. S. P., & Garg, R. D. (2017a, July). Surface water quality assessment of Ganga River basin, India using index mapping. In 2017 IEEE international geoscience and remote sensing symposium (IGARSS) (pp. 5609–5612). IEEE.
83. Abbas, F., Cai, Z., Shoaib, M., Iqbal, J., Ismail, M., Alrefaei, A. F., & Albeshr, M. F. (2024). Machine learning models for water quality prediction: A comprehensive analysis and uncertainty assessment in Mirpurkhas, Sindh, Pakistan. Water, 16(7), 941.
84. Shaaban, N. A., & Stevens, D. K. (2025). Transforming complex water quality monitoring data into water quality indices. Water Resources Management, 1–17.
85. Kassegne, A. B., & Leta, S. (2020). Assessment of physicochemical and bacteriological water quality of drinking water in Ankober district, Amhara region, Ethiopia. Cogent Environmental Science, 6(1), 1791461.

86. Dubey, P., Farooqui, A., Patel, A., & Srivastava, P. K. (2024). Microbial innovations in chromium remediation: Mechanistic insights and diverse applications. *World Journal of Microbiology and Biotechnology, 40*(5), 151.
87. Shukla, A. K., Ojha, C. S. P., Shukla, S., & Garg, R. D. (2021). Water quality challenges in Ganga River basin, India. *The Ganga River Basin: A Hydrometeorological Approach*, 1–19. https://doi.org/10.1007/978-3-030-60869-9_1
88. Karadeniz, S., Ustaoğlu, F., Aydın, H., & Yüksel, B. (2024). Toxicological risk assessment using spring water quality indices in plateaus of Giresun Province/Türkiye: A holistic hydrogeochemical data analysis. *Environmental Geochemistry and Health, 46*(8), 285.
89. Krishnakumar, A., Jose, J., Kaliraj, S., Aditya, S. K., & Krishnan, K. A. (2022). Assessment of the impact of flood on groundwater hydrochemistry and its suitability for drinking and irrigation in the river Periyar lower basin, India. *Environmental Science and Pollution Research, 29*, 1–40.
90. Hailu, H., & Haftu, S. (2023). Hydrogeochemical studies of groundwater in semi-arid areas of northern Ethiopia using geospatial methods and multivariate statistical analysis techniques. *Applied Water Science, 13*(3), 86.
91. Atumo Ante, A., Asefa Bogale, G., & Mohammed Adem, B. (2023). Bacteriological and physicochemical quality of drinking water and associated risk factors in Ethiopia. *Cogent Food & Agriculture, 9*(1), 2219473.
92. Sunkari, E. D., Hudu, A., Fosu, S., Gyimah, E., & Oppong, O. (2024). Hydrogeochemistry, sources, enrichment mechanism and human health risk assessment of groundwater fluoride in Saboba District in the Oti sub-basin of the Volta River basin, northern Ghana. *Groundwater for Sustainable Development, 25*, 101132.
93. Ling, Y., Podgorski, J., Sadiq, M., Rasheed, H., Eqani, S. A. M. A. S., & Berg, M. (2022). Monitoring and prediction of high fluoride concentrations in groundwater in Pakistan. *Science of the Total Environment, 839*, 156058.
94. Ubuoh, E. A., Nwogu, F. U., Ofoegbu, C. C., & Chikezie, P. C. (2023). Environmental pollution loads on surface water chemistry and potentially ecological risks of inland aquatic ecosystem in South-Eastern State, Nigeria. *Environmental Systems Research, 12*(1), 22.
95. Pant, M., & Singh, J. (2024). Seasonal variation of surface water quality and streamflow in Rispana: A tributary of Ganges river, India. *Environmental Quality Management, 34*(1), e22199.
96. Jahan, A., Khan, M. U., Rai, N., Kumar, S., & Dar, T. A. (2023). Geochemical characterization, its controlling factors, and comparison between the upstream and downstream segments of the Himalayan Satluj River basin, India. *Geochemistry, 83*(2), 125974.
97. Shukla, B. K., Gupta, L., Parashar, B., Sharma, P. K., Sihag, P., & Shukla, A. K. (2025). Integrative assessment of surface water contamination using GIS, WQI, and machine learning in urban–industrial confluence zones surrounding the National Capital Territory of the Republic of India. *Water, 17*(7), 1076. https://doi.org/10.3390/w17071076
98. Das, A. (2024a). Assessment of human health risk in Baitarani Basin, Odisha using water quality index (WQI), cluster analysis (CA), and geographic information systems (GIS). *Sustainable Development Using Geospatial Techniques*, 365–410.
99. Das, A. (2024b). Evaluation of surface water quality in Brahmani River basin, Odisha (India), for drinking purposes using GIS-based WQIs, multivariate statistical techniques and semivariogram models. *Innovative Infrastructure Solutions, 9*(12), 484.
100. Siddique, M. A. B., Islam, A. R. M. T., Hossain, M. S., Khan, R., Akbor, M. A., Hasanuzzaman, M., et al. (2022). Multivariate statistics and entropy theory for irrigation water quality and entropy-weighted index development in a subtropical urban river, Bangladesh. *Environmental Science and Pollution Research, 29*, 1–20.
101. Ansari, M. A., Kumar, U. S., Noble, J., Akhtar, N., Akhtar, M. A., & Deodhar, A. (2023). Isotope hydrology tools in the assessment of arsenic contamination in groundwater: An overview. *Chemosphere, 340*, 139898.

102. Shukla, A. K., Ojha, C. S. P., & Garg, R. D. (2017b). Application of overall index of pollution (OIP) for the assessment of the surface water quality in the upper ganga river basin, India. In *Development of water resources in India* (pp. 135–149). Springer International Publishing. https://doi.org/10.1007/978-3-319-55125-8_12
103. Edet, A., Emmanuel, C., & Kudamnya, E. A. (2023). Seasonal groundwater quality and associated health risk assessment in poorly organized coastal and inland towns, southern Nigeria. *International Journal of Energy and Water Resources*, 1–22.
104. Gani, A., Pathak, S., Hussain, A., Shukla, A. K., & Chand, S. (2024b). Emerging pollutant in surface water bodies: A review on monitoring, analysis, mitigation measures and removal technologies of micro-plastics. *Environmental Geochemistry and Health, 46*(7), 214. https://doi.org/10.1007/s10653-024-01992-7
105. Das, A. (2022, August). Water quality assessment using water quality index (WQI) under GIS framework in Brahmani Basin, Odisha. In *International conference innovation in smart and sustainable infrastructure* (pp. 131–150). Springer Nature Singapore.
106. Gupta, P., Samui, P., & Quaff, A. R. (2024). Prediction of water quality index using genetic programming. *Environmental Engineering & Management Journal (EEMJ), 23*(9), 1813.
107. Nisi, B., Vaselli, O., Taussi, M., Doveri, M., Menichini, M., Cabassi, J., et al. (2022). Hydrogeochemical surveys of shallow coastal aquifers: A conceptual model to set-up a monitoring network and increase the resilience of a strategic groundwater system to climate change and anthropogenic pressure. *Applied Geochemistry, 142*, 105350.
108. Vero, S., & Ascott, M. (2022). 3 Nitrate agricultural transport systems. In *Nitrate handbook: Environmental, agricultural, and health effects* (p. 45).
109. Hayes, E., Mullan, D., Geris, J., & Higgins, S. (2023). Novel tools for quantifying and managing sediment and nutrient losses from agricultural land. *Geography*.

Crop Health Monitoring and Yield Estimation Through Geospatial Technology

Ripninder Singh, Vishal Gautam, and Shray Pathak

1 Introduction

Since from the start, humans have tried to understand everything that has a relation with them or has an effect on them; so, understanding the crop patterns and calculating yield are small steps but also a big leap. It has more importance as now more and more farmers are using their lands for commercial purposes.

The need of crop monitoring can be understood by understanding the effects of low yield on the economy and the life of the everyday people and farmers. The effects of poor health are directly on the economy. The better we are at it, the more we can tackle these disasters of poor yield [1].

Yield production model is a necessary part of policy decision. The traditional methods are time consuming and also consume a lot of financial resources. The model uses images that are taken by the various types of sensors as aerial photographs and multispectral scanner images. These images can be of different resolutions in the spatial domain and spectral domain.

Pre-harvest crop yield model is essential for planning and making various policy decisions. Traditional methods are time consuming [2].

The prediction of crop yield is made better with the use of Leaf Area Index. It improves its prediction in the simulations when implemented operationally and at a certain level [3].

R. Singh (✉)
Department of Geography, Panjab University, Chandigarh, India

V. Gautam · S. Pathak
Department of Civil Engineering, Indian Institute of Technology Ropar, Rupnagar, Punjab, India
e-mail: 2023cem1016@iitrpr.ac.in; shray.pathak@iitrpr.ac.in

S. Pathak et al. (eds.), *Geo-Data Revolution*, Advances in Geographic Information Science, https://doi.org/10.1007/978-3-031-94078-1_14

We can access numerous models that are used for crop yield estimation. The extensive use of indices is for understanding vegetations and predicting drought. To understand crop yield, we need to work further with the indices and use additional data.

The yield estimation and crop health monitoring can be of two types: one of them uses biophysical measurements, and another one uses the remote sensing model to calculate the yield. USDA, known as the United States Department of Agriculture, is known to use both of the above methods [4].

The website of USDA provides data of the yield. We look upon this data for comparison. Once we understand the characteristics of the area we are working on, it is important to have a high density of corn and soybean cultivation to ensure accurate and reliable calculations. We need not only recent data but also old data to justify the model, which is fulfilled by this website.

In 1987, the Corn Belt of the USA contributed 80% of the total production of the entire country. It also had a 36% share in the global maize production in the same year [4]. Corn Belt is an area in the mid-western region of the USA. It covers the states of Indiana, Illinois, Iowa, Missouri, and eastern Nebraska. The dominant crops of the area are corn and soybeans [5].

The more digital this world is becoming, the more we need to stress on remote sensing components and digital work rather than the traditional approaches. The accuracy of the model's results heavily relies on the quality of input data. As the saying goes, "Garbage in, Garbage out," data must be as precise as possible [6].

2 Indiana

2.1 Introduction

Indiana is located in the mid-western region of the USA. It is surrounded by Lake Michigan and Michigan state in the north; it is the 38th largest state and 17th by population; the capital of Indiana is Indianapolis [5]. It is located at the center of the state; it was named Indiana due to high presence of Red Indians at the early times of the USA. Its name generally means "the land of the Indians"; Indiana was added to the USA on December 11, 1816, as the 19th state of the union.

This state lies in the Corn Belt region of the USA; it is known for its primary activities mostly. The northwestern region is much different as it is much dependent on industries and has a dependency on Chicago for its economic, industrial, and social continuum. In the south, the industrialization reduces and is taken over by agriculture.

2.2 Economic Activities

The state is dominant in the field of agriculture, with the most profiting being corn and soybean. Around 80% of the total area is devoted to agriculture and forest. Out of every ten jobs, eight are related to agriculture.

The top resources of the state are the two crops mentioned above; in a typical year, 50% of the cropland of Indiana is dominated by corn. This is followed by soybeans, which are used for poultry and livestock. These crops are used to make biofuels and fiber materials such as ethanol and carpet.

The corn alone accounts for $3.28 billion in sales for the state. These are used to feed the animals of the farms in the state. Almost 20% of popcorn supply in the USA is from Indiana itself [7].

The sale of soybean accounts for $3.08 billion. Its product contributes to half of Indiana products in the export sector. The farmers of Indiana have planted 5.9 million acres of soybeans [7].

2.3 Relief

The land of Indiana was modified by glacial actions, which has left much of the fertile soil that is present today [8]. Apart from soil, it has accumulated loess, sand, gravel, and glacial till. The south of the state was much eroded, so it gave birth to more fertile plains that are extensively used for agriculture [9]. At the time of glacial retreat, these were glacial lakes and moraine. Much of the Indiana state lies at the elevation of 150–300 m [10].

The slope of the Indiana state is towards south and southwest. In the north, we can see many small glacial lakes with many being the state's largest. The most part of the state is plain. So, these are pulling factors for extensive agriculture [11].

A high percentage of the land in Indiana is forested and also owned by private organizations or individual families. The same applies to the farmlands, as these estates are family-owned and operated privately. Families that farm here are not native to the land but who have migrated from the south in the early times of the USA.

As we can understand, the relief of Indiana is relatively flat and beneficial for agriculture. Fertile soils and ample water resources all make Indiana a major agriculture-dominant state.

2.4 Climate

The climate of Indiana falls in the temperate region; it distinctively has a four-season cycle. In the south, the temperature revolves around −6 °C to 4 °C. Coming to the north, the temperature ranges from −9 °C to −1 °C. In the summer season, temperature for both north and south ranges between 28 °C and 32 °C [12].

Annual precipitation is little more in the south-central state as compared to the north. Snow usually falls in a period of 6 months annually. The cities in the northern part receive more rainfall since they are located at the periphery of the lakes.

Indiana is mostly influenced by its location near the lakes as they provide a moderating influence as well as aid in precipitation in the area. It also amplifies snowfall in the region. This also makes the temperature cooler in spring and warmer in autumn. Indiana is part of the mid-western states where there is an unusually large frequency of severe storms; spring is generally the month of erratic storms that sweep across the state. This also makes tornados frequent in the state [12].

3 Dataset Used

Data is the raw information that we use and arrange in an order, so we can get the desired information. The data should always be taken from a trusted source and should also be verified by the concerned individual. A more precise dataset is preferred ,as demonstrated in (Table 1) [13].

3.1 Satellite Images

The first dataset to be used is a satellite image. These images are in the raster format (pixel grid). Each pixel has a specified value. The more the pixels, the better the resolution, the more the sharpness, and the more the information we can obtain. We need to understand our work purpose, so we can easily define our need of different resolutions like spectral, temporal, and spatial resolution.

Table 1 Datasets and their sources

Dataset	Source	Data type	Date
Satellite images (Landsat 7)	USGS	Raster	1 August–10 September 2023
NDVI	Calculated (USGS)	Raster	1 August–10 September 2023
Land surface temperature	Calculated (USGS)	Raster	1 August–10 September 2023
Soil moisture (10 cm)	Geovanni	Raster	August 2023
Annual rainfall	Geovanni	Raster	August 2023

The overview of the area is provided by these satellite images, which also provide all the desired band [14]. We need these images for further NDVI calculation and calculation of land surface temperature. Moreover, these images do give us a general picture of the area [15].

3.2 Normalized Difference Vegetation Index (NDVI)

Normalized difference vegetation index is used to understand vegetation of a particular area [16]. This is produced by a combination of bands that are near infrared and red in a way that they will highlight the vegetation.

NDVI does not give a distinct boundary for each land cover but a general idea of that land cover as shown in Fig. 1. For example, if we are having a value that is negative, it is most likely a water body; if we have a value closer to +1, it is most probably a dense vegetation; and if we get a value that is closer to zero, it is an urbanized and less dense area of vegetation. As shown in the image, greener area reflects the vegetation, and violet shows built-up area and water. NDVI can be categorized as an image that takes less space because of the usage of 2 bands, but it provides us relatively large information when compared.

$$\mathrm{NDVI} = \frac{(\mathrm{NIR} - \mathrm{Red})}{(\mathrm{NIR} + \mathrm{Red})} \tag{1}$$

3.3 Soil Moisture

Soil moisture is one of the important parts in the crop yield prediction model as it signifies an important preferable condition of crops, that is, the level of moisture held in the soil for the plant to thrive [17]. Soil moisture can be of many types, but the one that is used in the model is 4 in. or 10 cm. It is preferred because soils have roots only in the upper layers, which is perfect for the model and measuring the yield.

3.4 Land Surface Temperature

This mode of measuring the temperature is different from the air temperature warnings we get from day-to-day life. We calculate the land surface temperature because it is more suitable for the model. Crops are more affected by the ground surface temperature, and thus a suitable surface temperature helps the crop to increase its yield and keeps the green zone healthy. This even helps farmers in many ways, from understanding the need of the hour to the need of the future [18].

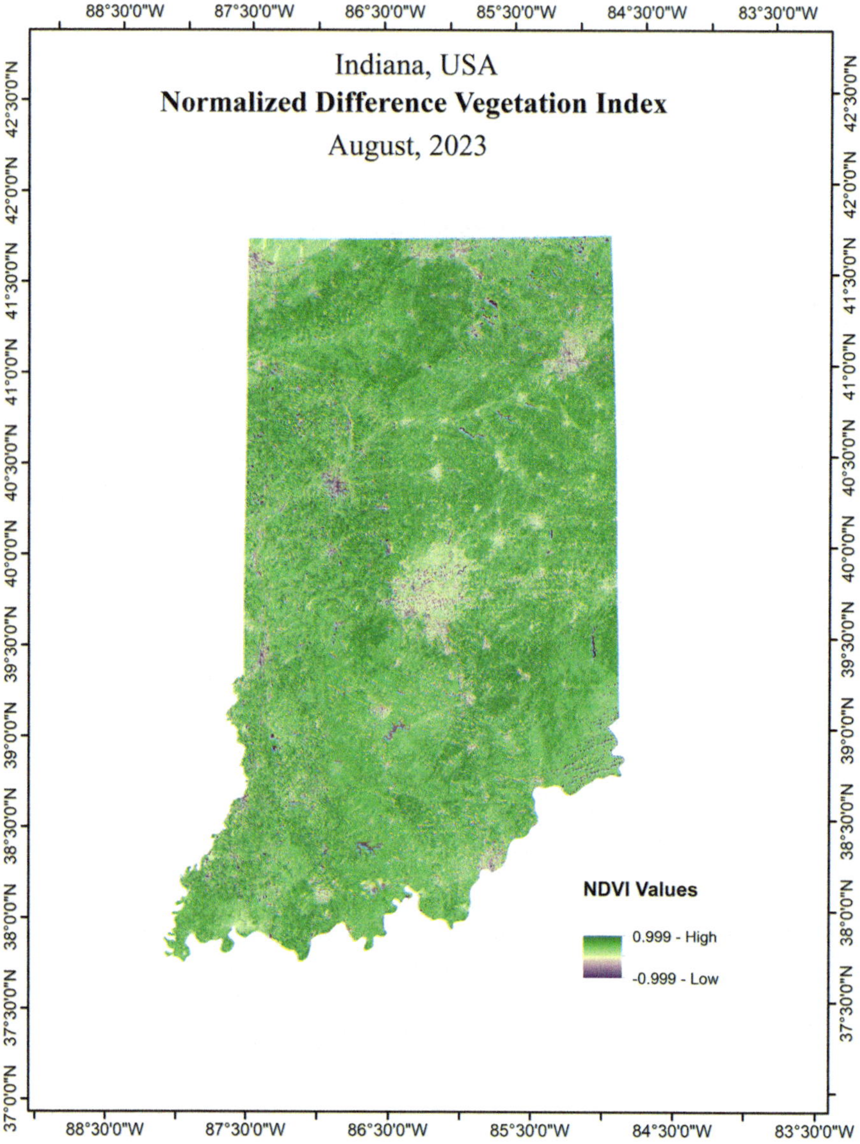

Fig. 1 NDVI of Indiana

Farmers that are indulged in commercial agriculture use the maps of temperature to evaluate the need of watering the crops during intense heat waves. In turn, during the season of shivering cold, these can help farmers to have the information of plants that can be prone to frost (Fig. 2).

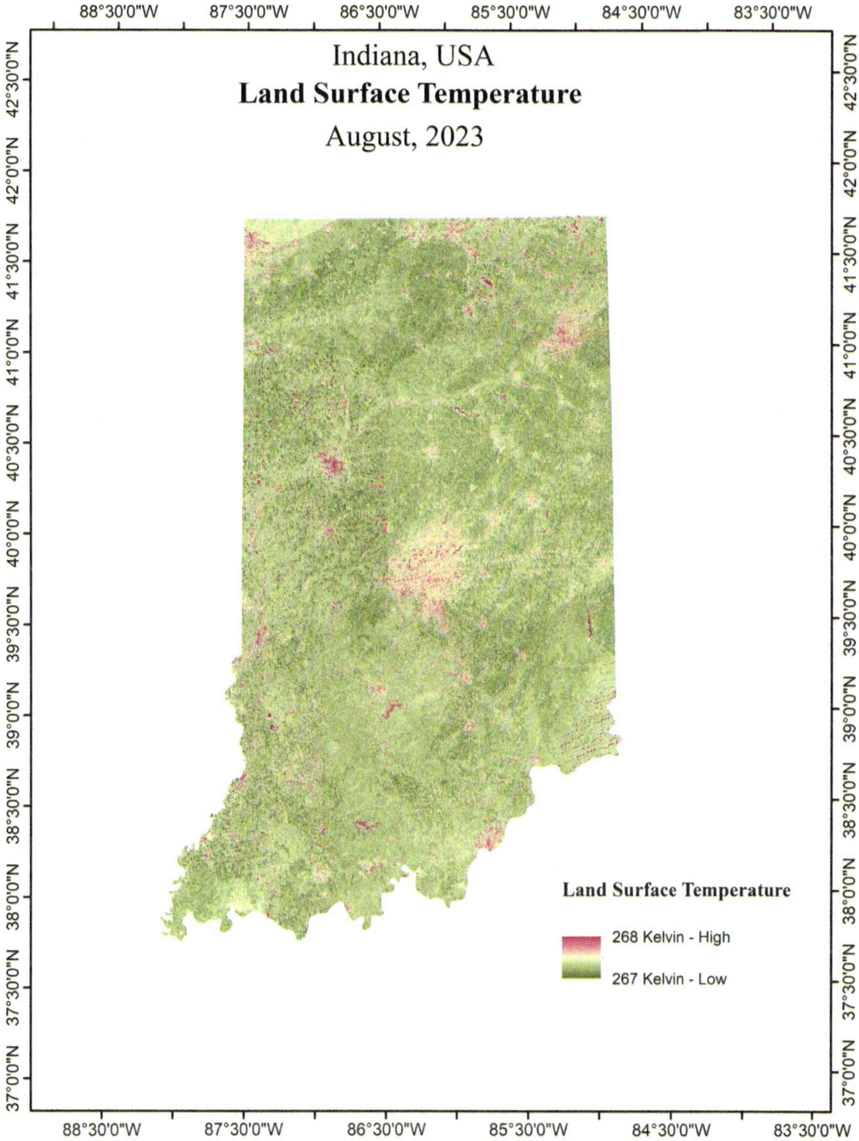

Fig. 2 Land surface temperature of Indiana

3.5 *Precipitation*

Another important factor for yield estimation is precipitation as water is a beneficial resource for the increase in the yield and thus plays a major role in the process of crop yield estimation. Rainfall data that we need is average daily accumulated rainfall. This gives us a general idea, so we prefer this more over other different ways of precipitation measurement (Fig. 3).

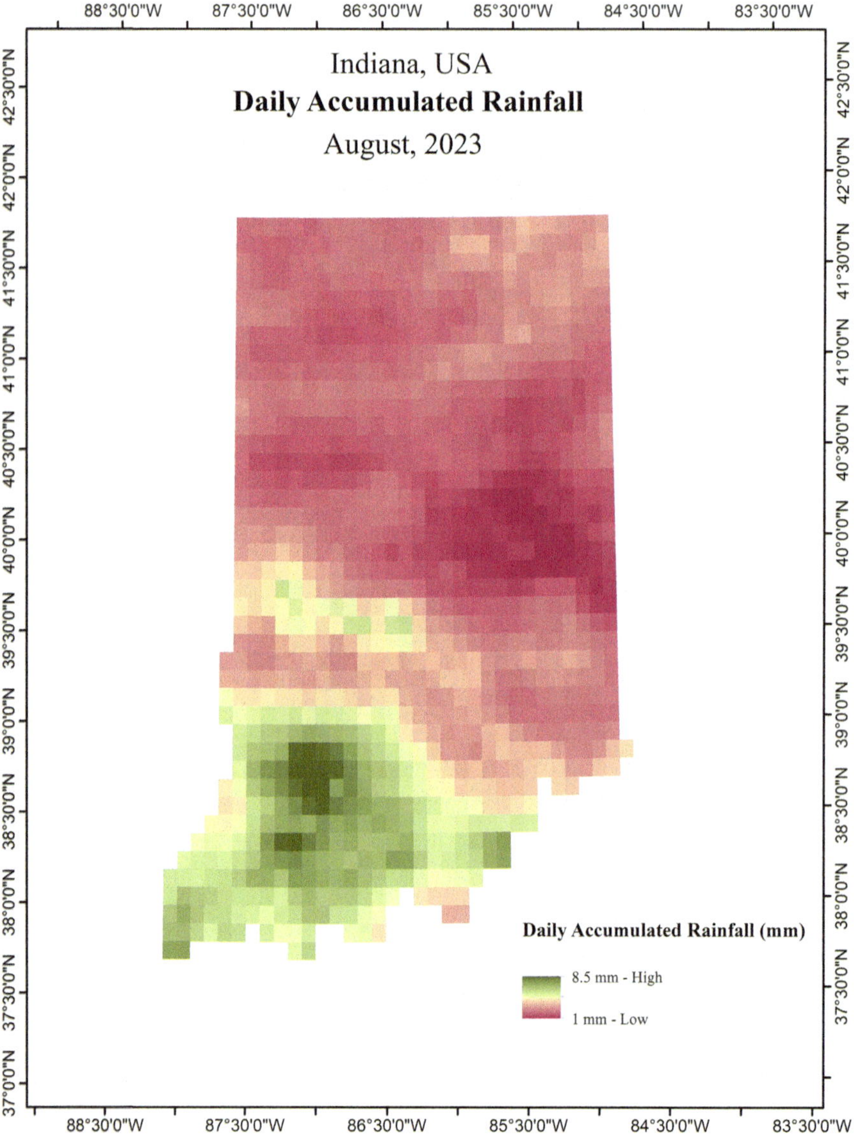

Fig. 3 Rainfall in Indiana

4 Methodology

The model that is to be used has been given by R.P. Singh. The model works with some predefined coefficients. Coefficients of empirical equation have been obtained using this method. Breakpoint (m) is the mean of the number of years we are considering for the model. Corn or soybean crop yield of Indiana [19]:

$$\begin{aligned}&\text{Crop Yield} = \left(c_1 + \left(a_1 \times \text{NDVI}\right) + \left(a_2 \times \text{SM}\right) + \left(a_3 \times \text{ST}\right) + \left(\vec{a}_4 \times \text{rf}\right)\right)\\&\qquad\qquad\left\{\text{for crop yield} < \text{breakpoint}\, m\right\}\\&\text{Or } \left(c + \left(b_1 \times \text{NDVI}\right) + \left(b_2 \times \text{SM}\right) + \left(b_3 \times \text{ST}\right) + \left(b_4 \times \text{RF}\right)\right)\\&\qquad\qquad\left\{\text{for crop yield} > \text{breakpoint}\, m\right\}\end{aligned} \tag{2}$$

In rare cases, if breakpoint (m) becomes equal to crop yield, condition of crop yield > breakpoint should be applied. The model empirical equations for corn and soybean crops thus obtained with coefficients are given below that are to be used in the equation [19].

The model is used to provide closest result to the actual data of the yield; it provides us a very closer estimate of the actual result in reality. The controlling factors for the result are precipitation, NDVI, soil moisture, and land surface temperature. The result is a combined value with the use of above coefficients and factors.

The factors that have been used above were used from a predefined model for Illinois, and its accuracy had to be confirmed for other states as well.

5 Results

At the end, we need to compare the predicted results with the actual ones. This helps us to test the accuracy of the model as well as the accuracy of the work performed. If there is no assessment, the work is a mere failure even if it is correct because assessing does give us insight into the small errors that occurred even if everything is performed within the rules of the book.

Here, the accuracy that we get is in bushel per acre that is a unit used to define the output of the model. So, the initial output of the model was 290–224 compared to the actual output of 203 bushels per acre for corn.

With regard to soybean, the output yield of the model was 56–45 compared to the actual yield of 61 bushels per acre. Both results can be seen in Fig. 4.

While the prediction or estimation results are expected to show only minor deviations, it is important to acknowledge the inherent uncertainties associated with

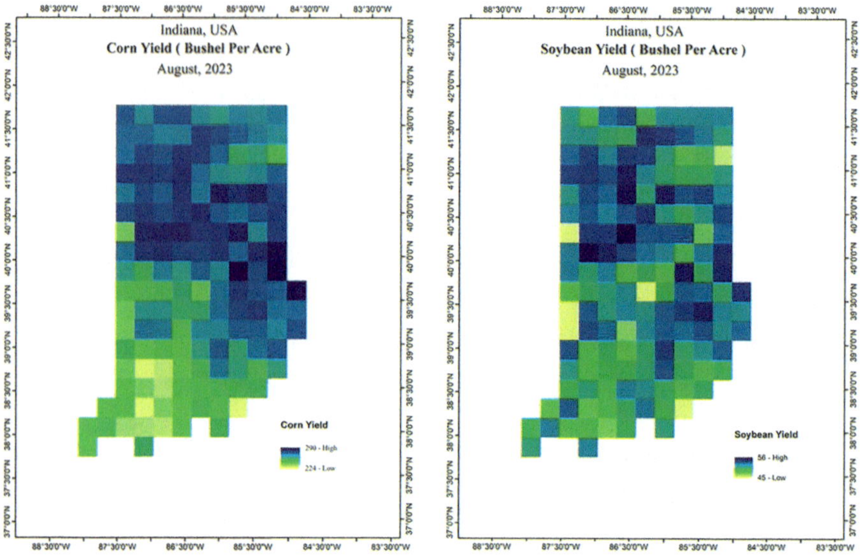

Fig. 4 Results of crop yield

environmental variability. Agricultural systems operate within highly dynamic environments where meteorological fluctuations—such as temperature extremes, precipitation variability, or storm events—can significantly impact crop outcomes. These changes may arise from both predictable seasonal patterns and unexpected disturbances, including convective storms or tornadoes. Additionally, the proximity of the region to large water bodies, such as lakes, increases the likelihood of sudden and localized precipitation events, further influencing model performance and observed variability." Please let us know if any further clarification is needed.

6 Conclusion

Crop yield prediction model is a big leap in remote sensing; it enhances the ways a farmer can contribute more towards world hunger. It gives us valid and reasonable results and does let us know about the areas that need the attention of the hour. It is calculated by considering all the factors that influence crops. But these should also not be trusted blindly as these are not the end result; these are just a mere result, and thus, this can be used to bring a desirable change in the yield if needed. The crop model has been a drastic element; more and more models are made suitable for various conditions and various areas; in addition, models are made accurate each year thinning the line of actual yield and predicted yield. But for now, we need to understand that uncertainties occur in all of the models, and thus, this model is not an exception.

The model was initially made for the state of Illinois, but its accuracy needed to be assessed as well to check its compatibility with the conditions of the other states. Now with the above result, we can confirm that the model is suitable for the state of Indiana as the difference in the yield output of the model and actual yield is only 10%.

The study demonstrates the transformative potential of geospatial tools in modern agriculture. By integrating remote sensing, GIS, and advanced data analytics, this approach provides precise, real-time insights into crop health and yield predictions. This not only enhances agricultural productivity and sustainability but also enables proactive decision-making for farmers, ensuring efficient resource management and improved crop outcomes. The study underscores the importance of continued innovation and adoption of geospatial technology to address the challenges of global food security and agricultural resilience.

References

1. Keating, B. A., & Thorburn, P. J. (2018). Modelling crops and cropping systems—Evolving purpose, practice and prospects. *European Journal of Agronomy, 100*, 163–176. https://doi.org/10.1016/j.eja.2018.04.007
2. Ali, A. M., Abouelghar, M., Belal, A., Saleh, N., Yones, M., Selim, A. I., Amin, M. E., Elwesemy, A., Kucher, D. E., Maginan, S., & Savin, I. (2022). Crop yield prediction using multi sensors remote sensing (review article). *The Egyptian Journal of Remote Sensing and Space Science, 25*(3), 711–716. https://doi.org/10.1016/j.ejrs.2022.04.006
3. Bandaru, V., Yaramasu, R., Jones, C., César Izaurralde, R., Reddy, A., Sedano, F., Daughtry, C. S., Becker-Reshef, I., & Justice, C. (2022). Geo-cropsim: A geo-spatial crop simulation modeling framework for regional scale crop yield and water use assessment. *ISPRS Journal of Photogrammetry and Remote Sensing, 183*, 3453. https://doi.org/10.1016/j.isprsjprs.2021.10.024
4. Retrieved from www.usda.gov
5. Retrieved from www.britannica.com
6. Retrieved from www.in.gov
7. US Department of Agriculture, National Agricultural Statistics Service. (2021). *Quick stats.* USDA-NASS: Washington, DC, USA.
8. Larson, G., & Schaetzl, R. (2001). Origin and evolution of the Great Lakes. *Journal of Great Lakes Research, 27*(4), 518–546. https://doi.org/10.1016/s0380-1330(01)70665-x
9. Indiana Glacial Boundaries. (n.d.). *Indiana geological & water survey*. Retrieved from https://legacy.igws.indiana.edu/Surficial/IndBoundaries
10. Indiana.edu.
11. wikipedia.org
12. statesummaries.ncics.org
13. Datamation.com
14. Wagner, K. (2018). Geographic information systems and glacial environments. *Past Glacial Environments*, 503–536. https://doi.org/10.1016/b978-0-08-100524-8.00015-4
15. Panjala, P., Gumma, M. K., Mesapam, S., Shukla, A. K., & Pignatta, G. (2025). Geospatial analysis of crop residue burn areas and their dates for emission mitigation strategies. *Sustainability, 17*(6), 2508. https://doi.org/10.3390/su17062508
16. gisgeography.com

17. Holzman, M., Rivas, R., & Piccolo, M. (2014). Estimating soil moisture and the relationship with crop yield using surface temperature and vegetation index. *International Journal of Applied Earth Observation and Geoinformation, 28*, 181–192. https://doi.org/10.1016/j.jag.2013.12.006
18. Li, Z., Tang, B., Wu, H., Ren, H., Yan, G., Wan, Z., Trigo, I. F., & Sobrino, J. A. (2013). Satellite-derived land surface temperature: Current status and perspectives. *Remote Sensing of Environment, 131*, 14–37. https://doi.org/10.1016/j.rse.2012.12.008
19. Prasad, A. K., Chai, L., Singh, R. P., & Kafatos, M. (2006). Crop yield estimation model for Iowa using remote sensing and surface parameters. *International Journal of Applied Earth Observation and Geoinformation, 8*(1), 26–33. https://doi.org/10.1016/j.jag.2005.06.002

Comprehensive Water Quality Assessment and GIS-Based Analysis in Kaithal District: Integrating Digital Technologies and Statistical Methods

Anamika Kumari Gautam, Gaurav Singh, Bhupender Parashar, Lokesh Gupta, Nitin Bhardwaj, Pushpendra Kumar Sharma, and Bishnu Kant Shukla

1 Introduction

Water serves as the foundation for sustainable environmental practices because it influences both natural ecosystems and economic operations and public health systems [1–3]. The water resources of Kaithal District in Haryana face substantial impacts from agronomy production together with domestic consumption and small industrial operations in rural areas. The introduction of pollutants and contaminants through these activities disrupts the natural water cycle and damages the quality of water sources [4]. The main occupation of Kaithal District relies on agriculture, while its irrigation depends on surface water collected from ponds, lakes, and canals.

A. K. Gautam · G. Singh
Department of Civil Engineering, JSS Academy of Technical Education, Noida, India

B. Parashar
Department of Mathematics, JSS Academy of Technical Education, Noida, India
e-mail: parashar_b@jssaten.ac.in

L. Gupta
College of Pharmacy, JSS Academy of Technical Education, Noida, India
e-mail: lokeshgupta@jssaten.ac.in

N. Bhardwaj
Department of Mathematics & Statistics, Ch. Charan Singh Haryana Agricultural University, Hisar, India
e-mail: nitinbhardwaj.mdu@hau.ac.in

P. K. Sharma
School of Civil Engineering, Lovely Professional University, Phagwara, India

B. K. Shukla (✉)
Department of Civil Engineering, JSS Academy of Technical Education, Noida, India

School of Civil Engineering, Lovely Professional University, Phagwara, India
e-mail: bishnukantshukla@gmail.com

S. Pathak et al. (eds.), *Geo-Data Revolution*, Advances in Geographic Information Science, https://doi.org/10.1007/978-3-031-94078-1_15

The local population depends on the water quality of these water bodies for drinking water, agricultural irrigation, and domestic uses. Throughout history, this area has dealt with poor water quality because of rising dissolved solid content, fertilizer-derived nutrient pollution, and microbial contamination. The ecological equilibrium together with human health remains endangered because of these environmental issues [5–7]. The last few years have brought about a shift toward executing detailed research, which includes chemical and physical characteristics, microbial analysis, and spatial distribution patterns to justify water quality monitoring [8–10].

Monitoring the quality of drinking water is of great importance to ensure safety and constitutes a considerable public health issue. Nitrites, coupled with microbial infections and heavy metals through drinking, can cause serious health challenges that provoke digestive and neurological issues and chronic diseases over a considerable number of years [11–15]. Monitoring systems continuously monitor for harmful sources, enabling prompt identification and action. Irrigation is significant in agriculture since irrigation water directly influences both soil productivity and soil fertility; thus, water quality assessment in agriculture continues to be important [16–18]. Degradation of soil composition, along with levels of salt concentration of harmful substances, leads to reduction of agricultural yield. The periodic assessment of water quality allows farmers to enhance their irrigation approaches and develop improved farming practices, which support sustainability. Ecological balance needs continuous water quality research. The changes in pH values, dissolved oxygen levels, and nutrient concentrations have a substantial impact on aquatic ecosystems. Water pollution causes three main environmental problems, which include habitat destruction together with species extinction and disruptions to ecosystem functions [19–21]. Regular monitoring of these factors allows the protection and regulation of aquatic ecosystems together with their ability to withstand environmental instabilities. The implementation of effective legislation and resource management depends on obtaining exact and detailed water quality data. Water management policies, regulatory frameworks, and conservation programs receive guidance from the data, which enables their effective development and implementation by policymakers [22, 23]. The identification of specific locations requiring resource allocation and intervention enables better water management practices.

The main focus of this study is to assess water quality in Kaithal District, Haryana, using a multidisciplinary approach. The research investigates the physical and chemical properties of surface water through measurements of turbidity, total suspended solids (TSSs), total solids (TSs), total dissolved solids (TDSs), electrical conductivity (EC), ultrasonic velocity, and refractive index. The research investigates essential chemical parameters through measurements of pH acidity levels, chemical oxygen demand (COD), biological oxygen demand (BOD), and dissolved oxygen (DO) content. The research evaluates microbial contamination through total coliform and faecal coliform bacteria concentration analysis in addition to physical and chemical assessments. The research implements GIS mapping to display water quality parameter distribution patterns and detect possible contamination zones for enhanced understanding. The research employs descriptive statistics to determine the maximum and minimum values of each parameter while also calculating standard deviation and mean values across different ranges and assessing skewness and kurtosis. The research studies various water quality measurement relationships while creating a correlation matrix from the data. The research study calculates Water Quality Index (WQI) through ICMR methods to evaluate overall water quality by comparing results to established standards. The research project develops recommendations for water quality enhancement along with risk reduction based on its findings.

The research evaluates water quality at 20 distinct locations across Kaithal District. The research gathers water samples from ponds, rivers, lakes, and canal beds to guarantee full district coverage. The research approach includes physio-chemical tests to evaluate turbidity, TS, TSS, TDS parameters, ultrasonic velocity, electrical conductivity, and refractive index. The analysis examines water chemical properties including pH, COD, BOD, DO, acidity, nitrate, phosphate, hardness, and chloride to determine organic composition and potential contaminants. The analysis checks for microbial contaminants including total coliform and faecal coliform bacteria. GIS maps have been developed using advanced geographic information systems to show the spatial distribution of water quality parameters within a district. The research also employs descriptive statistics and correlation analysis to show the relationships between the different water quality metrics. The Water Quality Index (WQI) is a calculated comprehensive measure of water quality that can be used for comparison and interpretation. This study aims at incorporating these factors and coming up with a detailed understanding of water quality in Kaithal District, which would help in the formulation of informed decisions and effective water management plans.

2 Literature Review

2.1 Overview of Water Quality Assessment

Water quality assessment represents an essential process, which examines drinking water properties through physical, chemical, and biological testing for multiple uses including drinking supply, irrigation, recreation, and industrial applications. The evaluation produces essential data, which supports both water resource conservation and management and ensures that water resources meet fundamental human and environmental health and safety standards [24]. The assessment process starts with water sample collection from different sources followed by laboratory testing with standardized procedures and ends with result evaluation against established water quality standards. The evaluation includes multiple parameters, which provide essential information about water quality components. Water exists in its current state because of physical elements such as temperature and turbidity. The chemical parameters of water include pH, dissolved oxygen (DO), and chemical oxygen demand (COD), which show its pH level and potential pollution. The presence of coliform bacteria serves as a biological contamination indicator. The need to monitor water quality arose from urbanization and industrial and agricultural activities to detect pollution patterns and implement corrective measures [25].

The escalating problems with water contamination and limited access make it essential to evaluate water quality. The combination of urbanization, industrialization, and farming practices has resulted in higher toxin emissions into water bodies. Regular observation together with analysis of water bodies enables the detection of pollution patterns, which allows for suitable remediation actions [26]. Advanced technologies including geographic information system (GIS) and remote sensing have enhanced our capability to track water quality across extensive geographic areas and time periods. The collection of extensive information has become possible, which enables effective water resource management [27].

2.2 *Physio-Chemical Parameters*

Water quality receives its definition mainly from physiological-chemical parameters. The physical and chemical characteristics of these parameters have a significant impact on the suitability of water bodies for diverse applications. Additionally, their overall condition is important. This study examines physio-chemical parameters, which include turbidity, TSS, TS, and TDS. The study, additionally, examines electrical conductivity, ultrasonic velocity, refractive index, pH, chemical oxygen demand (COD), biochemical oxygen demand, dissolved oxygen (DO), acidity, nitrate, phosphate, hardness, and chloride [28]. Water's turbidity refers to the degree to which suspended particles in it reduce its appearance or cause cloudiness. Higher levels of turbidity can hinder light penetration, which has negative effects on the growth of aquatic plants and the overall health or well-being of the aquatic ecosystem. Particulate matter that is suspended in water and dissolved in water is measured as total suspended solids (TSSs) and total diluted solids (TDSs). Elevated concentrations of TSS and TDS can indicate pollution and can negatively impact water's purity, taste, and fitness for diverse applications [29].

An aqueous solution's electrical conductivity (conductivity) is directly proportional to the concentration of dissolved ionic species, including salt and other substances. Accordingly, this feature functions as an alternative gauge of the overall dissolved electrolyte quantity and the general salinity of water samples. The propagation speed of ultrasonic waves within the water matrix is influenced by the specific molecular composition, mass density, and temperature. This property is known as ultrasonic velocity. Thus, the determination of ultrasonic velocity facilitates the understanding of the basic physical properties of water, thereby contributing significantly to its comprehensive assessment of quality [30]. Depending on the constituent composition and the general purity of water sample, the refractive index is an optical property that measures deflection of the light beam at either end of air-to-water interface. The detection of foreign substances and quantitative determination of dissolved solutes can be made using this measurement. The hydrogen ion activity, expressed as pH, is a critical parameter defining the acidic or alkaline nature of water, with a neutral condition defined at pH 7. Significant deviations from the circumneutral pH range can induce adverse physiological effects on aquatic biota and profoundly influence the speciation, solubility, and, ultimately, potential toxicity of various chemical constituents present within the aquatic environment [31]. Notably, the study documented in the sources revealed pH values between 3.9 and 4.6 in the Kaithal District, falling considerably below the permissible limit of 6.5–8.5 outlined in IS 10500, thus indicating highly acidic and corrosive conditions.

Chemical oxygen demand (COD) and biochemical oxygen demand (BOD) are fundamental parameters employed to quantify the oxygen required for the oxidation of organic and inorganic constituents present within an aqueous matrix. Elevated concentrations of both COD and BOD serve as indicative markers of organic pollution, suggesting the presence of substantial biodegradable and non-biodegradable organic matter. Such organic loading can lead to a critical depletion of dissolved oxygen (DO) levels, consequently imposing significant stress on the survival and overall well–being of aquatic organisms. Dissolved oxygen (DO) concentration, representing the amount of molecular oxygen present in the water column, is an indispensable requirement for the respiration of most aquatic species. Elevated levels of organic pollutants, as reflected by high BOD, typically result in diminished DO concentrations, thereby creating hypoxic conditions that can severely impair aquatic ecosystems [32]. The study in Kaithal District revealed that mean DO values significantly below the optimal minimum help in sustaining aquatic life. Acidity, expressed as the concentration of hydrogen ions (H^+), dictates the corrosive potential of water and its suitability for various applications. The documented pH range of 3.9–4.6 in the Kaithal District signifies highly acidic conditions. Nitrate and phosphate are essential nutrients; however, their presence in excessive concentrations can trigger eutrophication, an ecological imbalance characterized by the rapid proliferation of algal biomass and subsequent oxygen depletion upon algal decomposition. Hardness, a measure of the total concentration of divalent cations, primarily calcium (Ca^{2+}) and magnesium (Mg^{2+}) ions, influences the suitability of water for both domestic (e.g., soap lathering) and industrial processes (e.g., scale formation) [33]. The study in Kaithal District found hardness levels generally within the permissible range. Chloride concentration acts as an indicator of saline intrusion into freshwater sources or contamination from anthropogenic activities. The study found chloride concentrations below the permissible limit in the Kaithal District.

2.3 *Microbial Indicators*

Microbial indicators are essential for evaluating the biological quality of water resources and for monitoring the possible presence of pathogenic microorganisms that might represent serious risks for public health. One of the key indicators is to detect faecal contamination and then assess its potential pathogenic bacteria, viruses, and protozoa. Two of these types are faecal coliforms and total coliforms

[34]. Total coliforms are a group of related bacteria found in the environment (including soil, vegetation, and surface waters). Marine organisms in this locality are experiencing increasing levels of waterborne pathogens, and although their presence may not directly define pathogenicity, they have a sentience to be aware of potential waterborne pathogens. The detection of coliforms in the total volume of water sample indicates a condition of hygiene compromised in the form of not suitable water for potable use without appropriate cleaning methods [35]. Thermotolerant faecal coliforms, thermotolerant subgroups of total coliform organisms (generally termed as "thermotolerant" in the absence of airborne heat and humidity), have a more specific association with faecal material from all homeothermic animals, humans included. Enterococci, faecal coliforms, and *E. coli* are confirmed as better indicators that faecal contamination has occurred recently and therefore pathogens may be present (*E. coli*, *Salmonella* spp.) or for the introduction of other pathogenic organisms into the water supply. Intake of contaminated water can cause diarrhoea and other systemic health problems, including infections or digestive problems [36]. The results revealed that mean faecal coliform counts ranged significantly above acceptable levels in Kaithal District.

Microbiological water quality assessment is done for identifying the potential pathogens and their health risk through indicator organisms. Sampling is performed at systematic intervals, and the samples undergo rigorous laboratory analysis using well–known methodologies such as membrane filtration, multiple–tube fermentation process, or enzyme substrate assays. To this end, microbiologic analysis is critical. These methodologies allow the quantification and detection of coliform bacteria, which proves to be fundamental regarding the microbiological characteristics of the water body [37]. In the case of the study conducted in Kaithal District, total coliforms were enumerated using a multiple–tube fermentation method (Indian Standard (IS) 1622) and the faecal–added Friendzone using a membrane filtering method (IS 5401). Since drinking and recreational water quality is dependent on the presence and levels of freshwater indicator bacteria, their identification and quantification are critical. The use of microbial indicators is valuable for the early identification of contamination points, thus allowing point–specific intervention to safeguard human health both now and in the future.

Water Resources and Sustainable Pest Management: The sustainable management of water resources and their microbiological safety are significant in eliminating waterborne diseases. ↑ High level of coliform counts has been seen in Kaithal District, highlighting the need for its surveillance and remedial measures.

2.4 GIS and Remote Sensing in Water Quality Studies

Geographic information systems (GISs) and remote sensing technologies have transformed the field of water quality research, providing novel analytical tools for tracking and analysing the quality of water over broad spatial and temporal scales.

Climate and ocean sensors have been improved with advanced biomimicry to identify sources of pollution, monitor changes in the environment, and manage water resources with evidence–based practices. GIS is a powerful integrative technology that combines spatial data with associated attribute information and can facilitate exhaustive cartographic visualization and sophisticated analysis of water quality drivers [38]. Its advantages are substantial. The ability to layer these datasets one on top of the other and then parse the information provides for a graphical interpretation of complicated topographical patterns that explain the interrelationships of various environmental measurements. This analytical approach allows to provide tailored measure of maximum contamination spread, to precisely trace space-time origin of input of infectious agents, and to quantitatively assess complex impacts of land-use pressure and anthropogenic activities on water bodies' integrity. As exemplified in this study on Kaithal District, QGIS software was employed for spatial analysis, integrating GPS-derived sampling locations with water quality attribute data. Spatial interpolation techniques, such as inverse distance weighting (IDW), were applied to generate continuous spatial representations of key water quality parameters, effectively highlighting areas of concentrated contamination and regions necessitating focused management interventions [39]. Remote sensing, conversely, involves the acquisition of data pertaining to water bodies from distal platforms, such as satellite or aerial-borne sensors. These sensors are equipped to record electromagnetic radiation across multiple spectral bands, enabling the quantitative assessment of crucial water quality characteristics, including chlorophyll-a concentrations (an indicator of algal biomass), turbidity levels (a measure of water clarity), and surface water temperature. The principal advantage of remote sensing lies in its capacity to conduct surveillance over vast and often inaccessible geographical regions, offering a synoptic and longitudinal perspective on the dynamic variations in water quality over extended periods [40]. While the current study on Kaithal District primarily utilized GIS for spatial analysis of collected samples, the synergistic integration of remote sensing data could further enhance the temporal and spatial comprehensiveness of future water quality assessments in the region.

The synergistic integration of geographic information systems (GISs) and remote sensing technologies significantly enhances the capacity for comprehensive observation and informed management of water resources. Specifically, satellite-based multispectral imagery provides a valuable tool for the remote identification and spatial delineation of critical water quality phenomena, such as the proliferation of algal blooms (indicating potential eutrophication), the dynamic tracking of alterations in land utilization patterns (which can have profound implications for non-point source pollution), and the thorough evaluation of the impacts of natural calamitous events on the ecological integrity of aquatic ecosystems. And GIS platforms provide a highly dynamic environment for the complex investigation of spatially referenced ecological variables. These capabilities utilize the ability to combine different datasets, develop predictive models to predict trends in water quality over time, and ultimately support informed decisions about allocating resources and managing water resources. We thus are able to conclude, as shown in the large-scale analysis of water quality in Kaithal District, that GIS software (QGIS) was instrumental in mapping the spatial distribution of various physico-chemical and microbial parameters, identifying potential contamination hotspots through spatial interpolation techniques, and elucidating spatial relationships among different water quality indicators [41].

The use of GIS and remote sensing in water quality studies offers many some obvious benefits including the ability to perform large-scale environmental assessments over wide geographic areas, to improve the accuracy and spatial specificity of water quality assessments, and to increase the overall efficiency of both data collection and subsequent analysis. Moreover, these technological approaches provide a low-cost and time-efficient methodology for the synoptic evaluation of water quality conditions, thus representing an essential contribution for the planning and management of sustainable and adaptive water resource management plans [42].

2.5 Water Quality Index (WQI) and its Applications

As a tool for environmental assessment and communication, the Water Quality Index (WQI) has profound advantages. Importantly, it presents a clear and easily interpretable summary of complex water quality datasets that can be used to compare across differing regimes and geographic areas. Moreover, the WQI can greatly assist in evidencing and prioritizing the allocation of resources and environmental management intervention planning [11]. In addition, it provides an essential vehicle for communicating important water quality information to a large audience of both the general public and selected stakeholder groups while also increasing awareness of and engagement in water quality issues and encouraging responsible water resource management. Nevertheless, it is important to point out the intrinsic limitations of the WQI methodology. These include the potential for subjectivity in the selection of relevant parameters, the assignment of appropriate weighting factors, and the establishment of meaningful scoring systems [7]. Moreover, the shaping of a multi-variable index inherently bears a risk of oversimplification of complex water quality dynamics and may obscure specific pollutants or vital biological interactions. Thus, a more prudent approach recommends the application of the WQI in addition to complementary monitoring approaches and location-specific in-depth assessments, in order to give a more detailed and comprehensive picture of the water quality situation [2].

Previous studies from other regions have provided valuable knowledge about water quality drivers, success of different types of best management practices, and ongoing stressors challenging the persistence and improvement of the ecological status of aquatic resources [9]. These studies highlight the need for regular and rigorous monitoring regimes and the need for integrated assessment frameworks. However, it is a well-known fact that water quality index (WQI) is potentially an effective tool to minimize the complexity of water data and also allows a quick overview of water bodies' quality for comparison, prioritization of resources, and enlightenment of the common man with a knowledge about the water quality problems [1]. But the parameter selection, weighting, and scoring system introduce subjectivity, and while the WQI provides a thumb on the scale of baseline water quality, it can also simplify data and miss detailed problems. Accordingly, it is used by experts along with the other monitoring methods to fully assess water quality. This demand is supported by previous research that showed the importance of comprehensive and ongoing assessments of water quality.

3 Methodology

pH: The pH, which indicates acidity or alkalinity of water, was determined using a pH meter using IS 3025 (Part 11). The meter was calibrated with standard buffer solutions prior to taking any measurements, to ensure precision.

Chemical oxygen demand (COD): COD was estimated by the dichromate reflux method as per the IS 3025 (Part 58), which represents the mass of oxygen required to chemically oxidize both organic and inorganic materials in a quantified water sample. The material was digested with sulphuric acid and potassium dichromate; then the unconsumed dichromate was titrated to calculate the COD.

Biological oxygen demand (BOD): BOD, which stands for biochemical oxygen demand, was measured using a 5-day BOD test under IS 3025 (Part 44), which measures the amount of oxygen required by microorganisms to decompose organic material in water. BOD was determined by using the difference between DO before and after incubation. Water samples were incubated at 20 °C for 5 days.

Dissolved oxygen (DO): The Winkler titration method was used for DO testing as per IS 3025 (Part 38), because it is essential for aquatic life. The sample was treated with reagents that immobilized the oxygen, and the oxygen content was determined from titration with sodium thiosulfate.

Acidity: The acidity of water, which reflects its capacity to neutralize bases, was determined by titrating a standard sodium hydroxide solution, according to IS 3025 (Part 22) guidelines. The endpoint was found using methyl orange as an indicator.

Nitrate: Assessments of nitrate, a marker of nutrient pollution levels, were performed in conformation with IS 3025 (Part 34) by a UV spectrophotometric screening method. Filtration and absorbance at 220 nm were used to determine the concentration of nitrate in the sample.

Phosphate: Following IS 3025 (Part 31), the ascorbic acid method was used to determine the concentration of phosphate, a key nutrient. The sample was treated with ascorbic acid and ammonium molybdate, generating a blue complex that was detected by spectrophotometry.

Hardness: According to IS 3025 (Part 21), EDTA titrimetric method was followed to determine the concentration of calcium and magnesium ions that indicate the hardness of water. Eriochrome Black T was employed as an indicator to titrate the sample with EDTA solution.

Chloride: Chloride, an indicator of saline incursion or pollution, was measured by argentometric method based on IS 3025 (Part 32). Potassium chromate was employed as an indicator during the titration to locate the endpoint.

3.1 Microbial Parameter Analysis

Total coliform: Total coliform was detected using the multiple-tube fermentation method, according to IS 1622, which specifies methods for detection of total coliform bacteria, which indicates faecal contamination. Coliform count was quantified by incubating water samples with lactose broth and measuring the gas output created during this process.

Faecal coliform: The levels of faecal coliform, which is an indicator of pooch pollution with warm-blooded animals, were quantified using the membrane filtering method as per IS 5401. For faecal coliform concentration assessment, samples were filtered through a membrane and incubated at 44.5 °C, per m-FC agar, where the blue colonies were counted.

3.2 GIS Mapping and Analysis

Software and tools used: QGIS software, a potent open-source tool for spatial analysis, was used to undertake geographic information system (GIS) mapping. Satellite imagery data and GPS devices for precise location tracking were additional tools.

GIS mapping methods: Some main steps are required in GIS mapping. First, the locations of the sampling were collected using GPS and imported into QGIS for further processing. The physical coordinates were then linked to the attributes of water quality factor data. Inverse distance weighting (IDW) and other spatial interpolation methods were employed to generate continuous surfaces for the water quality measures. These maps reflected the areas where various water quality parameters were concentrated and dispersed in Kaithal District, which helped identify the potential hotspots of contamination as well as those vulnerable regions requiring greater attention. Maps were subjected to further analysis to identify spatial patterns and correlation between different indicators, thus providing a detailed understanding of the water quality indices in the area under investigation.

3.3 Statistical Analysis

Statistical analysis was used to interpret data of water quality of 20 sampling sites in Kaithal District. Analysis included descriptive statistics, correlation, and multivariate analyses to explore interrelations between water quality parameters. Descriptive statistics were summarized for each parameter by calculating average mean, median, standard deviation, minimum, maximum, skewness, and kurtosis to determine data patterns and variability between locations.

3.3.1 Correlation Analysis

Pearson correlation coefficients were calculated to quantify the linear correlations between pairs of water quality measures. This study helped identify significant correlations and potential interdependencies across factors. A correlation matrix was created to visualize these interactions and indicate the parameters that tend to increase or decrease in tandem with the others. Robust associations may indicate shared origins of contamination or analogous ecological mechanisms impacting these variables.

3.4 Water Quality Index Calculation

At each of the 20 sampling sites in Kaithal District, the overall water quality was evaluated using the Water Quality Index (WQI). To make interpretation simple, the WQI combines several criteria into a single score. The selection of parameters was based on the significance of each parameter for both human health and water quality, including pH, DO, BOD, COD, TDS, nitrate, phosphate, and turbidity.

Normalization of parameters: The following formula (Eq. 1) was used to normalize the parameters to a scale from 0 to 100:

$$\text{Normalized value} = \frac{\text{Actual value} - \text{Minimum standard}}{\text{Maximum standard} - \text{Minimum standard}} \times 100 \tag{1}$$

Weight assignment: Each parameter was given a weight according to its significance, and the total of all weights was set at 1.

Calculation of sub-indices: Sub-indices were determined by multiplying the normalized value of each parameter by its weight.

Aggregation of sub-indices: The sub-indices were added up to determine the WQI (Eq. 2) for every location:

$$\text{WQI} = \sum(\text{Normalized value} \times \text{Weight}) \tag{2}$$

Classification of water quality: WQI values were classified as:

- 0–25: Very poor
- 26–50: Poor
- 51–70: Fair
- 71–90: Good
- 91–100: Excellent.

Interpretation and visualization: To illustrate the distribution of water quality throughout the district and pinpoint regions that require attention, WQI values were plotted and charted. In order to support efficient communication and water management strategies, this approach offered a succinct, comprehensive evaluation of the quality of water.

3.5 *Sampling Locations of Kaithal*

For this study, 20 sampling sites were selected in Kaithal District, Haryana, India (Fig. 1), to gather surface water. These were sites where the water quality of different types of water bodies was measured extensively, including ponds and canals. Water sampling was performed by the Indian Standard Code (IS 3025) methods to ensure uniformity and consistency of sampling and analysis. Sampling locations

Fig. 1 Kaithal district map

Fig. 2 Source of surface water

were mapped, and the coordinates were noted through GPS. Locations of two of the sampling sites in the district are shown in Fig. 2, from where the surface water comes.

Table 1 provides spatial reference for each site, including the 20 sampling sites with latitude and longitude for each occurrence. Such a combined visualization as well as tabular overview of study area is able to systematically inform for the differentiated water management plans and to aid in understanding the spatial linkage of water quality.

4 Results and Discussion

4.1 Physical Test Results

Table 2 provides a statistical summary of the physio-chemical parameters of water samples collected from Kaithal District. Total solids (TSs), total dissolved solids (TDSs), total suspended solids (TSSs), electrical conductivity (EC), turbidity, refractive index, and ultrasonic velocity are the parameters that were analysed. Significant variability is observed in the values across the sampling locations. For example, the concentration of TS varied from 1500 to 9000 mg/L, while TDS ranged from 135 to 1367 mg/L and TSS ranged from 1062 to 8040 mg/L. TS, TDS, and TSS have mean values of 5690 mg/L, 520.45 mg/L, and 5169.55 mg/L, respectively, suggesting moderate levels of contamination. Significant variation across various sites is indicated by standard deviation values, including 2517.72 mg/L for TS and 320 mg/L for TDS. The asymmetry and distribution characteristics of these

Table 1 Detailed list of water sampling locations in Kaithal District, accompanied by their latitude and longitude

Location	Location ID	Latitude	Longitude
Chandana	K1	29.750653	76.401292
Bidkyar Lake	K2	29.800027	76.399028
Sikka market	K3	29.793729	76.395497
Mohalla Sundrao	K4	29.793668	76.390477
Neem sahib	K5	29.803996	76.386295
Patti Khot	K6	29.800942	76.384198
Surya Kund	K7	29.778414	76.421488
Pundari	K8	29.77747	76.521572
Mator	K9	29.623876	76.266051
Simla	K10	29.635119	76.220423
Chausala	K11	29.659733	76.32139
Johad	K12	29.785495	76.68016
Keorak	K13	29.871292	76.472676
Bakal	K14	29.640932	76.577028
Kab Sarover	K15	29.653385	76.550097
Arnoli water pond	K16	30.174762	76.412366
Gaggarpur	K17	30.132713	76.408225
Jajanpur	K18	29.903512	76.605102
Ramthali	K19	29.921125	76.225306
Mahmoodpur	K20	30.12202	76.229967

Table 2 Statistical analysis of physical parameter results

Statistical parameter	Total solid	TDS	TSS	Conductivity	Turbidity (NTU)	Refractive index	Ultrasonic velocity
Min	1500	135	1062	210	1	1.355	1449
Max	9000	1367	8040	2734	75	1.44	1564
Mean	5690	520.45	5169.55	817.8	13.55	1.39	1512.55
Standard deviation	2517.72	320	2294.41	666.60	15.37	0.03	37.78
Skewness	−0.39	0.65	−0.48	1.27	3.12	−0.01	−0.34
Kurtosis	−1.19	0.55	−1.07	1.66	13.21	−1.53	−1.04

parameters are further indicated by the skewness and kurtosis values. Turbidity exhibits a significant positive skewness of 3.12, which indicates a higher incidence of extreme values.

The spatial insights into the distribution of these physical parameters across the district are provided by GIS mapping, as illustrated in Figs. 3, 4, and 5. Figure 3 emphasizes the spatial variability in refractive index and turbidity, with regions of elevated turbidity that may necessitate intervention. Figure 4 illustrates regions with increased concentrations of solid particles and ionic content through GIS mapping for total dissolved solids (TDSs), total suspended solids (TSSs), electrical conductivity (EC), and total solids (TSs). The velocity of ultrasonic waves through water is

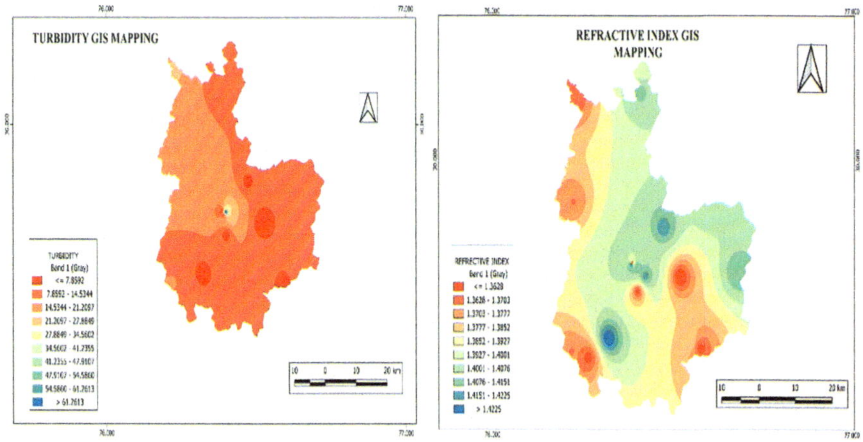

Fig. 3 GIS mapping for (**a**) Turbidity; (**b**) refractive Index

the primary focus of Fig. 5, which provides a visual representation of the physical properties of water. These figures collectively improve comprehension of the spatial variability in water quality parameters, thereby facilitating the identification of specific regions that may necessitate targeted water management strategies.

4.2 Chemical Parameter Results

Table 3 provides a statistical summary of the significant chemical characteristics identified in water samples collected from Kaithal District. The parameters consist of pH, biological oxygen demand (BOD), chemical oxygen demand (COD), dissolved oxygen (DO), hardness, chloride, total acidity, nitrate (NO_3), and total phosphates (PO_4). The pH values, which ranged from 3.9 to 4.6 with a mean of 4.24, suggest corrosive conditions. This value is below the permissible limit of 6.5–8.5 as per IS 10500. The average BOD level was 11.96 mg/L, which exceeded the acceptable limit of 3 mg/L and indicated substantial organic pollution. The BOD levels ranged from 6.8 to 20.2 mg/L. The concentration of organic pollutants was moderate, with a mean of 15.78 mg/L and a range of 2–28.48 mg/L for COD. The average DO value was 3.97 mg/L, with a range of 0.9–7.79 mg/L. This value is below the optimal minimum of 5 mg/L for the maintenance of aquatic life. Hardness levels were within the permissible range of 200–600 mg/L, with a mean of 175 mg/L and a range of 105–215 mg/L. The chloride concentrations were below the permissible limit of 250 mg/L, with an average of 191.22 mg/L and a range of 79.98–240.05 mg/L. The mean total acidity was 323.25 mg/L, with a range of 105–480 mg/L. Nitrate levels were within the permissible limit of 45 mg/L, with an average of 2.92 mg/L, ranging from 1.5 to 4.7 mg/L. The mean total phosphorus concentration was 0.63 mg/L, with a range of 0.44–0.9 mg/L.

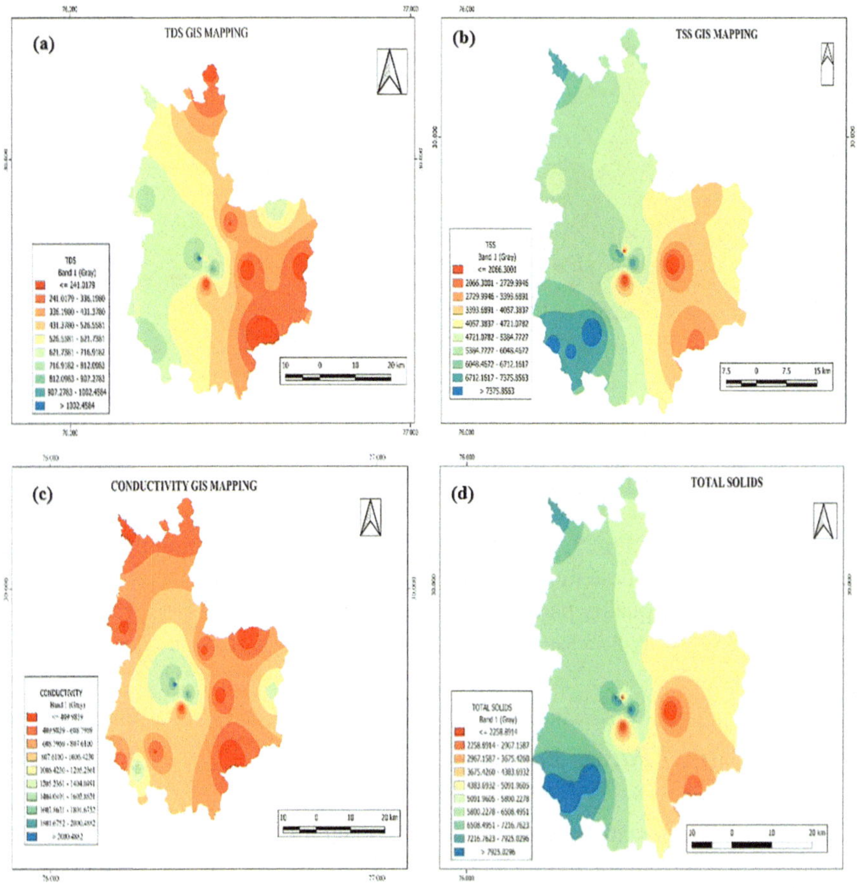

Fig. 4 GIS mapping for **(a)** total dissolved solids; **(b)** total suspended solids; **(c)** electrical conductivity; and **(d)** total solids

This statistical summary provides a more thorough understanding of the district's water quality by highlighting each indicator's variability, central tendency, and range. The results of this analysis highlight the most extreme values and provide insight into the general distribution of chemical characteristics of water samples. GIS mapping was performed to visualize the spatial distribution of these chemical parameters across Kaithal District. The spatial differences in these important water quality parameters are shown in Fig. 6, where the GIS mapping for pH, BOD, COD, and DO is shown. This type of map allows to highlight possible staging areas during pollution monitoring by identifying zones contributing significantly to the overall deviations from mean values. Total phosphate is represented by the GIS mapping in Fig. 7, which shows this distribution in the district. Figure 8 shows the GIS mapping of hardness, chloride, total acidity, and nitrate, offering wide coverage on the spatial variability of these parameters. These maps work in tandem to help target the spatial

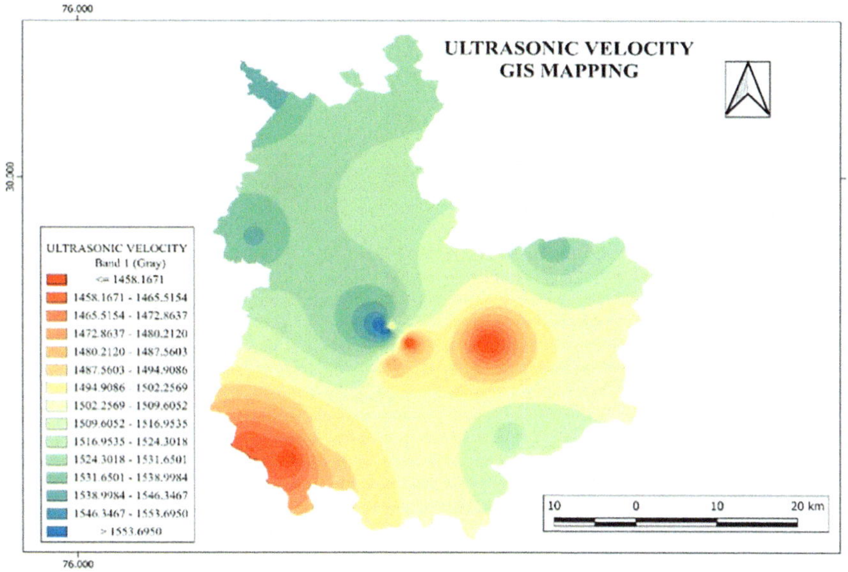

Fig. 5 GIS mapping for ultrasonic velocit

Table 3 Statistical analysis of chemical parameters across the sampling locations

Statistical parameter	pH	BOD (mg/L)	COD (mg/L)	DO (mg/L)	Hardness (mg/L)	Chloride (mg/L)	Total acidity (mg/L)	Nitrate (NO_3) (mg/L)	Total phosphates (PO_4) (mg/L)
Min	3.9	6.8	2	0.9	105	79.98	105	1.5	0.44
Max	4.6	20.2	28.48	7.79	215	240.05	480	4.7	0.9
Mean	4.24	11.96	15.78	3.97	175	191.22	323.25	2.92	0.63
Standard Deviation	0.17	3.78	8.57	2.01	25.35	43.20	89.68	1.03	0.15
Skewness	0.20	0.34	−0.33	0.35	−0.73	−1.33	−0.76	−0.76	0.30
Kurtosis	−0.111	−0.77	−1.03	−0.50	1.36	1.73	0.77	−1.46	−1.47

distribution of water quality for specific interventions and guide general water management strategies.

4.3 *Analysis of Biological Parameters*

The microbiological parameters in water samples collected from Kaithal District, including total coliforms and faecal coliforms, are statistically summarized in Table 4. It also lists a few crucial statistical parameters, such as the minimum, which signifies the lowest detected concentration of coliforms; the maximum, which

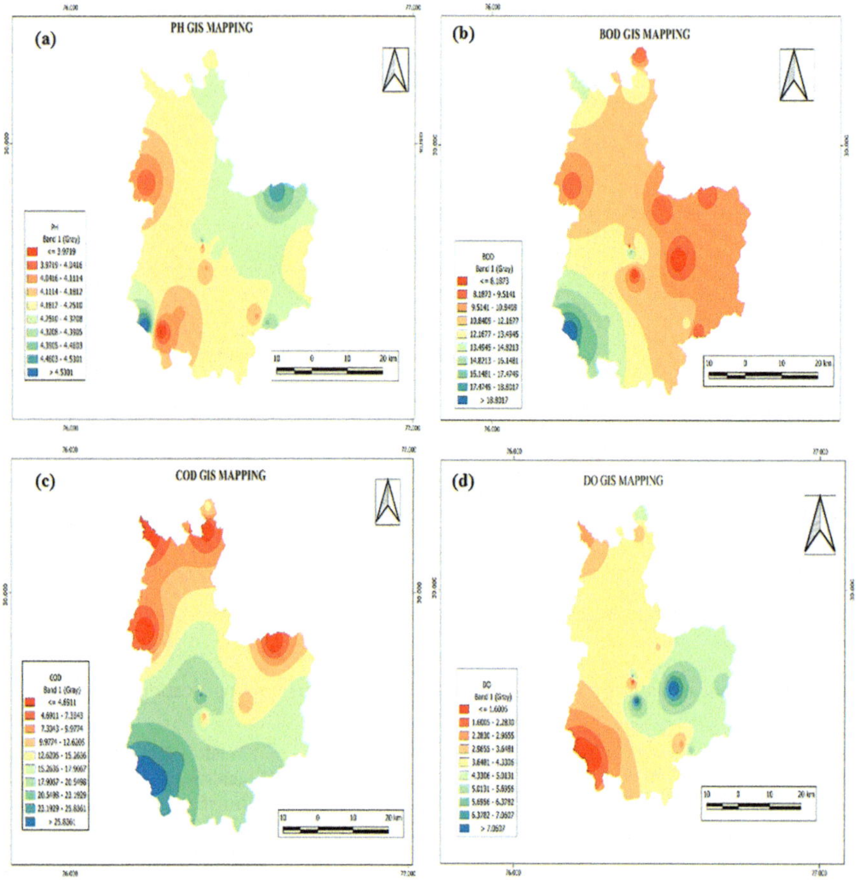

Fig. 6 GIS mapping for **(a)** pH; **(b)** biochemical oxygen demand; **(c)** chemical oxygen demand; and **(d)** dissolved oxygen

implies the highest concentration recorded; the mean, showing the average concentration among all samples; the standard deviation, reflecting how much coliform concentrations vary from the mean; the skewness, indicating the asymmetry of the coliform concentration distribution; and finally, the kurtosis, measuring the degree of "tailedness" or outlier presence in the coliform concentration distribution.

Table 4 illustrates that the minimum value for total coliforms is 12.9 MPN/100 mL, while the maximum value is 542.2 MPN/100 mL. The mean value is 197.875 MPN/100 mL, with a standard deviation of 138.812. The mean value for faecal coliforms is 65.14 MPN/100 mL, with a standard deviation of 30.94. The minimum value is 20.1 MPN/100 mL, and the maximum value is 110.2 MPN/100 mL. The observed values are considerably higher than the permissible limits for coliform bacteria in drinking water, as per IS 10500:2012. This means that serious contamination problems do exist. The statistical summary highlights the mean, standard

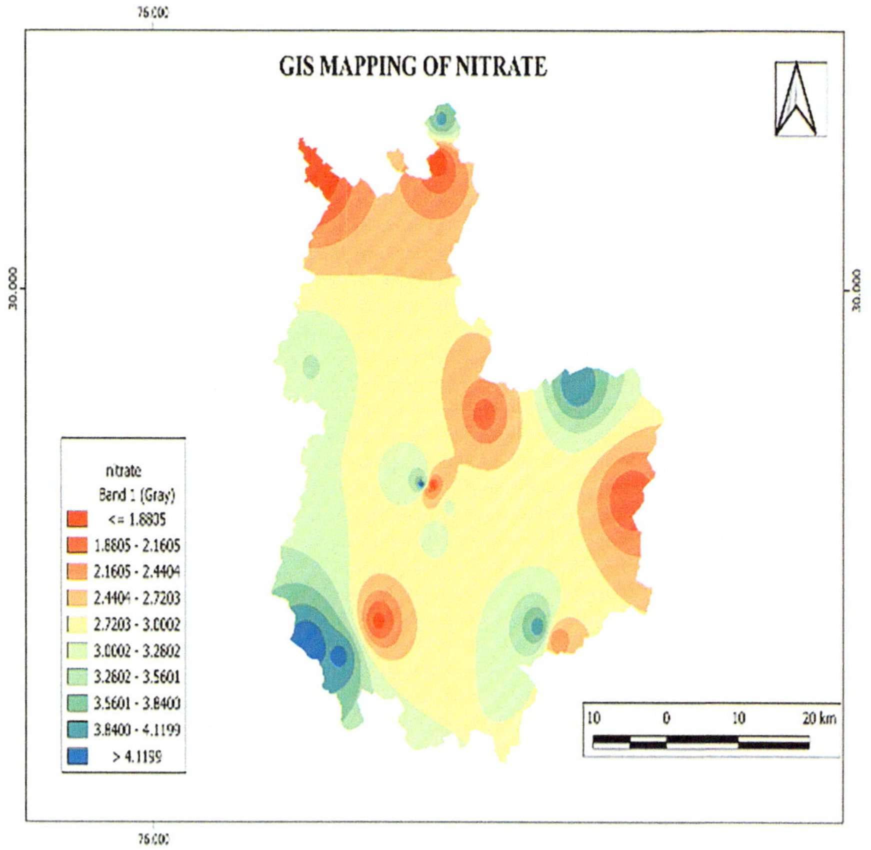

Fig. 7 GIS Mapping for Total Phosphat

deviation, and range of the coliform counts, thus contributing significantly to understanding the microbiological status of district water. Assessing contamination levels and trends is essential for evaluating the potential risks of human exposure to such pathogens and the human health risk associated with drinking contaminated water.

GIS mapping was used to visualize the spatial distribution of microbiological parameters over Kaithal District. Differences of these key microbiological parameters in space are shown in Fig. 9, which shows the GIS mapping for total coliforms and faecal coliforms. Coliform presence can tell us where to spend our money to protect public health and environment, and these maps are the first step—each dot tells us a place of significant coliform contamination that should be investigated further and possibly remediated.

As shown in Fig. 10, the media transfer process through a Nichrome wire, a technique often used by researchers, reflects an experiential approach that helps one to understand the challenges of water quality beyond bookish knowledge, which assists in assessing parameters of water samples for microbes. Statistical analysis

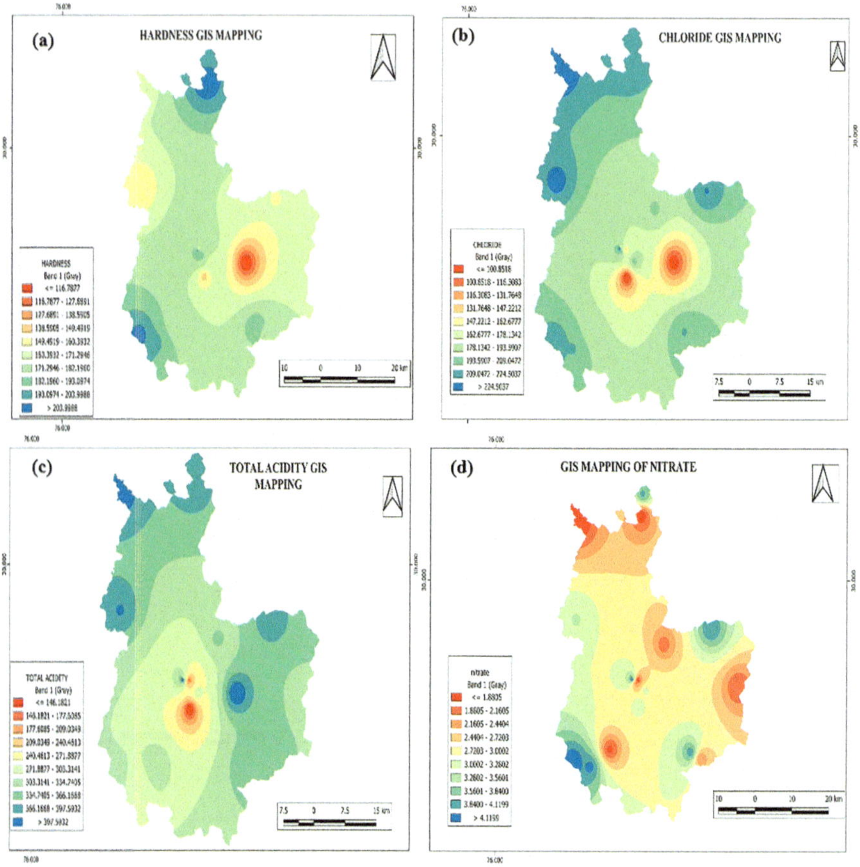

Fig. 8 GIS Mapping for (**a**) Hardness; (**b**) chloride; (**c**) total acidity; (**d**) nitrate

Table 4 Statistical analysis of biological parameters across the sampling locations

Statistical parameter	Total coliforms	Faecal coliforms
Min	12.9	20.1
Max	542.2	110.2
Mean	197.88	65.14
SD	138.81	30.93
Skewness	1.66	0.07
Kurtosis	2.18	−1.29

combined with GIS mapping provides an integrated approach for effective strategies in the protection of public health and management of water, which offers a comprehensive understanding of the microbiological quality of water in Kaithal District.

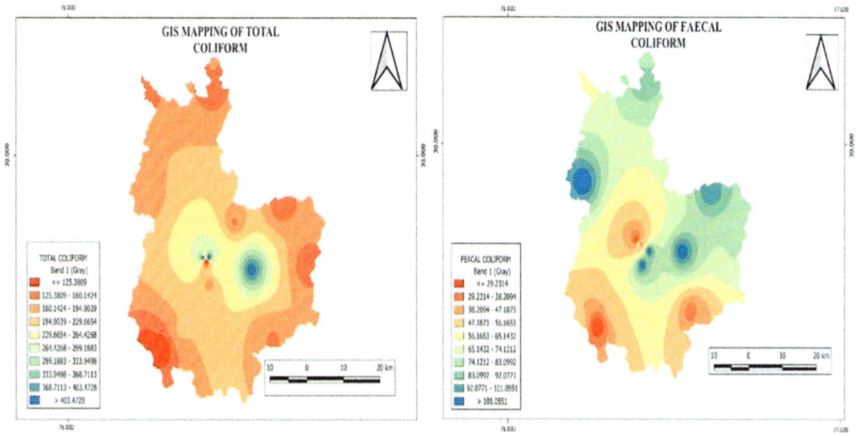

Fig. 9 GIS mapping of **(a)** total coliform and **(b)** faecal coliform

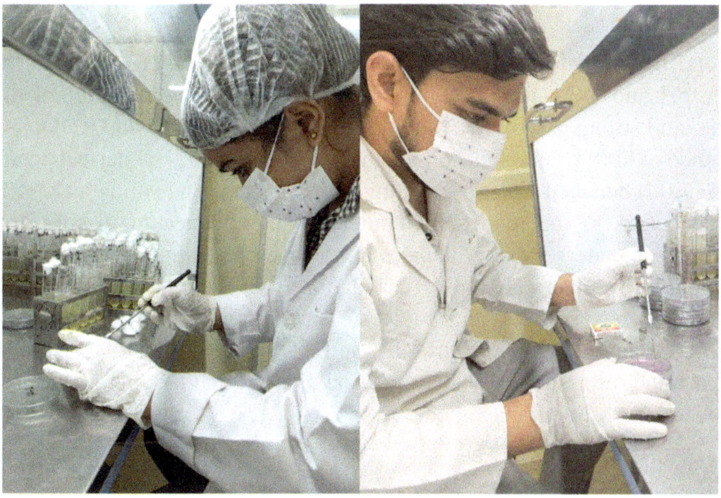

Fig. 10 Media transfer using Nichrome wire

4.4 *Correlation Analysis*

In the correlation table of Kaithal District, the variety of water quality metrics displays considerable correlations. Table 5 shows the correlations between parameters such as pH, biological oxygen demand (BOD), chemical oxygen demand (COD), dissolved oxygen (DO), total hardness (HN), chloride (CHL), total acidity (TA), nitrate (NO_3), and phosphate (PO_4).

Table 5 **Correlation analysis of evaluated parameters**

Positive Correlation (+)								Strong (0.8-1)			Moderate (0.4-0.8)			Weak (0-0.4)		
	TS	TDS	TSS	EC	TUR	RI	UV	pH	BOD	COD	DO	HN	CHL	TA	NO_3	PO_4
TS	1.															
TDS	0.73	1.														
TSS	1.00	0.49	1.													
EC	0.49	-0.16	0.44	1.												
TUR	-0.16	0.36	-0.21	0.14	1											
RI	0.36	0.24	0.34	0.47	0.16	1										
UV	0.24	-0.17	0.24	0.12	-0.14	0.23	1									
pH	-0.17	0.71	-0.18	-0.19	0.04	-0.06	-0.05	1								
BOD	0.71	0.38	0.70	0.38	0.14	0.27	-0.04	-0.16	1							
COD	0.38	-0.72	0.37	0.52	0.02	0.24	-0.26	-0.16	0.61	1						
DO	-0.72	0.48	-0.71	-0.29	-0.09	-0.35	-0.12	0.29	-0.90	-0.50	1					
HN	0.48	0.58	0.49	0.27	0.11	0.38	0.24	-0.03	0.67	0.42	-0.73	1				
CHL	0.58	0.22	0.57	0.16	0.02	0.27	0.38	0.01	0.45	-0.03	-0.66	0.61	1			
TA	0.22	0.21	0.22	-0.02	-0.34	0.09	0.21	0.07	0.03	-0.27	-0.19	0.04	0.60	1		
NO_3	0.21	0.18	0.17	0.20	-0.25	-0.20	-0.09	0.08	0.10	0.21	-0.11	0.16	0.23	0.25	1	
PO_4	0.18	0.40	0.14	0.21	-0.17	-0.13	-0.27	0.00	0.13	0.20	-0.14	0.11	0.22	0.31	0.95	1
Negative Correlation (-)								Strong(0.8-1)			Moderate (0.4-0.8)			Weak (0-0.4)		

Among the most notable findings are the moderate positive correlation (0.29) between pH and DO—i.e., the higher the pH, the greater the value of dissolved oxygen—and strong positive correlation (0.61) between BOD and COD, indicating the reliance of both components on organic contamination. The perfect positive correlation (1.00) between TS and TSS suggests that dispersed solids make a substantial contribution to the total solid content. The influence of dissolved ions on conductivity is underscored by the strong positive correlation (0.73) between TDS and TS and the moderate positive correlation (0.49) between EC and TDS. The impact of dispersed particles and acidity on water clarity is reflected in the moderate positive correlations between turbidity and TDS (0.36) and total acidity (0.22). These correlations assist in the identification of areas that require targeted interventions for water quality management.

4.5 *Water Quality Index*

Table 6 presents the Water Quality Index (WQI) values for the 20 sampling sites in Kaithal District, which highlights the quality of water at these locations. Important metrics that are used to calculate the WQI are the pH, DO, BOD, COD, TDS, nitrate (NO_3), phosphate (PO_4), and turbidity. Each location's WQI was computed by summing the values for each parameter after normalization and respective weighting based on their importance. And here are the WQI scores: 0–50 (excellent), 50–100 (good), 100–200 (poor), 200–300 (very poor), and > 300 (unsuitable). The water quality of places like Chandana (79.86), Bidkyar Lake (181.02), and Sikka Market (210.97) was found to be good, poor, and very poor, respectively. It is important to note that Patti Khot (359.56) and Surya Kund (267.49) are classified as unsuitable

Table 6 Water quality index for water sources

Location	Sample	Latitude	Longitude	WQI
Chandana	K1	29.750653	76.401292	79.86
Bidkyar Lake	K2	29.800027	76.399028	181.02
Sikka market	K3	29.793729	76.395497	210.97
Mohalla Sundrao	K4	29.793668	76.390477	219.93
Neem sahib	K5	29.803996	76.386295	150.62
Patti Khot	K6	29.800942	76.384198	359.56
Surya Kund	K7	29.778414	76.421488	267.49
Pundari	K8	29.77747	76.521572	129.83
Mator	K9	29.623876	76.266051	242.05
Simla	K10	29.635119	76.220423	270.42
Chausala	K11	29.659733	76.32139	233.79
Johad	K12	29.785495	76.68016	143.77
Keorak	K13	29.871292	76.472676	133.48
Bakal	K14	29.640932	76.577028	95.35
Kab Sarover	K15	29.653385	76.550097	136.11
Arnoli water pond	K16	30.174762	76.412366	164.55
Gaggarpur	K17	30.132713	76.408225	174.48
Jajanpur	K18	29.903512	76.605102	155.67
Ramthali	K19	29.921125	76.225306	175.09
Mahmoodpur	K20	30.12202	76.229967	210.09
Excellent (<50)	Good (50–100)	Poor (100–200)	Very poor (200–300)	Unsuitable (>300)

and very poor, respectively. These classifications suggest that there is a need for substantial intervention to enhance the quality of water. This table provides geographic insights that indicate the spatial distribution of water quality throughout the district, identifying areas that require immediate attention.

This WQI table can be employed by environmental administrators and policymakers to assess and efficiently manage water quality in Kaithal District. For example, locations with WQI scores below 50 are considered to have excellent water quality, while scores above 200 indicate that urgent remediation is necessary. A targeted approach to environmental management is ensured by this detailed analysis, which assists in the prioritization of areas for water quality enhancement initiatives.

5 Conclusion

The exhaustive analysis of water quality in Kaithal District, Haryana, demonstrates substantial variability in both physio-chemical and microbial parameters. Water quality issues, particularly acidity and high levels of organic pollution, are indicated by high levels of total solids (TSs: 1500–9000 mg/L), total dissolved solids (TDSs:

135–1367 mg/L), and low pH values (3.9–4.6). This is supported by BOD (6.8–20.2 mg/L) and COD (2–28.48 mg/L). The permissible limits for total coliforms (12.9–542.2 MPN/100 mL) and faecal coliforms (20.1–110.2 MPN/100 mL) have been significantly exceeded, suggesting severe microbial contamination. Several critical sites, such as Patti Khot (359.56) and Surya Kund (267.49), were identified as inadequate by the Water Quality Index (WQI), requiring immediate remediation effort. The necessity of integrated water management strategies is underscored by correlation analysis, which reveals significant interdependencies, including a strong positive correlation (0.61) between BOD and COD and a strong negative correlation (−0.90) between DO and BOD. A spatial overview was provided by GIS mapping of water quality parameters, which identified concentrations of contamination and guided targeted interventions. To safeguard public health and guarantee the availability of safe water in Kaithal District, it is crucial to implement comprehensive and systematic water quality monitoring and management. These findings underscore the significance of this approach. The findings highlight the importance of employing advanced digital tools for sustainable water resource management, addressing both contamination and conservation in the district. These results are critical for policymakers and environmental managers aiming to improve water quality and ensure public health in the region.

References

1. Shukla, B. K., Bhowmik, A. R., Raj, R. B., & Sharma, P. K. (2019). Physio-chemical parameters and status of ground water pollution in Jalandhar-Phagwara region. *Journal of Green Engineering, 9*(2), 212–223.
2. Shukla, B. K., Gupta, A., Sharma, P. K., & Bhowmik, A. R. (2020). Pollution status and water quality assessment in pre-monsoon season: A case study of rural villages in Allahabad district, Uttar Pradesh, India. *Materials Today: Proceedings, 32*, 824–830.
3. Hussain, A., Deshwal, A., Priyadarshi, M., Pathak, S., Sambandam, G., Chand, S., & Shukla, A. K. (2024). *Landfill leachate analysis from selected landfill sites and its impact on groundwater quality, New Delhi* (pp. 1–26). India. Environment.
4. Shenoy, A., Bansal, V., & Shukla, B. K. (2022). Treatability of effluent from small scale dye shop using water hyacinth. *Materials Today: Proceedings, 61*(02), 579–586.
5. Gupta, S., Bhatnagar, M., & Jain, R. (2003). Physio-chemical characteristics and analysis of Fe and Zn in tube well water and sewage water of Bikaner City. *Asian Journal of Chemistry, 15*(2), 727.
6. Rajagopal, T., Thangamani, A., Sevarkodiyone, S. P., Sekar, M., & Archunan, G. (2010). Zooplankton diversity and physio-chemical conditions in three perennial ponds of Virudhunagar district, Tamil Nadu. *Journal of Environmental Biology, 31*(3), 265–272.
7. Shukla, B. K., Bashir, M., Sharma, P. K., & Abhimanyu. (2021). An analytical investigation of surface water quality and pollution status in Srinagar, Jammu and Kashmir, India. *Journal of Green Engineering, 11*(1), 952–962.
8. Shukla, B. K., Goel, A., & Sharma, P. K. (2020). An experimental study on oxygenation performance of rectangular shaped solid jet aerator having rounded edges. *Journal of Green Engineering, 10*(06), 2728–2745.

9. Singh, S., Singh, V. P., Garg, P., & Shukla, B. K. (2023, February). Physico-chemical analysis of groundwater in Noida-Ghaziabad region. In *IOP conference series: Earth and environmental science* (Vol. 1110, No. 1, p. 012028). IOP Publishing.
10. Imtiyaz, I., Krishnakant Shukla, B. K., Varadharajan, S., & Bharti, G. (2023). A comprehensive review of mass transfer phenomenon in gas-liquid phase flow during aeration in wastewater treatment. *Lecture Notes in Civil Engineering, 281*, 127–135.
11. Shukla, B. K., Teeli, M. A., Shukla, S. K., Chandra, R., Bharti, N., & Singh, U. (2023). A comprehensive overview of vital water quality parameters. *Lecture Notes in Civil Engineering, 439*, 1–20.
12. Shukla, B. K., Singh, R. D., Singh, N., Singh, S., Burman, Y. K., & Singh, R. (2023). Application of nanomaterials in water purification: A thematic review. *Lecture Notes in Civil Engineering, 439*, 97–113.
13. Agarwal, A., & Saxena, M. (2011). Assessment of pollution by physicochemical water parameters using regression analysis: A case study of Gagan River at Moradabad-India. *Advances in Applied Science Research, 2*(2), 185–189.
14. Gorlach-Lira, K., Pacheco, C., Carvalho, L. C. T., Melo Júnior, H. N., & Crispim, M. C. (2013). The influence of fish culture in floating net cages on microbial indicators of water quality. *Brazilian Journal of Biology, 73*, 457–463.
15. Singh, S., Singh, V. P., Yadav, S., Kushwaha, V., Kumar, A., & Shukla, B. K. (2023, September). A compendious study on groundwater quality evaluation and monitoring. In *AIP conference proceedings* (Vol. 2800, No. 1). AIP Publishing.
16. Josefsson, H. (2025). The water framework directive and transnational situations: A call to rethink transboundary water management? *Journal for European Environmental & Planning Law, 22*(1–2), 168–186.
17. Kostanda, M., Spriņģe, G., Ozoliņš, D., Skuja, A., & Purmalis, O. (2025). Longitudinal distribution of benthic macroinvertebrates related to river ecological quality. *Limnological Review, 25*(1), 8.
18. Gani, A., Hussain, A., Pathak, S., & Banerjee, A. (2024). An empirical investigation on the elimination of heavy metals using bioremediation method for selected plant species. *Physics and Chemistry of the Earth, 134*, 103568.
19. Flory, E., Gloyne-Phillips, I., Veal, A. J., & Milner, A. M. (2024). Role of riparian vegetation in colonization and succession of stream macroinvertebrates in Glacier Bay, Alaska. In *Advances in ecological research* (Vol. 70, pp. 67–97). Academic Press.
20. Gelis, M. M. N., Canino, A., Bouchez, A., Domaizon, I., Laplace-Treyture, C., Rimet, F., & Alric, B. (2024). Assessing the relevance of DNA metabarcoding compared to morphological identification for lake phytoplankton monitoring. *Science of the Total Environment, 914*, 169774.
21. Kleinteich, J., Frassl, M. A., Schulz, M., & Fischer, H. (2024). Climate change triggered planktonic cyanobacterial blooms in a regulated temperate river. *Scientific Reports, 14*(1), 16298.
22. Al Haffar, M., Fajloun, Z., Azar, S., Sabatier, J. M., & Abi Khattar, Z. (2024). Lesser-known cyanotoxins: A comprehensive review of their health and environmental impacts. *Toxins, 16*(12), 551.
23. Pathak, S., Gupta, S., & Ojha, C. S. P. (2021). Assessment of groundwater vulnerability to contamination with modified DRASTIC index: A case study in Haridwar, Uttarakhand, India. *ASCE Journal of Hazardous, Toxic, and Radioactive Waste, 25*(2), 04020081.
24. Gani, A., Singh, M., Pathak, S., & Hussain, A. (2024). Groundwater quality index development using ANN model of Delhi Metropolitan City, India. *Environmental Science and Pollution Research*, 1–16.
25. Kamel Elshaarawy, M., & Eltarabily, M. G. (2024). Machine learning models for predicting water quality index: Optimization and performance analysis for El Moghra, Egypt. *Water Supply, 24*(9), 3269–3294.
26. Tugrul, S., & Yalcin, M. G. (2014). Water quality assessment using water quality index and multivariate analysis in Saraydüzü dam Lake, Turkey. *Ecological Indicators, 36*, 672–679.

27. Shukla, B. K., Sharma, P. K., & Goel, A. (2020, May). Study on oxygenation performance of solid jet aerator having circular opening corresponding to variable jet length and flow area. In *Journal of physics: Conference series* (Vol. 1531, No. 1, p. 012117). IOP Publishing.
28. Gani, A., Pathak, S., Hussain, A., Shukla, A. K., & Chand, S. (2024). Emerging pollutant in surface water bodies: A review on monitoring, analysis, mitigation measures and removal technologies of micro-plastics. *Environmental Geochemistry and Health, 46*(7), 214. https://doi.org/10.1007/s10653-024-01992-7
29. Shukla, A. K., Ojha, C. S. P., & Garg, R. D. (2017). Application of overall index of pollution (OIP) for the assessment of the surface water quality in the upper ganga river basin, India. In *Development of water resources in India* (pp. 135–149). Springer International Publishing. https://doi.org/10.1007/978-3-319-55125-8_12
30. Mishra, S., & Mishra, D. R. (2018). Remote sensing of water quality parameters: A review and analysis. *Environmental Monitoring and Assessment, 190*(3), 134.
31. Zhu, Z., Qu, Y., & Yu, P. (2020). Recent advances in sensor networks for water quality monitoring: A review. *Journal of Sensors, 2020*, 1–18.
32. Tripathy, S., Nayak, G., & Mishra, S. (2021). Machine learning-based water quality assessment: A systematic review. *Water, 13*(9), 1249.
33. Hussain, A., Deshwal, A., Priyadarshi, M., Pathak, S., Sambandam, G., Chand, S., & Shukla, A. K. (2024). Landfill leachate analysis from selected landfill sites and its impact on groundwater quality. New Delhi, India. *Environment, Development and Sustainability*, 1–26. https://doi.org/10.1007/s10668-023-04403-6
34. Shukla, A. K., Ojha, C. S. P., Mijic, A., Buytaert, W., Pathak, S., Garg, R. D., & Shukla, S. (2018). Population growth, land use and land cover transformations, and water quality nexus in the upper Ganga River basin. *Hydrology and Earth System Sciences, 22*(9), 4745–4770. https://doi.org/10.5194/hess-22-4745-2018
35. Shukla, B. K., Gupta, L., Parashar, B., Sharma, P. K., Sihag, P., & Shukla, A. K. (2025). Integrative assessment of surface water contamination using GIS, WQI, and machine learning in urban–industrial confluence zones surrounding the national capital territory of the Republic of India. *Water, 17*(7), 1076. https://doi.org/10.3390/w17071076
36. Yaseen, Z. M., & Afan, H. A. (2020). Water quality modelling: Artificial intelligence models. In *Water quality monitoring and management* (pp. 173–202). Elsevier.
37. Barzegar, R., & Alvisi, S. (2021). Artificial intelligence in water quality monitoring and management in artificial.
38. Shukla, A. K., Shukla, S., Surampalli, R. Y., Zhang, T. C., Yu, Y. L., & Kao, C. M. (2023). Modeling microconstituents based on remote sensing and GIS techniques. *Microconstituents in the Environment: Occurrence, Fate, Removal and Management*, 227–246. https://doi.org/10.1002/9781119825289.ch10
39. Shirk, J. L., Ballard, H. L., Wilderman, C. C., Phillips, T., Wiggins, A., Jordan, R., McCallie, E., Minarchek, M., Lewenstein, B. V., Krasny, M. E., & Bonney, R. (2012). Public participation in scientific research: A framework for deliberate design. *Ecology and Society, 17*(2), 29.
40. Shukla, A. K., Ojha, C. S. P., Shukla, S., & Garg, R. D. (2021). Water quality challenges in Ganga River basin, India. *The Ganga River Basin: A Hydrometeorological Approach*, 1–19. https://doi.org/10.1007/978-3-030-60869-9_1
41. IPCC. (2014). Climate change 2014: Synthesis report. In *Contribution of working groups I, II, and III to the fifth assessment report of the intergovernmental panel on climate change*. IPCC.
42. Pahl-Wostl, C., Arthington, A., Bogardi, J., Bunn, S., Hoff, H., Lebel, L., Nikitina, E., Palmer, M., Poff, L. N., Richards, K., Schlüter, M., Schulze, R., St-Hilaire, A., & Tharme, R. (2013). Environmental flows and water governance: Managing sustainable water uses. *Current Opinion in Environmental Sustainability, 5*(3–4), 341–351.

Illegal Foothill Settlements in Visakhapatnam, Andhra Pradesh: A Case Study

Peetala Pavan Kumar and Aditya Anand

1 Introduction

The rapid development of Visakhapatnam city and the establishment of numerous industries have spurred a significant influx of people migrating from surrounding rural areas [1]. Due to the scarcity and high cost of land within the city limits, residents have gradually encroached upon the delicate foothill ecosystems of Visakhapatnam. This encroachment has resulted in adverse effects on the residents' quality of life, and it is also a serious cause of concern in terms of ecological and disaster management perspective [2].

Over the past two decades, the escalating encroachment on hillsides and unregulated construction activities have substantially diminished green spaces and altered the geological integrity of Visakhapatnam's hills. This phenomenon is observable in neighborhoods such as Gajuwaka, Marripalem, Kancharapalem, Thatichetlapalem, Simhachalam, Hanumanthuwaka, Pendurthi, Arilova, and others throughout the city.

The haphazard construction of dwellings on hillsides, coupled with subsequent deforestation, has rendered these areas highly susceptible to landslides. Factors such as wedge failure of rock slopes, deforestation, and alterations in topography further exacerbate the risk of soil erosion and rockfall [3].

The limited accessibility of these hillside communities poses significant challenges during evacuations and other emergency situations, particularly for vulnerable groups such as the elderly, disabled, hospitalized individuals, and children [4].

These concerns are particularly pronounced during cyclones like Hudhud [5], where the ferocious winds wreak havoc on kutcha and semi-pukka structures [6]. Houses constructed on steep slopes without adequate support from retaining walls

P. P. Kumar (✉) · A. Anand
Amity School of Architecture and Planning, Amity University,
Gwalior, Madhya Pradesh, India

S. Pathak et al. (eds.), *Geo-Data Revolution*, Advances in Geographic
Information Science, https://doi.org/10.1007/978-3-031-94078-1_16

face the imminent threat of landslides, compounding the risk to residents' safety and property [7].

This study aims to examine the historical spatial growth patterns of illegal foothill settlements within the city, evaluate the ecological ramifications of these unauthorized settlements, and assess their vulnerability to natural and man-made disasters, including cyclones, landslides, and urban fires. To achieve these objectives, the development of built areas, settlement characteristics, movement patterns, and fluctuations in green cover will be analyzed across the years 2005, 2010, 2015, and 2020 in the selected locations. Furthermore, an exploration of the policies implemented by the local government, which contribute to the persistence of such illegal foothill settlements, will be undertaken.

The insights gleaned from this study shed light on how ecological degradation contributes to the evolution of unsafe spaces, ultimately rendering them susceptible to disasters.

2 Methodology and Data Collection

The study of illegal foothill settlements in Visakhapatnam started with the study of the evolution and characteristics of foothill settlements in the city's core area near Hanumanthuwaka and Arilova. Four case studies have been selected to understand the various characteristics of the illegal settlements in the foothills in Visakhapatnam. The different aspects that are studied to analyze and make recommendations for illegal foothill settlements in the city include demography, socioeconomic conditions, housing typologies, development control regulations, policies, programs and schemes, various aspects of land and its utility contributing to development in these foothills, material and technology, master plan, organization, and legislative framework. Primary and secondary data were collected from GVMC, VMRDA, and APTIDCO.

The report exhibits a detailed study of the existing scenario in the foothill settlements in GVMC area so that recommendations for the foothills of the city can be drawn according to the requirements [8]. In the data collection stage, data was collected from primary and secondary sources, and extensive research work was carried out. The findings of the study serve to address the key issues and potentials of the foothill settlements with respect to various aspects such as ecological perspective, disaster management perspective, and social and physical infrastructure perspective. Based on the findings; to address the issues in the existing illegal foothill settlements, suitable recommendations are provided based on tenability of the foothills. To protect the other foothills in the city from future encroachments illegals are suggested in recommendations from the understandings achieved from the study of the evolution of the illegal foothill settlements.

2.1 Study Area

Visakhapatnam, situated in the Indian state of Andhra Pradesh, holds significant financial influence. It serves as the state headquarters for the Indian Coast Guard and is strategically located between the Eastern Ghats and the Bay of Bengal coastline [9]. With a population of 2,035,922 as of 2011 (represented in Table 1) [10], it is the most populous city in the state, ranking as the 14th largest city in the country and the fourth most populous in South India. Visakhapatnam, commonly known as Vizag, is the second largest urban agglomeration in Andhra Pradesh, covering a total area of 53,000 hectares under the Greater Visakhapatnam Municipal Corporation (GVMC). Visakhapatnam boasts the oldest shipyard in India and possesses the sole natural harbor along the east coast [11].

Visakhapatnam features a rapidly expanding port, currently ranking as the fifth busiest cargo port in India [12]. It serves as the headquarters for the Indian Navy's Eastern Command and the recently established South Coast Railway Zone. Renowned for its picturesque beaches, Visakhapatnam is a popular tourist destination. It goes by several monikers, including "The City of Destiny" and "The Jewel of the East Coast." Additionally, it has been chosen as one of the cities in India to undergo development as a smart city under the Smart Cities Mission [13]. Figure 1 indicates the location map of study area.

The population of Visakhapatnam witnessed a notable increase due to the expansion of its municipal area. In 2005, the jurisdiction of the municipal area was

Table 1 Demography data (Census 2011 [10])

Level	Location	Population (in lakhs)
State	Andhra Pradesh	494
District	Visakhapatnam	42.9
Corporation	GVMC	18.8

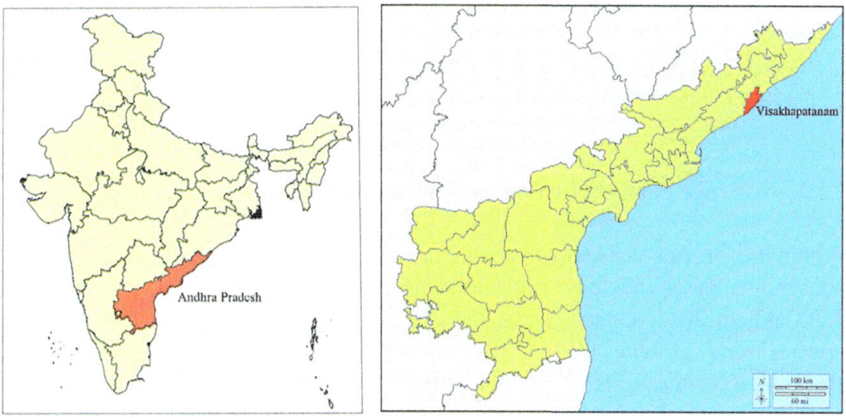

Fig. 1 Location in India and Andhra Pradesh map

extended to encompass the Gajuwaka municipality and 32 villages within the Greater Visakhapatnam Municipal Corporation (GVMC). Subsequently, in 2013, the boundaries of GVMC underwent further expansion to include the Anakapalli and Bheemili zones. These expansions played a significant role in the population growth of Visakhapatnam [14].

Evolution of the City:

- 1700s: The small fishing settlement of Vizagapatam experienced growth as a result of the presence of the British, French, and Dutch. The establishment of the old city municipality in 1858 marked the beginning of urban development in the area.
- 1900–1950: The settlement witnessed significant growth during this period with the development of road networks and the introduction of railways in 1904. This led to a fourfold increase in the population, transforming Vizagapatam into a thriving urban center. The establishment of the Eastern naval command in 1947 and Hindustan Shipyard Limited in 1949 provided improved transportation facilities and employment opportunities.
- 1960–1990: The city experienced further prosperity with the establishment of major industries such as BHPV in 1960, Vizag Port Trust in 1964, CORIL-HPCL in 1978, and Vizag Steel Plant in 1982. As a result, the suburbs also witnessed development, facilitated by the expansion of road networks. The concentration of industries in the southern part of the city led to its growth. To accommodate this shift, the Town Planning Trust (TPT) was transformed into the Visakhapatnam Urban Development Authority (VUDA) in 1975, and the municipality was upgraded to a corporation in 1979.
- 2005: The Municipal Corporation of Visakhapatnam was restructured into the Greater Visakhapatnam Municipal Corporation (GVMC) by incorporating the areas of MCV, 32 merged villages, and the former Gajuwaka Municipality.
- 2017: The Greater Visakhapatnam Municipal Corporation (GVMC) underwent further expansion, incorporating parts of the Anakapalli and Bheemili regions.
- 2018: The Visakhapatnam Urban Development Authority was dissolved and transformed into the Visakhapatnam Metropolitan Region Development Authority. The boundaries of the authority were extended northwards, encompassing areas towards Vizianagaram [15].

2.1.1 Average Household Size and Sex Ratio

Within the Greater Visakhapatnam Municipal Corporation (GVMC), the number of households has witnessed a substantial rise, increasing from 2.24 lakh in 2001 to 4.75 lakh in 2011. Concurrently, the average household size has also shown an upward trend, growing from 3.9 in 2001 to 4.4 in 2011. These figures reflect the changing dynamics of residential patterns within the GVMC.

Furthermore, the sex ratio in GVMC has experienced notable improvement over the decades. In 1961, the sex ratio stood at 921, while in 2011, it increased to 982.

This positive shift indicates progress towards achieving a healthier gender balance within the city.

2.1.2 Economy of the City

Visakhapatnam, the largest city in Andhra Pradesh, holds the distinction of being the ninth richest city in India, boasting a GDP of $43.5 billion [16]. The city's economy is primarily characterized by its thriving industrial sector. What was once traditional fishing town has transformed into a significant industrial hub in southern India. Visakhapatnam's economic success can be attributed to various factors, including its strong fishing industry, well-developed road and rail connectivity, and presence of numerous heavy industries. Key industrial entities in the city include Hindustan Petroleum Corporation Ltd., Visakhapatnam Steel Plant, Hindustan Shipyard Limited, Visakhapatnam Port Trust, Bharat Heavy Electricals Ltd., Naval Science and Technological Laboratory, Naval Dockyard, Dredging Corporation of India, Strategic Petroleum Reserve, NMDC, CONCOR, and Andhra Pradesh Medtech Zone Limited, among others. These industries contribute significantly to the city's overall economic growth and prosperity.

The emergence of Visakhapatnam as an industrial hub can be attributed to the presence of major private sector companies such as Coromandel International, Ferro Alloys Corporation, Gangavaram Port, and others. These influential private sector entities have played a pivotal role in shaping the industrial landscape of the city. They have significantly contributed to the growth and development of Visakhapatnam as a thriving industrial center [17].

2.1.3 Economic Base and Workforce Contribution

Approximately 4% of the overall population is involved in the primary sector, while 31% is engaged in the secondary sector, and the remaining 65% is employed in the tertiary and service sectors. Visakhapatnam is quickly becoming a hub for IT industries, with over half of the workforce concentrated in the tertiary sector. The primary sector primarily consists of fishing and agriculture, serving as the main sources of employment. Around one-third of the workforce participates in the secondary sector, indicating a robust manufacturing sector, which includes steel and thermal power plants.

2.1.4 Growth Drivers of the City

The city is expanding its growth towards the north of the IT hub towards the proposed airport. The south of the city is dominated by the industrial zones, which are facilitated by the government. Many PSUs are situated to the south of the port area, which helps them to receive their imports and reduces the transportation charges.

The city core is focused around the port area, the collector's offices, and the KGH hospital, which has become very congested.

With the thrust of development in the city and setting up of various industries, many people migrated here from nearby rural areas. On account of the scarce and costly land within the city, the fragile foothill ecosystem of Visakhapatnam was gradually encroached by the people. It is estimated that around 40,000 households live on various locations of foothills illegally in the city.

2.1.5 Growth Direction

GVMC area occupies an area of 545 square km. It is characterized by hilly terrain with moderate valleys and coastal plains coming under Western Ghats.

Due to the constraints, the growth of the city happens linearly. The urban centers like Visakhapatnam city center, Gajuwaka, and Madhurawada can be connected linearly, and the foothills near these areas are encroached. The future growth of the city tends to move in a linear direction due to new proposals of Bhogapuram Airport and the upcoming VCIC industrial corridor. The hills along this line can be seen to be slowly populating, and the danger of illegal foothill settlements to fully grow in these areas is very high [16].

2.1.6 Hills in Visakhapatnam

Peddakonda area is a reserve forest area, which has comparatively less urban dominance. Same can be seen in the Nallakonda forest area. Kambalakonda forest area is designated as a forest area, but its contiguous hillocks beside Simhachalam hills are not protected enough to protect them from encroachments. Also, the Simhachalam hills which are under the temple trust have seen encroachment when core city started developing during 1970s. Rushikonda area belt along the city core to airport has lucrative land prices. Yarada hill near the port has seen encroachments due to PSU and industrial activity.

Rushikonda has seen development after the new airport has been proposed making the belt along the city core to airport to have lucrative land prices.

2.1.7 Growth and Hill Encroachment Relation

Due to migration, industrialization, and urbanization, hill ranges are affected by deforestation. Asbestos or even thatched roof houses are being constructed in foothill ranges without adhering to building norms and topographical conditions, Jointed khondalite hillocks, porous and permeable area, and weathered zones are the most probable zones for landslides. Normal soil bearing capacity in these areas of foothills ranges from 28 to 40 t/m sq. Apart from SBC, terrain characteristics are to be taken into account while constructing here. Terraced or split-level foundation with adequate retaining wall support is necessary to withstand hill slope conditions, but illegal housing does not adhere to their requirements [16].

2.1.8 Simhachalam Devasthanam Hill Encroachments

The Sri Varaha Lakshmi Narasimha Swamy temple at Simhachalam is a popular pilgrimage destination located in Visakhapatnam. However, there have been encroachments on the temple land over the years mostly on the foot of the hill.

In 2019, the temple authorities launched a massive drive to reclaim the encroached land, which had been occupied by individuals and private organizations. The temple authorities identified around 360 acres of land, which were encroached upon, and started taking steps to remove the encroachments.

Simhachalam Devasthanam records tell that there are 10,000 encroachments in various places of its land under survey no. 275 across 9000 acres. The temple authorities sought the help of the district administration and the police to carry out the eviction drive. The temple authorities issued notices to the encroachers to vacate the land, and those who did not comply with the notices were forcefully evicted [18].

The temple authorities also took measures to prevent further encroachments, such as constructing boundary walls and deploying security personnel to guard the temple land [16].

2.1.9 Assigned Land D-Form Patta

The term "Assigned Land" is defined under Sect. 2(1) of the AP Assigned Lands (Prohibition of Transfers) Act, 1977.

Assigned land is unutilized government land given to the landless poor for agriculture purposes or as house site under the rules for the time being in force subject to the condition of non-alienation and includes land allotted or transferred to the landless poor persons before 3 January 2007.

In Andhra Pradesh, the Assigned Land D-Form Patta is a legal document that confirms temporary ownership of assigned lands, which are lands that have been granted to landless poor and other eligible beneficiaries under various government schemes and programs. Once they get approval from the joint collector, the tahsildar will issue patta to them. The farmers can enjoy the rights to stay on such lands but do not have the right to sell or transfer it to other persons. Assigned lands are heritable but not alienable.

The assigned person may get complete ownership after 30 years of continued usage. After 30 years, once the no objection certificate is issued from the joint collector, then the land could be transferred or sold.

Landless poor persons mean a person who owns an extent of land not more than 2 and 1/2 acres of wetland and 5 acres of dryland or such other extent of land has been or may be specified by the government in this behalf from time to time and who has no other means of livelihood.

"Transfer" means any sale, gift, exchange, mortgage with or without possession, lease, or any other transactions with assigned lands.

"Assigned Houses" means houses provided by the government to homeless people.

2.1.10 Application Process for Assigned Land D-Form Patta

The process of obtaining an Assigned Land D-Form Patta in Andhra Pradesh involves the following steps:

- Identify the eligible beneficiaries who have been granted assigned lands under various government schemes and programs.
- Collect necessary documents such as identity proof, address proof, and proof of assignment of land from the beneficiaries.
- Verify the eligibility of the beneficiaries through a committee appointed by the state government.
- Once the eligibility is confirmed, the beneficiaries can apply for an Assigned Land D-Form Patta by applying to the local tahsildar's office.
- After the approval of the collector, the tahsildar will process the application and issue the Assigned Land D-Form Patta after verification.

The Assigned Land D-Form Patta typically contains info such as the name of the owner, the extent of the land, and the type of land.

The disbursement of D-form pattas as temporary ownership rights showcases that vote bank politics is playing a crucial role in encouraging the encroachments. This further encourages new foothills to get further encroachment.

2.2 Issues and Challenges

Lack of Adequate Housing:

- High land rates of land making them unaffordable.
- Poverty making their affordability less and making them unable to buy or rent houses.
- Political patronage received by encroachers encouraging them to further encroach.

The Green Hills, Which Are the Lungs of the City, Are Encroached:

- City being prone to cyclones and high-velocity winds can create havoc on hills.

Lack of Any Framework on Arresting the Growth of Illegal Foothill Settlements:

- No foothill-specific building norms framed to be followed on foothills.
- No enforcement mechanism to curb encroachments on foothills.
- No mechanism to oversee the safety of the people living in foothills.
- No delineation of no-development zones on foothills.

Lack of Schools, Health Centers, and Anganwadis Within the Vicinity:

- There are no designated open spaces, for it being an encroached land.
- Vehicular access is present only at the bottom of the foothill. With no designated parking, the streets are choked by vehicles parked.
- The only way to reach the upper level of the settlement is by staircase only, making it riskier during emergencies.
- With trees being cut, unscientific digging carried out in certain areas, and retaining walls either not built or built weak, there is a high probability of landslide when heavy rain pours down.
- There are dangerous horizontal pathways in the upper part of the settlement.
- The existing street lighting is either installed but not working or not present at all in the upper parts of the settlement.
- With staircase making it difficult to access, the doorstep collection of waste is not being possible. They dump the waste in one location, and the garbage collection people collect once in 3–4 days.
- There are no drainage connection and toilets in some houses in the upper part.
- It is hard to take material to the upper parts or use concrete mixer there for slabs. Hence, most of the roofs in the upper parts are made of GI sheets, PVC sheets, etc., which is also affordable compared to concrete. These roofs are very much prone to risk under high-velocity winds of cyclones.
- People are not able to avail other alternatives of housing as the current house is under their name via D-form patta and are being excluded from beneficiaries list.
- Local politicians support encroachments due to vote bank politics.
- They are not vigilant enough to stop the encroachments in the early stage and later not sensitive enough to take some action on populated and untenable foothill encroachments.

3 Result and Discussion

Exploring PMAY Potential to Provide Livable and Affordable Housing to All:

- Helps to contribute their share to the city's growth and development.
- Protects and rejuvenates the green lungs.
- Can make the city disaster resilient.
- A common policy framework for foothills encompassing disaster mitigation, ecological fragility, byelaws, and standards.
- Opportunity to address social infra gap.
- Demarcation of open spaces.
- Potential to address parking issue.
- Creating universal accessibility.
- Creating disaster resilience.
- Securing the safety of people.
- Making the area more sanitary in accordance with Swachh Bharat Mission.

- Potential to reduce the disaster risk.
- Expanding the scope of PMAY for safe and affordable housing.
- Potential to make foothills free from politics.
- Potential to make the governance better and to make officials more vigilant.
- Integrating with PMAY—making housing more affordable by integrating PMAY with the rehabilitation component for untenable and tenable components separately.
- City-level mock drills for disaster response—increasing better coordination between departments for better response during disasters by organizing mock drills. Designating relief shelters for people living on the foothills at the bottom of the hills when hit by cyclones. Devising local-level response on evacuation routes specific to each foothill.
- Reforestation on foothills—rejuvenating greens by reforestation on the hills where the trees are cleared by the people illegally for them to encroach. Making sure that the greens on the retaining walls are not cut so that the soil is held by the roots of the trees and probability of landslides is reduced.
- Legislation to protect the greens and restrict encroachments—coming up with a legislation with detailed work division of departments to protect the greens of the foothills and at the same time restrict and regulate the growth of encroachments.
- Collaborative interactions to sensitize people with the help of NGOs—undertaking workshops and collaborative interactions with the local people to sensitize them of the dangers and issues of encroaching foothills with the help of NGOs.
- Mandating bimonthly safety audits by a special committee of experts to check the tenability of populated areas of foothills—mandating safety audits for foothill settlements for every 6 months to check the tenability of existing encroachments.
- Surveying to set up no-habitation zones on hills—with the help of GSI and GVMC SRU, setting no-habitation zones on hills in the city by surveying probable encroachable areas to prevent future illegal growth.
- Untenable areas are to be relocated.
- After the safety audit is undertaken by the special committee of experts, the inhabitable zones are to be identified.
- Under PMAY, the HHs in the designated area are to be relocated and given homes.
- Skill training and self-employment opportunities are to be done in relocated area for the promotion of their livelihood.
- Tenable areas to be developed phase-wise.
- Open spaces are to be demarcated, and dilapidated buildings are to be replaced with open spaces in future.
- A vehicular access connecting to the upper portion of the settlement for emergencies is to be constructed.
- Horizontal pathways are to be constructed where the pavements are very unsafe.
- Retaining walls are to be repaired and strengthened.
- Improving social infrastructure phase-wise.
- Mandating future redevelopment with minimum social infra facilities such as schools, clinics, anganwadis, and community centers.

- Improving physical infrastructure phase-wise.
- Establishing garbage collection points and ensuring frequent collection of segregated waste.
- Gradual upgradation of physical infra facilities such as drainage connection until the upper tenable parts.
- Street lighting is to be improved in the upper tenable areas where they are lagging.
- Repairing dilapidated pavements.
- Awareness campaigns and sensitizing people.
- Awareness of government housing schemes.
- Enabling them by enhancing their representability to make them capable of reaching for redressal mechanism.
- Sensitizing people of the dangers of living on the foothills and tuning them to avail government housing schemes.
- Enforcing legislation to regulate politics.
- Strict legislation which is hard to bypass through loopholes, which usually gives way for vote bank politics.
- Building capacity, imbibing impartiality, and enforcing anti-corruption measures.
- Strict vigilance over the future encroachments to stop them from arising in the beginning itself.
- Anti-corruption cell to look after the corrupt practices, which give rise to illegal foothill settlements.

4 Recommendations

Ideally, illegal foothill settlement should not be allowed. But it evolved into a subsystem which enjoys political patronage. Almost constituting about 10% of the city, it cannot be simply pushed away.

To address this issue, twofold recommendation is provided as an integrated approach focusing on all the alternatives:

1. The first one is to provide recommendation on addressing the issue in existing foothill settlements.
2. The second is to provide recommendations to ensure that no further new settlements evolve on non-encroached hills.

Again, the first one can be further divided into recommendations separately with respect to the tenability considering slope, landslide probability, and cyclone impact proneness.

- For tenable location on the foothill settlement, recommendations are to be in a long-term planning perspective to upgrade the housing conditions in phase-wise manner and through preparation of redevelopment plan at neighborhood level, specific to each foothill.

- For untenable locations on the foothill settlement, there is short-term temporary measure to relocate the families from there and rehabilitate them to mitigate from potential disaster and a long-term permanent measure to provide them with housing which does not impact their livelihood.

4.1 Addressing the Issue in Existing Foothill Settlements

(a) **For Tenable Locations on Foothills:**

- The lower parts of the foothills, which are mostly tenable but without sufficient physical and social infrastructure, are to be developed in the future with a phase-wise plan for development of infrastructure.
- The future redevelopment can be mandated in accordance with neighborhood-level plans designating open and green spaces, specifying design norms (min room sizes, setbacks, etc.) and proposals for street widening.
- Transferable Development Rights (TDR) can be used as a bargaining tool to make the residents give the land (part or full) which they are residing under the D-form patta ownership without any resistance for development of common spaces (opens, greens, community halls, health and wellness centers, etc.) and for street widening.
- The upper tenable areas are to be mandated with structural safety norms (foundation, walls, roof, etc.) and design norms (min room size, setbacks, etc.) in case they want to redevelop their plot areas.
- PMAY support for beneficiary-led construction components for plots in the tenable areas of the foothills is to be made conditional that the financial support will be given only if the norms set for the foothill settlement as mentioned above are adhered.
- PPP model of housing is to be proposed in the lower part of the foothill settlement in the redevelopment plan under government financing from in situ slum rehabilitation component of PMAY.
- The height restrictions are to be eased for PPP affordable apartments proposed at the bottom of the foothill to accommodate more floors and more flats, which can be provided to the encroaching families who surrender land on the foothill who have D-form pattas.
- Stringent measures such as revocation of allotment of flats or D-form patta allotted need to be taken if the flats given via this scheme or D-form pattas are transferred later to others by the beneficiaries.
- The retaining walls at each level and stair access maintenance or reconstruction are to be undertaken by the local body.
- At least one access road till the upper part of the settlement has to be developed for emergency vehicle passage (ambulances, fire vehicles) with a minimum width of 6 meters.

- The upper greens are to be rejuvenated on the foothill, which were deforested due to encroachment. Greens near the retaining walls are to be focused to hold the soil to mitigate landslides.
- Cyclone relief centers are to be designated in the foothill, and local-level response plan is to be devised.
- Boundary wall on the upper part of the foothill settlement is to be constructed taking proper structural measures to stop people from encroaching further up the settlement in future.

(b) **For untenable locations on foothills:**

- Families in these locations are to be immediately relocated to camps designated as a temporary measure until they are allotted flats as a permanent measure under PMAY through in situ rehabilitation or PPP apartment model back in the same locality to sustain their livelihood as it is intricately linked to the location.
- If enough houses are not being able to be provided within the location, then lottery basis is to be followed for allotment of houses within the location and the rest are to be relocated to places where the government land is available.
- People who are to be relocated to distant places from the location where they used to live are to be provided with skill upgradation platforms through programs such as Kaushal Vikas Yojana and are to be given a handhold to make them self-employed; and finances for it should be given through Mudra loans such that their lives and livelihoods are normalized.

4.2 Ensuring no Further New Settlements Evolve on Non-encroached Hills

- Strict vigilance on tree clearance on foothills, as it is the first step for encroachment. Immediate action needs to be taken when encroachers or anyone tries to cut down trees and clear land on the foothills. Stringent punishment through fines or imprisonment is to be backed by law for trying to encroach green hills.
- Hills, where the bottom of the foothill is freshly encroached overnight, are to be cleared on an immediate basis, and boundary walls have to be constructed with security posted there to stop the encroachment.

5 Conclusion

Visakhapatnam's unlawful foothill settlement problem is a complicated one with broad ramifications for both natural sustainability and human existence. Rapid population expansion, industrialization, and urbanization have all pushed encroachments onto the fragile foothill ecosystems throughout time, with unfavorable outcomes like increased susceptibility to disasters, ecological degradation, and deficits in social infrastructure. There pertains an intricate relationship that exists between the demands of urbanization and the need for environmental preservation initiatives. In addition to offering vital biological functions, the foothills around the city act as natural barriers against landslides and floods as well as catchment regions for bodies of water. Unchecked settlement growth in these places jeopardizes these essential services and increases hazards to the environment and human populations.

The study emphasizes how important it is to have a diversified approach to handle complex problems. Strict enforcement of current restrictions and implementation of foothill-specific policies that take ecological fragility, catastrophe resilience, and urban planning standards into account are essential. It is also essential to raise public awareness and encourage community involvement to educate locals about the dangers of illegal settlements and to promote adherence to legal frameworks.

Tenable foothill areas need to be given priority in sustainable redevelopment plans, which need to have emergency access roads, health facilities, and any other basic amenities. Government agencies, nongovernmental organizations, and local communities must work together to prevent further encroachments and to rehabilitate those who are already living in unlivable conditions. Promising results can be obtained from integrating programs such as the Pradhan Mantri Awas Yojana (PMAY) into frameworks for urban foothill development and its disaster management policies. This holistic approach aims to uncover sustainable management and development of illegal foothill settlements in Visakhapatnam. By utilizing these initiatives to maintain and revitalize green spaces alongside offering safe and cheap housing options, Visakhapatnam can aim to become a leader in sustainable urban development and catastrophe resilience.

References

1. Reddy, M. G., & Kumar, K. A. (2010). *Political economy of tribal development: A case study of Andhra Pradesh* (p. 13). Centre for Economic and Social Studies.
2. Reddy, V. R., Rao, M. R., & Venkateswamy, M. (2010). *Centre for economic and social studies. Begumpet, Hyderabad-500016, research unit for livelihoods and natural resources (supported by Jamsetji Tata trust)*. Centre for Economic and Social Studies, Begumpet, Hyderabad.
3. Khan, S., Rathore, D., Singh, A., Kumari, R., & Malaviya, P. (2024). Socio-economic and environmental vulnerability of urban slums: A case study of slums at Jammu (India). *Environmental Science and Pollution Research*, 18074–18099.
4. Bahera, P. K., & Barik, K. Recently discovered middle Palaeolithic foothill site at Kundakhai in the southern Bargarh upland of Odisha: A preliminary report. *Ancient Asia*.
5. Ramuje, K., & Rao, B. N. M. (2015). Hudhud cyclone—A severe disaster in Visakhapatnam. *International Journal of Research Engineering & Technology*, 156–163.
6. Vivek, G., & Srinivasa, K. T. (2015). Impact assessment of tropical cyclone Hud Hud on coastal region of Visakhapatnam, Andhra Pradesh, India. In *ISPRS annals of the photogrammetry, remote sensing and spatial information sciences* (pp. 123–130).

7. Das, S., & Dsouza, N. M. (2020). Identifying the local factors of resilience during cyclone Hudhud and Phailin on the east coast of India. *Ambio, 49*, 950–961.
8. Anuradha, & Gupta, S. (2022). AHP-based multi-criteria decision-making for forest sustainability of lower Himalayan foothills in northern circle, India—A case study. *Environmental Monitoring and Assessment, 849.*
9. Ravi, S. S., Bhagat, S., & Rai, R. (2024). Smart city initiatives entangled in political priorities: Case-based analysis on Visakhapatnam, Andhra Pradesh. In *Smart cities: Innovations, challenges and future perspectives* (pp. 279–299). Cham.
10. Chandramouli, C., General, R., & Census of India. (2011). *Rural urban distribution of population, provisional population total.* Office of the Registrar General and Census Commissioner, India.
11. Prasad, B., Rao, P. R., & Tigga, A. (2021). Groundwater quality assessment using the weighted arithmetic index method in the selected villages of Butchayyapeta Mandal, Visakhapatnam, Andhra Pradesh, India. *Environmental Monitoring and Assessment*, 1–17.
12. Gupta, A., & Shukla, A. K. (2024). Optimal approaches in global warming mitigation and adaptation strategies at city scale. *Discover Sustainability, 5*(1), 272. https://doi.org/10.1007/s43621-024-00497-8
13. Alizadeh, T., Kurian, L., Bansal, C., & Prasad, D. (2023). Smart cities mission in the face of COVID: Scope and scale of 'smart' COVID responses in India. *International Journal of Environmental Research and Public Health, 7036.*
14. Ravi, S., Tomar, A., Bhatia, A., & Kane, J. (2016). *Building smart cities in India: Allahabad, Ajmer and Visakhapatnam.*
15. Kumar, P. S., Rao, G. V., & Nitish, S. (2020). A study on land use changes in southern part of Visakhapatnam city due to urbanization: A case study. *PalArch's Journal of Archaeology of Egypt/Egyptology*, 7318–7333.
16. Barodia, R. A. (2021). *How adversely were the socio-economic conditions of the people affected due to Covid-19 in the areas of Andhra Pradesh and Telangana (doctoral dissertation.* SRM University, SRM.
17. Abraham, M. (2023). Perceptions of the tribal population on grama Sachivalayam (village secretariat) concept and its services in the tribal areas of Visakhapatnam. An empirical study. *Journal of Contemporary Research Social Sciences, 14*, 14–26.
18. Malakar, K. D., Kuzur, G., Maity, D., Yadav, M., & Roy, S. (2022). A socio-ecological study of population, migration, urbanization, and socio-climate variation in Andhra Pradesh and Telangana, India. *Journal of Geography, Environment and Earth Science International*, 1–33.

Modelling of Urban Ecosystem Services in Chandigarh Using InVEST Modelling

Sudhir Rohilla, Neha G. Paswan, and Shray Pathak

1 Introduction

Urbanism is a way of living centered around the city [1]. Urbanization reflects the transformation of social and physical spaces due to an increase in population growth and economic ventures in cities [2]. Urbanization in the recent decade has become a major factor causing environmental degradation impacting the health of urban dwellers. Urbanization in developing countries is at a faster pace compared to developed countries. Urbanization is projected to increase from 55% to 68% by 2050. Future growth in urbanization is concentrated in a few countries like China, India, and Nigeria. These three countries account for 35% of the estimated urban growth by 2050. According to projections, 416 million people will be added to urban areas by 2050 [3]. About 70% of global carbon emissions originate from urban areas, serving as an essential reminder of Sustainable Development Goal 11—"Sustainable Cities and Communities" established by UN-HABITAT. The aim of SDG 11 is to make human settlements and urban areas more environmentally friendly, safe, sustainable, and resilient to disaster and climate change [4].

For the last few decades in India, migration from rural to urban areas has been increasing at a tremendous level. According to the 2011 census, India has 31.6% urbanization. It is projected to increase by 40% between now and 2030 [3]. Among all states and union territories in the country, NCT Delhi and Chandigarh are the most urbanized [5].

S. Rohilla (✉)
Department of Geography, Panjab University, Chandigarh, India

N. G. Paswan · S. Pathak
Department of Civil Engineering, Indian Institute of Technology Ropar, Rupnagar, Punjab, India
e-mail: neha.23cez0014@iitrpr.ac.in; shray.pathak@iitrpr.ac.in

S. Pathak et al. (eds.), *Geo-Data Revolution*, Advances in Geographic Information Science, https://doi.org/10.1007/978-3-031-94078-1_17

Ecosystem services (ES) are the benefits that human populations derive, directly or indirectly, from ecosystem functions [6]. They provide an effective framework for assessing the quality of life of urban dwellers and the environmental impacts of urbanization. Human action is degrading natural resources and putting the upcoming generation at risk, as depletion of the environment is a major impedance to achieving sustainable goals, and it has become necessary to collectively seek a sustainable manner to maintain or sustain ecosystem services [7].

Green infrastructure in the city is important for urban dwellers in both ecological and social contexts. It provides microclimate regulation and provides social space for the residents and is essential for the ecological balance of the city. Urban green spaces include parks, trees, green roofs and wetland regulate the temperature of the city or we can say it reduces the effect of urban heat island in the city and reduce the pollutant in environment that affect the health of the society [8, 9].

Land use change affects ES negatively, and converting green infrastructure into built-up areas leads to a decline in ecosystem service value in the long term [10, 11]. The blue landscape in the city is of vital importance in the context of ES, which includes waterbodies, rivers, lakes, and wetlands, which provide a home for the diverse fauna and flora, along with fertile soil, carbon sequestration, and purification of water [12]. Assessment of ecosystem services is important for decision making and urban planning that make the city more sustainable and resilient. It is also important for resource management [13]. Recreational benefits provided by natural ecosystems, e.g., green space in urban areas, are a way to reconnect people with nature and contribute to society's well-being [14].

2 Study Area

Chandigarh is the Union Territory (UT) and capital of Haryana and Punjab, located in the foothills of Shivalik Hills (Fig. 1). It is a planned post-independence city, the master plan of which was prepared by Swiss French Architect Le Corbusier. The approximate area of the UT is 114 square kilometers and is located 260 km north of Delhi. Chandigarh is located at 30.7333° N latitude and 76.7794° E longitude. The mean sea-level elevation of the city is 321 m. It lies under Koeppen climate classification CWg, i.e., monsoon type with dry winter. Hence, the city experiences a hot summer season and cold, dry winters [15]. The city is known for its green space and beauty, which is why it is referred to as "City Beautiful." But rapid urbanization and influx of population in the city stressed the ecosystem of the city, and the increase in vehicle and pollution concentration in the city has degraded the environment, so it's become necessary to focus on the ecological planning and ecosystem services assessment in the city. The city was planned for 500,000 people, but this threshold was already crossed by 2011, when the population of Chandigarh reached 1,054,686 with a population density of 9252 persons per square kilometer, ranking second highest in the country after the National Capital Territory of Delhi [16]. The projected population for 2031 is approximately 2,000,000 [17]. The air quality index in

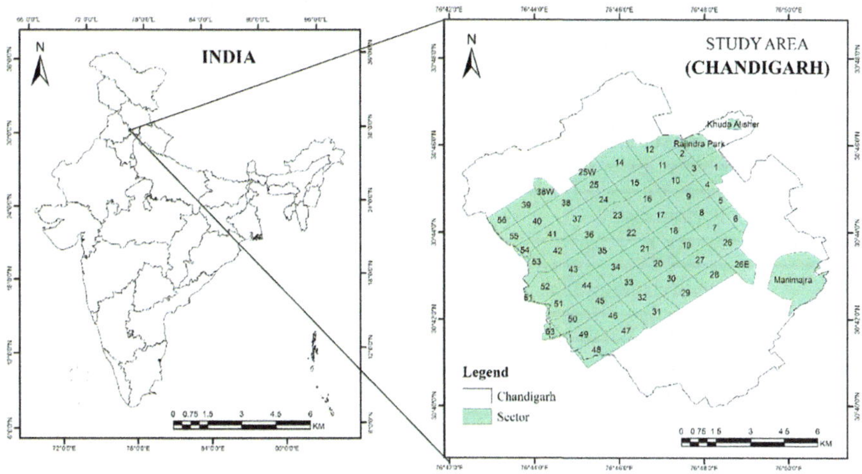

Fig. 1 Locational map of the Study area, Chandigarh, India

the UT is deteriorating, as current $PM_{2.5}$ is 12.5 times above the limit recommended by the WHO. The AQI of Chandigarh is a matter of concern, as it is unhealthy for urban dwellers and poses a threat to vulnerable populations [18].

3 Database and Methodology

For the classification of various land use categories in the study area, Landsat 8 OLI data from 2013 and 2023 were used. Landsat 8 imagery was obtained at 30 m resolution, and preprocessing was done before LULC. Supervised classification was used to classify the image, and the maximum likelihood classifier algorithm was selected. An administrative boundary shapefile of the study area was obtained from the Survey of India (Fig. 2; Table 1). All the spatial data is projected on WGS_1984 UTM ZONE 43 N in linear units of meters.

3.1 Land Use and Land Cover (LULC) Map Generation

Landsat 8 OLI images of 2013 and 2023 were used for land use and land cover (LULC). October was selected because it has the least cloud cover, and all the preprocessing was done with ERDAS Imagine software. Resolution merge is done with the 15-meter resolution panchromatic band, enhancing image resolution and interpretability, and is followed by supervised classification for LULC extraction. The maximum likelihood classifier algorithm was selected for the classification. The land cover of the study area is divided into five classes: built-up, green infrastructure, water, wasteland, and agriculture. Green infrastructure included all the urban

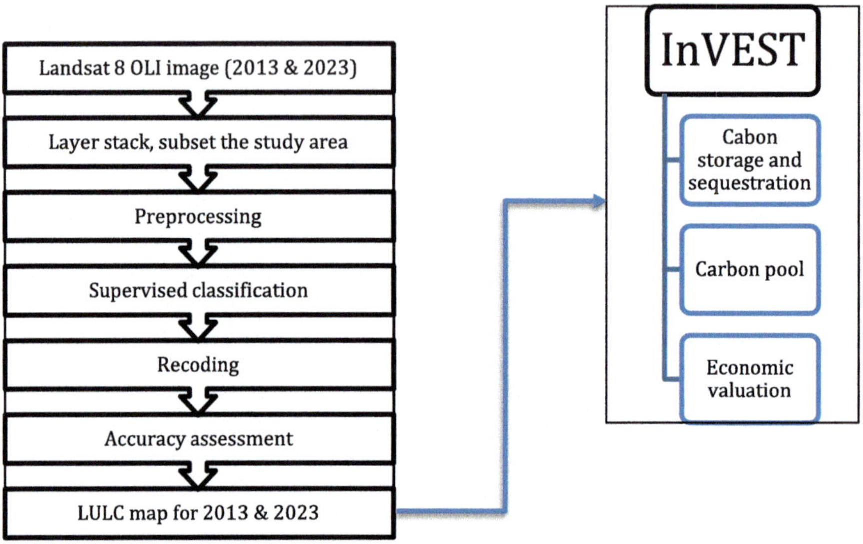

Fig. 2 Methodological flowchart

Table 1 Dataset used in this study

Satellite data used	Format	Product id	Row / path	Date	Source
Landsat 8 OLI	Raster	LC08_L1TP_147039_20131027_20200912_02_T1	39/147	2013	USGS
Landsat 8 OLI	Raster	LC08_L1TP_147039_20231007_20231011_02_T1	39/147	2023	USGS
Chandigarh shpfile	Vector				Survey of India

parks, trees, and forest areas within the administrative boundary. The water class includes water bodies and wetlands in the study area. A minimum of 20–25 pixels as used for each class of land cover. Reclassification corrected the merged classes, ensuring that agriculture and vegetation exhibit similar spectral reflectance, while built-up and bare land display similar reflectance. Accuracy assessment is performed by random sampling. We validated classification with Google Earth Pro for accuracy assessment.

3.2 Accuracy Assessment

Accuracy assessment is an essential part of the validation of classification. Simple random sampling was selected for accuracy assessment, and overall, 100 observations were taken for it. We use Google Earth for the reference check for observation by linking its view with the ERDAS Imagine. Overall accuracy was above 82% for both years, and the Kappa coefficient was above 0.75 for both years, which is acceptable for classification accuracy. It ranges from −1 to 1, and a value near 1 is an indication of a more accurate result. After preparing the LULC map, the area in hectares was calculated for each year, and a statistical table was prepared to analyze changes.

3.3 InVEST Modeling Framework

Integrated Valuation of Ecosystem Services and Tradeoff is a standalone tool developed by the Natural Capital Project and Stanford University for ecosystem services valuation and assessment by various models. InVEST 3.14.2 provides a total of 25 different models for various ecosystem services assessments, like urban cooling, sediment retention, urban nature access, carbon storage and sequestration, coastal vulnerability, etc. It evaluates ES with both economic and biophysical processes. The model runs as an independent tool for ES assessment, requiring biophysical input like LULC of the area. Other information (e.g., economic) and the output of this model are in a map format. Generally, ArcGIS is required for the analysis and visualization of the map output of Invest. We use a carbon storage and sequestration model of investment to investigate the carbon stock in the study area.

3.4 Carbon Storage and Sequestration Model

Carbon storage in the terrestrial ecosystem far exceeds what is stored in the atmosphere and is a vital component of the global carbon cycle. Vegetation, grassland, wetlands, and park (green) space are all parts of the terrestrial ecosystem playing major roles in absorbing CO_2 from the atmosphere and regulating the climate of an area.

Storage of carbon in terrestrial ecosystems is found in four pools as follows:

1. **Above ground biomass**: All the living flora above the surface is above ground biomass including trees, grass, and shrub land. Has the highest carbon storage on the earth.
2. **Below ground biomass**: All the plant root under the soil can store carbon and regulate the nutrient uptake of the plant.
3. **Soil organic carbon**: Comprises carbon stored in the soil due to plant decomposition, playing an important role in soil structure.
4. **Dead organic matter**: Comprises non-living organic material, i.e., fallen leaves, dead plants, debris, and litter [19].

Carbon storage helps in climate regulation, and terrestrial ecosystems act as carbon sinks as they absorb carbon during the photosynthesis process, accumulate it in plants and soil, and reduce the impact on climate change [20]. Carbon sequestration helps in the reduction of the concentration of GHGs in the atmosphere and helps in regulating climate and slowing the impact of global warming [21]. Biodiversity conservation in the form of habitat provisioning and ecosystem services is provided by carbon storage as it provides a habitat for diverse species, along with microclimate regulation, water purification, and reduction in temperature as an ecosystem service [7].

The terrestrial ecosystem functions as a resource provider providing timber, as well as medicinal and edible resources for livelihood. It is essential for sustainable development [22].

Carbon storage in a particular area depends on the size of four carbon pools: aboveground, belowground, soil, and dead organic matter. The model uses LULC data provided by the user to integrate carbon pool data with LULC classes, yielding the carbon stock in the land parcel. Carbon storage in a given pixel (*i, j*) for land use class *m* can be calculated using the formula presented in Eq. (1) [23].

$$C_{m,i,j} = A \times \left(\mathrm{Cam}_{I,j} + \mathrm{Cb}_{m,i,j} + \mathrm{Cs}_{m,i,j} + \mathrm{Cd}_{m,I,j} \right) \tag{1}$$

Where, A = Area of pixel in ha; Ca = Carbon stored above ground, Cb = Carbon stored below the ground, Cs = Carbon stored in soil, Cd = Carbon stored in dead matter.

The model requires generally two inputs—one is the LULC map, and the second one is the carbon pool (Table 2) that provides the carbon storage (in Mg/ha) for a particular class. For carbon sequestration, we need the current LULC as well as the future predicted LULC of the study area. In the present study, we take the year 2013.

Carbon sequestration for a parcel x can be calculated through the formula given below in Eq. (2) [24].

$$\text{Sequestration of parcel}(x) = V \frac{S_x}{q-p} \sum_{t=0}^{q-p-1} \frac{1}{\left(1+\frac{r}{100}\right)^t \times \left(1+\frac{c}{100}\right)^t} \tag{2}$$

Where,

V = Carbon price per Mg, S_x = Carbon amount in Mg that is stored in given pixel from atmosphere, q = Future year, p = Current year, r = Market discount rate per annum, c = Rate of change per year in the price of carbon.

Table 2 Carbon pool data for the Chandigarh City

Sr. no.	LULC Class	Carbon (ABG)	Carbon (BGB)	Carbon (SOC)	Carbon (DOM)
1	Water	0	0	0	0
2	Wasteland	0	0	0	0
3	Agriculture	7	2	35	1
4	Built-up	0	0	0	0
5	Green space	45	15	30	10

The carbon storage value in all four pools was taken only for the green space and agricultural class, and the value taken for all the other classes was zero. The value of the carbon pool was extracted from previous studies, and for the economic valuation we utilized the global carbon price, because in 2013, India did not have an official mechanism for carbon valuation or pricing. The social cost of carbon is based on the study of the Interagency Working Group (IWG) of the USA, which is based on the social cost of GHG [25]. So, for the economic evaluation of carbon we took carbon price, market discount rate, and annual price change. The social cost of CO_2, 38 per ton, was utilized for calculating the carbon price, i.e., $139, and the market discount rate was taken as 3%. The annual change in price of carbon was taken as zero [26].

4 Results

4.1 LULC Assessment

Five classes were taken for the study area, namely, built-up, vegetation (including open, urban, and dense), water, wasteland, and agriculture shown in Fig. 3. Table 3 summarizes changes that were detected in the area. A negative value indicates a decrease in a particular class area over a certain period of time, and a positive value in the change column indicates an increase for any particular class over a decade. In 2013, vegetation accounted for 45.31% of the total area, which remains 39.12% in the year 2023 with a 13% decadal decrease. This was primarily due to rapid urbanization and the encroachment of anthropogenic activities on green spaces, despite Chandigarh being is a planned city where, during its master planning, green spaces were included. In other words, linear tree plantations alongside roads and urban parks in each sector were established, population growth and housing demand are the main reasons for green space reduction in the city.

Built-up areas in the base year accounted for 40.78% of the total area, which increased to 49.55% in the year 2023, showing an increase of 21.51% since the base year. The past three decades proved to be a fast urbanization period because the standard of living and opportunities in Chandigarh are a pull factor for other areas. Studies also reveal an increase of 104% in the built-up class in Chandigarh from 1990 to 2020, which indicates double the amount from the base year [15]. Waterbodies in Chandigarh changed negatively over this period of time, as they decreased by 7.7% from the base year. The overall area of waterbodies in Chandigarh is very small in comparison to other classes, accounting for 159 hectares in 2013 and 147 hectares in 2023. Wasteland in the study area has small proportions, i.e., 1.7% in 2013 and 1.5% in 2023. This 11% decrease from the base year can be attributed to utilization of wasteland in built-up and other agricultural areas.

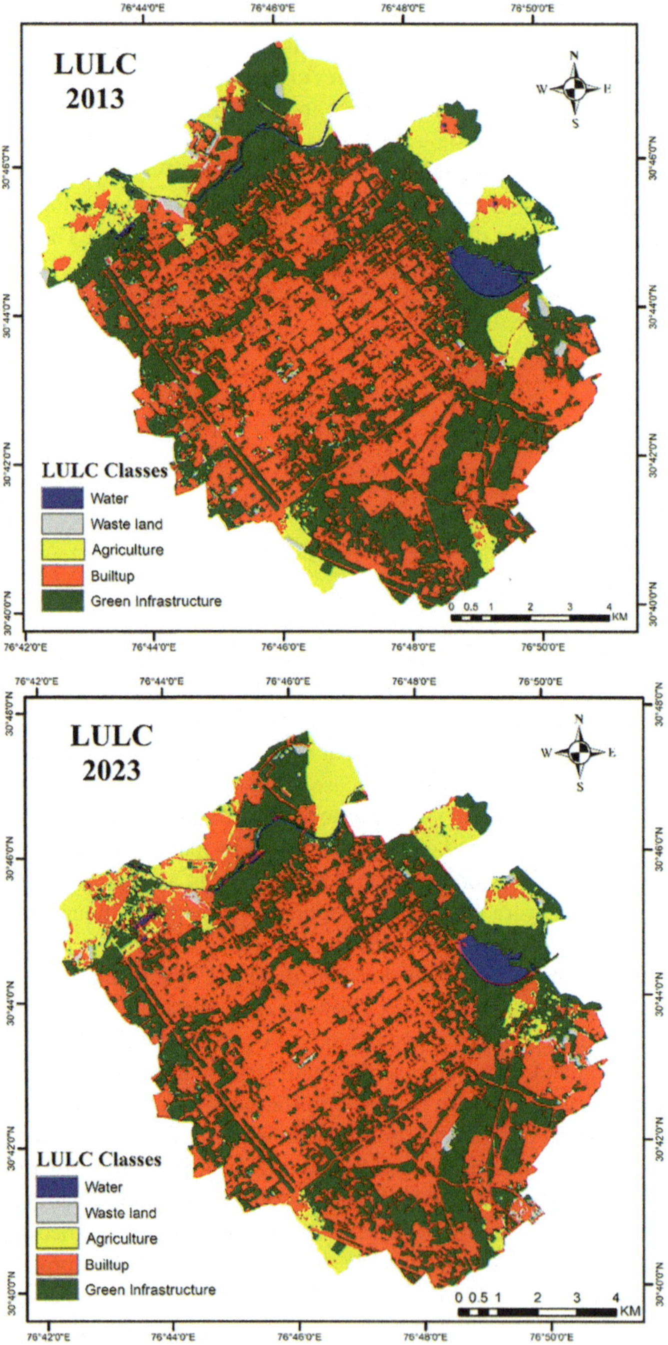

Fig. 3 LULC for the years 2013 and 2023 of the study area

Table 3 Change detection in study area from 2013 to 2023

	Year 2013		Year 2023		Change 2013–23	
Class	Area (hectare)	Area (percent)	Area (hectare)	Area (percent)	Area (hectare)	Area (percent)
Built-up	4678.63	40.78257	5685.39	49.55828	1006.76	21.51826
Vegetation	5198.92	45.31783	4488.66	39.12665	−710.26	−13.6617
Water	159.75	1.392505	147.42	1.285027	−12.33	−7.71831
Wasteland	204.73	1.784586	180.29	1.571548	−24.44	−11.9377
Agriculture	1230.1	10.72251	970.37	8.458499	−259.73	−21.1145
Total	11472.13	100	11472.13	100		

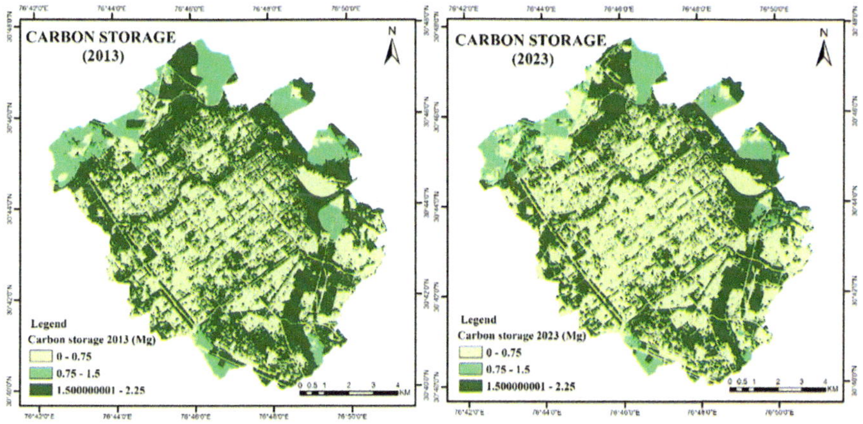

Fig. 4 Carbon storage in 2013 and 2023

4.2 *Analysis of Carbon Storage*

Total carbon is the amount of carbon being stored in Mg per pixel in the area. It accounts for all four carbon pools, i.e., aboveground, belowground, organic soil, and dead matter. Green space contains 1.5–2.25 Mg per pixel as shown in the map (Fig. 4), which is the highest, while the agricultural area contains 0.75–1.5 Mg carbon per pixel in the landscape and 0–0.75 Mg for the rest of the classes, i.e., built-up, water bodies, and wasteland. The total carbon stored in the area in 2013 was 5.8×10^5 Mg, and in 2023, it remained 4.9×10^5 Mg (Table 4).

The decrease in carbon storage could be due largely to the change in land use and land cover of the area. The amount in both the green space and agriculture classes was reduced substantially from 2013 to 2023. To validate the declining trend in carbon storage, we calculated the NDVI (Normalized Difference Vegetation Index) of the area for both years. We used the NIR (near-infrared) and red bands for this index, and the NDVI results indicate a decrease in green spaces due to rapid urbanization. As shown in Fig. 5, the highest value in both maps, 0.39 and 0.84,

Table 4 Carbon storage (Mg) in each LULC class for respective years

		2013		2023	
Sr. no.	Class name	Area (m^2)	Carbon storage (Mg)	Area in m^2	Carbon storage (Mg)
1	Water	1,597,500	0	1,474,425	0
2	Waste land	2,047,275	0	1,813,950	0
3	Agriculture	12,300,975	55354.39	9,799,425	44097.41
4	Built-up	46,786,275	0	57,025,800	0
5	Green space	51,989,175	519891.8	45,037,575	450375.8
6	Total		575246.1		494473.2

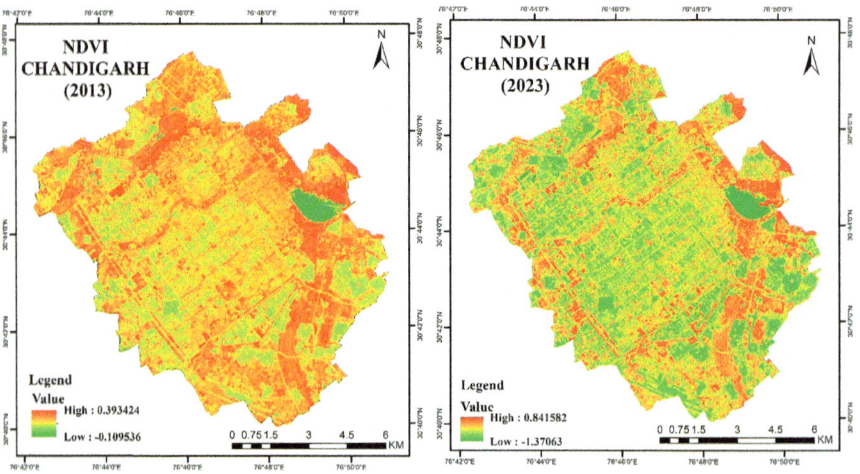

Fig. 5 Normalized difference vegetation index (NDVI) of study area in 2013 and 2023

respectively, shows the vegetation surface in red color. The total carbon stored in green spaces in 2023 was 4.5 × 10^5 Mg, which was a 13% decrease from the year 2013. And in the agricultural area, current carbon storage is 4.4 × 10^5 Mg, which was a 20% decrease from the base year. This shows a marked decrease in the agricultural area from the base year. The total carbon change in these 10 years was −8 × 10^4 Mg, which means carbon storage has decreased this amount from the base year to 2023.

4.3 *Assessment of Carbon Sequestration*

Sequestration of carbon is when carbon dioxide is captured from the atmosphere and stored in the terrestrial ecosystem. Reducing the proportion of CO_2 in the atmosphere can help to tackle the impact of climate change and maintain the average

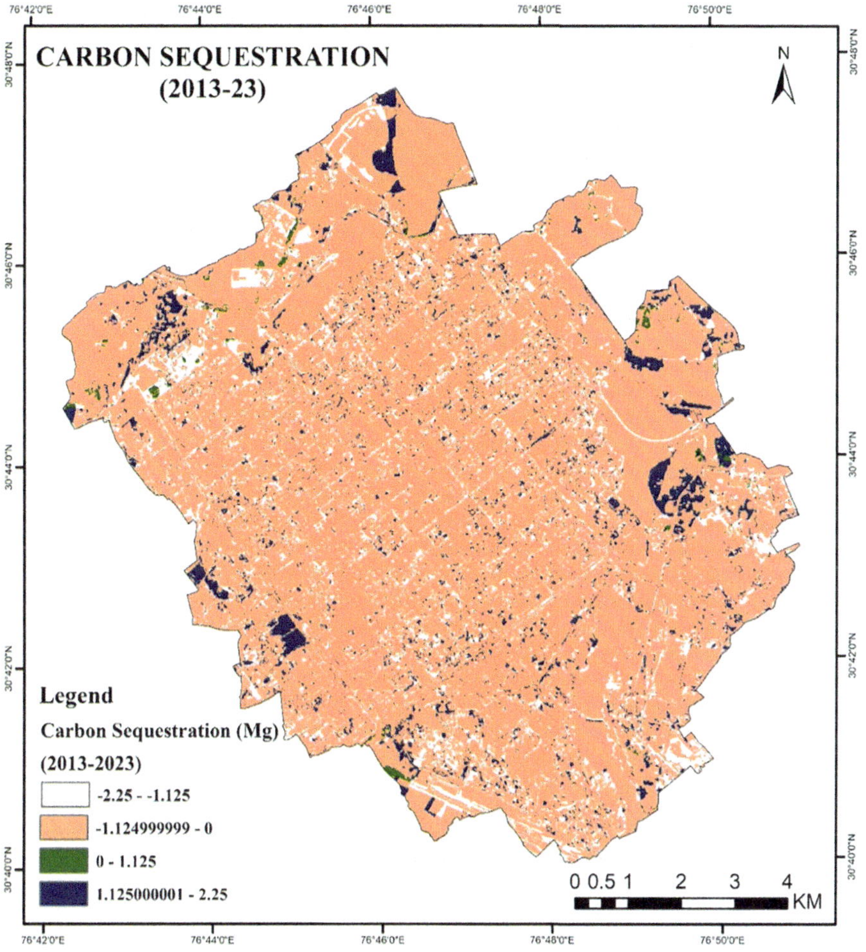

Fig. 6 Carbon sequestration from 2013 to 2023

global temperature to sustain the habitat on Earth [20]. The total amount of sequestered carbon from 2013 to 2023 is shown on the map in Fig. 6 and in Table 5. In the map of carbon sequestration, negative values show a loss of carbon in the atmosphere, and positive values show sequestration of carbon, i.e., capturing CO_2 from the atmosphere and investing it into terrestrial carbon sinks. The forests act as a carbon sink, as they are the largest reservoir of carbon. Total carbon sequestration is -8.2×10^4 Mg, which is negative, indicating that the storage of carbon dioxide from the atmosphere is less than the carbon emissions in the study area. The total emissions of carbon here are 1.3×10^5 Mg, and green spaces are the only class that sequestered the carbon account 4.8×10^4 Mg.

Table 5 Carbon sequestration in Mg from 2013 to 2023

Sr. No.	Class name	Carbon sequestration (2013–23) (Mg)
1	Water	−926.888
2	Waste land	−9532.46
3	Agriculture	−3151.46
4	Built-up	−117,907
5	Green space	48801.15
6	Total	−82716.9

4.4 Net Present Value (NPV)

The net present value (NPV) is a monetary term for carbon sequestration over two points in time, and it can be used for the cost-benefit assessment of any ecological or environmental project. NPV maps show the economic value in dollars per pixel with respect to carbon sequestration, showed by positive and negative values (Fig. 7). The total for the time period of 10 years from 2013 to 2023 was –$11.48 million.

5 Discussion

Carbon emission is the major source of global warming, as carbon dioxide is a major greenhouse gas, which causes an increase in global average temperature. The long-term effect of this process is significant climate change. All anthropogenic activities are culprits for CO_2 emissions, including fossil fuel burning, transportation, and a rapid growth in industrialization. The 1991 industrial revolution contributed significantly to increasing the proportion of GHGs in the atmosphere [27]. The International Panel on Climate Change (IPCC) of 2014 also reported an emphasis on climate regulation, the effect of CO_2 emissions on climate, and this observed increase in global warming. An increase in the concentration of CO_2 in the atmosphere also disrupts the carbon cycle [28]. Increases in carbon sinks by afforestation, reforestation, soil, and ocean, can store more atmospheric CO_2 and can play a major role in reducing the effect of long-term climate change, as well as in increasing global average temperature [29]. Rapid urbanization in the world, especially in developing countries, impacts the green infrastructure negatively, because urban expansion and built-up area expansion in any city happens at the cost of green cover. Studies also emphasize urbanization in Chandigarh in the past few decades and an increase in built-up areas at a fast rate, along with rapid population growth in the city, which converted the agricultural land into concrete surface. All infrastructural projects and residential building expand at the expense of green cover and agricultural area [30].

Urban growth and ecosystem services are negatively correlated, as urban ecosystem services decline with the expansion of the built-up land-use class, which

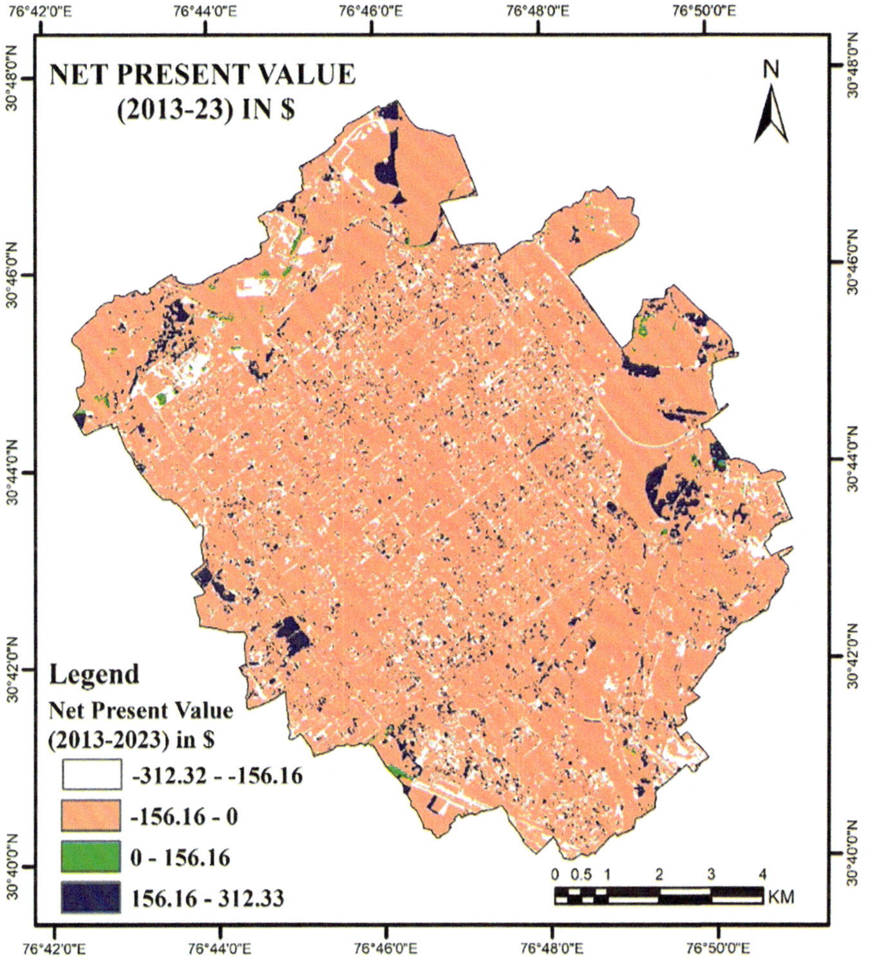

Fig. 7 Map showing Net present value ($) from 2013 to 2023

increases the urban heat island effect. The present study revealed a loss of green cover and agricultural land due to urbanization and a decrease in the carbon storage capacity of an ecosystem, which led to the deterioration of the habitat quality of the city. Natural habitat conversion to an impervious surface led to biodiversity loss and habitat fragmentation, which resulted in the endangerment of native species [31].

Through economic evaluation, we can get valuable insights into benefits the natural ecosystem provides to the welfare and stability of people in the domains of ecology and economics. Assigning monetary value and quantification of these services can help decision makers and urban planners to take robust steps for sustainability in society, as well as conservation [32].

6 Conclusion

This study focused on the storage and sequestration of carbon in the Union Territory (UT) of Chandigarh in the decadal period from 2013 to 2023 by utilizing the LULC map of Chandigarh, which showed the carbon stock potential from the base year to 2023. We took 2013 as the current year and assumed 2023 as the predicted LULC and quantified carbon sequestration, along with the monetary value of carbon. The outcome of the study showed a 14% decrease in total carbon storage from the base year. This decrease was attributed to a decrease in vegetation cover in the city. Land use and land cover (LULC) changes showed there was a significant amount of agricultural land and and green space converted into residential built-up areas. Built-up areas increased by up to 21% from 2013, while wasteland and waterbodies showed a decreasing trend. The total carbon sequestration was -8.2×10^4 Mg, which indicates the occurrence of more emission than sequestration of carbon. The total carbon emission was 1.3×10^5 Mg, which indicates the low proportion of green cover and low capacity for carbon sequestration during this time. The Net Present Value (NPV) of sequestration and carbon stock was \$11.48 million, where a negative value shows low carbon sequestration. As it stands, the current value of cost surpasses the current value of benefits. Based on the outcome of this study, there is more attention required in ecological planning and in the need for green conservation. This study can contribute to urban planning and decision-making.

References

1. Wirth, L. (1938). Urbanism as a way of life. *American Journal of Sociology, 44*(1), 1–24. https://doi.org/10.1086/217913
2. Friedmann, J. (1986). The world city hypothesis. *Development and Change, 17*(1), 69–83. https://doi.org/10.1111/j.1467-7660.1986.tb00231.x
3. United Nations. (2018). *World urbanization prospects: The 2018 revision*. Department of Economic and Social Affairs, Population Division.
4. UN-HABITAT.
5. Ministry of Housing and Urban Affair, Govt. of India.
6. Costanza, R., d'Arge, R., de Groot, R., Farber, S., Grasso, M., Hannon, B., et al. (1997). The value of the world's ecosystem services and natural capital. *Nature, 387*(6630), 253–260.
7. Millennium Ecosystem Assessment. (2005). *Ecosystems and human well-being: Synthesis*. Island Press.
8. Wang, C., Ren, Z., Dong, Y., Zhang, P., Guo, Y., Wang, W., & Bao, G. (2022). Efficient cooling of cities at global scale using urban green space to mitigate urban heat Island effects in different climatic regions. *Urban Forestry & Urban Greening, 74*, 127635. https://doi.org/10.1016/j.ufug.2022.127635
9. Kruize, H., Van der Vliet, N., Staatsen, B., Bell, R., Chiabai, A., Muiños, G., Higgins, S., Quiroga, S., Martinez-Juarez, P., Aberg Yngwe, M., Tsichlas, F., Karnaki, P., Lima, M. L., García de Jalón, S., Khan, M., Morris, G., & Stegeman, I. (2019). Urban green space: Creating a triple win for environmental sustainability, health, and health equity through behavior change. *International Journal of Environmental Research and Public Health, 16*(22), 4403. https://doi.org/10.3390/ijerph16224403

10. Gupta, L. K., Pandey, M., Raj, P. A., & Shukla, A. K. (2023). Fine sediment intrusion and its consequences for river ecosystems: A review. *Journal of Hazardous, Toxic, and Radioactive Waste, 27*(1), 04022036. https://doi.org/10.1061/(ASCE)HZ.2153-5515.0000729
11. Zhang, J., Lu, X., Qin, Y., Zhang, Y., & Yang, D. (2024). Can urbanization-driven land-use and land-cover change reduce ecosystem services? A case of coupling coordination relationship for contiguous poverty areas in China. *Land, 13*(1), 82. https://doi.org/10.3390/land13010082
12. Deeksha, S. A. K., Rama Devi, N., & Shukla, S. (2023). Remote sensing and its application on soil: An ecosystem services. In *River, sediment and hydrological extremes: Causes, impacts and management* (pp. 307–319). Springer Nature Singapore. https://doi.org/10.1007/978-981-99-4811-6_17
13. Arcidiacono, A., Ronchi, S., & Salata, S. (2015). Ecosystem services assessment using invest as a tool to support decision making process: Critical issues and opportunities. *Computational Science and Its Applications—ICCSA, 2015*, 35–49. https://doi.org/10.1007/978-3-319-21410-8_3
14. Hermes, J., Van Berkel, D., Burkhard, B., Plieninger, T., Fagerholm, N., Von Haaren, C., & Albert, C. (2018). Assessment and valuation of recreational ecosystem services of landscapes. *Ecosystem Services, 31*, 289–295. https://doi.org/10.1016/j.ecoser.2018.04.011
15. Nayak, D., Shukla, A. K., & Devi, N. R. (2023, July). Effects of urbanization on urban ecosystem services (UES S)-A geospatial approach. In *IGARSS 2023–2023 IEEE international geoscience and remote sensing symposium* (pp. 6216–6219). IEEE.
16. Census of India, 2011.
17. Department of urban planning, UT Chandigarh 'Draft master plan Chandigarh-2031'.
18. IQAIR, 2024.
19. Harmon, M., Franklin, J., Swanson, F., Sollins, P., Gregory, S., Lattin, J., Anderson, N., Cline, S., Aumen, N., Sedell, J., Lienkaemper, G., Cromack, K., & Cummins, K. (1986). Ecology of coarse Woody debris in temperate ecosystems. *Advances in Ecological Research*, 133–302. https://doi.org/10.1016/s0065-2504(08)60121-x
20. IPCC. (2014). *Climate change 2014: Synthesis report. Contribution of working groups I, II and III to the fifth assessment report of the intergovernmental panel on climate change*. IPCC.
21. Lal, R. (2004). Soil carbon sequestration impacts on global climate change and food security. *Science, 304*(5677), 1623–1627. https://doi.org/10.1126/science.1097396
22. FAO. (2010). *Global Forest Resources Assessment 2010. FAO forestry paper 163*. Rome: Food and Agriculture Organization of the United Nations.
23. Nayak, D., Shukla, A. K., & Devi, N. R. (2024). Decadal changes in land use and land cover: Impacts and their influence on urban ecosystem services. *AQUA—Water Infrastructure, Ecosystems and Society, 73*(1), 57–72. https://doi.org/10.2166/aqua.2024.211
24. Carbon storage and sequestration—InVEST® documentation. Retrieved from https://storage.googleapis.com/releases.naturalcapitalproject.org/invest-userguide/latest/en/carbonstorage.html
25. Interagency Working Group on Social Cost of Carbon, United States Government (2013). *Technical support document: Technical update of the social cost of carbon for regulatory impact analysis—Under executive order 12866*. Link to document.
26. Babbar, D., Areendran, G., Sahana, M., Sarma, K., Raj, K., & Sivadas, A. (2021). Assessment and prediction of carbon sequestration using Markov chain and invest model in Sariska tiger reserve, India. *Journal of Cleaner Production, 278*, 123333. https://doi.org/10.1016/j.jclepro.2020.123333
27. Le Quéré, C., Andrew, R. M., Friedlingstein, P., Sitch, S., Pongratz, J., Manning, A. C., Korsbakken, J. I., Peters, G. P., Canadell, J. G., Jackson, R. B., Boden, T. A., Tans, P. P., Andrews, O. D., Arora, V. K., Bakker, D. C., Barbero, L., Becker, M., Betts, R. A., Bopp, L., et al. (2018). Global carbon budget 2017. *Earth System Science Data, 10*(1), 405–448. https://doi.org/10.5194/essd-10-405-2018
28. Friedlingstein, P., Jones, M. W., O'Sullivan, M., Andrew, R. M., Hauck, J., Peters, G. P., Peters, W., Pongratz, J., Sitch, S., Le Quéré, C., Bakker, D. C., Canadell, J. G., Ciais, P., Jackson,

R. B., Anthoni, P., Barbero, L., Bastos, A., Bastrikov, V., Becker, M., et al. (2019). Global carbon budget 2019. *Earth System Science Data, 11*(4), 1783–1838. https://doi.org/10.5194/essd-11-1783-2019

29. Nayak, D., Shukla, A. K., & Devi, N. R. (2022, November). Effects of urbanization on urban ecosystem services (UESS)—A framework. In *International conference on Mediterranean geosciences union* (pp. 149–153). Springer Nature Switzerland. https://doi.org/10.1007/978-3-031-48715-6_33
30. Jat, M. K., Garg, P., & Khare, D. (2008). Monitoring and modelling of urban sprawl using remote sensing and GIS techniques. *International Journal of Applied Earth Observation and Geoinformation, 10*(1), 26–43. https://doi.org/10.1016/j.jag.2007.04.002
31. Deeksha, & Shukla, A. K. (2022). Ecosystem services: A systematic literature review and future dimension in freshwater ecosystems. *Applied Sciences, 12*(17), 8518. https://doi.org/10.3390/app12178518
32. Costanza, R., D'Arge, R., De Groot, R., Farber, S., Grasso, M., Hannon, B., Limburg, K., Naeem, S., O'Neill, R. V., Paruelo, J., Raskin, R. G., Sutton, P., & Van den Belt, M. (1997). The value of the world's ecosystem services and natural capital. *Nature, 387*(6630), 253–260. https://doi.org/10.1038/387253a0

Integrating Multi-Criteria Decision Making and Advanced Technologies in Water Quality Index Development

Bishnu Kant Shukla, Lokesh Gupta, Hans Kumar Singh, Abhinav Malik, Ayushi Karanwal, and Aditi Dhama

1 Introduction

The Water Quality Index (WQI) is a significant and well-recognised indicator that provides an aggregated assessment of the state of water quality in various aquatic habitats. WQI enables understanding of water quality and makes it easier for the general public, policy makers, and water resource managers to interpret water quality [1] by combining various water quality metrics into a single numerical score. The Water Quality Index (WQI), which dates back to the 1960s, has significantly changed through time in order to accommodate various water quality factors and evaluation techniques [2]. In the past, many models have been created to more accurately depict the WQI, with regional variations being customised to the unique needs and environmental conditions of particular geographic regions [3]. WQI models, for instance, have proved useful in evaluating groundwater quality monitoring networks, as evidenced in Daughney et al.'s research, demonstrating the model's adaptability in various aquatic situations [1]. Accurate and understandable data on water quality are essential for efficient management of water resources. The WQI assumes top priority in this context. WQI enables stakeholders to make informed decisions, whether it is for drinking, agricultural, industrial, or recreational uses by offering a concise yet complete assessment of the water quality [4]. It acts as a fundamental parameter in the processes of allocating, treating, and utilising water resources. Additionally, merging WQI with contemporary analytical procedures, such Multi-Criteria Decision Making (MCDM) techniques, provides for increased objectivity and precision in evaluating water quality [5]. The repeated use of WQIs

B. K. Shukla (✉) · H. K. Singh · A. Malik · A. Karanwal · A. Dhama
Department of Civil Engineering, JSS Academy of Technical Education, Noida, India

L. Gupta
College of Pharmacy, JSS Academy of Technical Education, Noida, India

S. Pathak et al. (eds.), *Geo-Data Revolution*, Advances in Geographic Information Science, https://doi.org/10.1007/978-3-031-94078-1_18

in international research, from the assessment of groundwater quality in Tunisia to the evaluation of drinking water in Southern India, further emphasises its applicability and necessity in water resources engineering [2, 6].

2 Historical Perspective on the Water Quality Index

Since its early stages, water quality evaluation has advanced significantly, giving rise to the development of strict procedures and frameworks. With its historical roots and current significance, the Water Quality Index (WQI) is a prime example of this change.

2.1 Evolutionary Progress of WQI Models

The Water Quality Index (WQI) was first created in the middle of the twentieth century, according to historical records. These early models were created with the intention of reducing the complex web of water quality characteristics to a format that the average person could understand [7]. The substantial effects of water quality on public health, ecological balance, and the sustainability of the ecosystem as a whole, however, came to light as societies developed [8]. This highlighted the necessity for an expanded, comprehensive, and all-encompassing Water Quality Index.

A complete chronology of WQI models from 1960 to 2020 is shown in Fig. 1 [9]. The image intriguingly highlights how these models have multiplied and changed over the course of these six decades. Given the global push for environmental

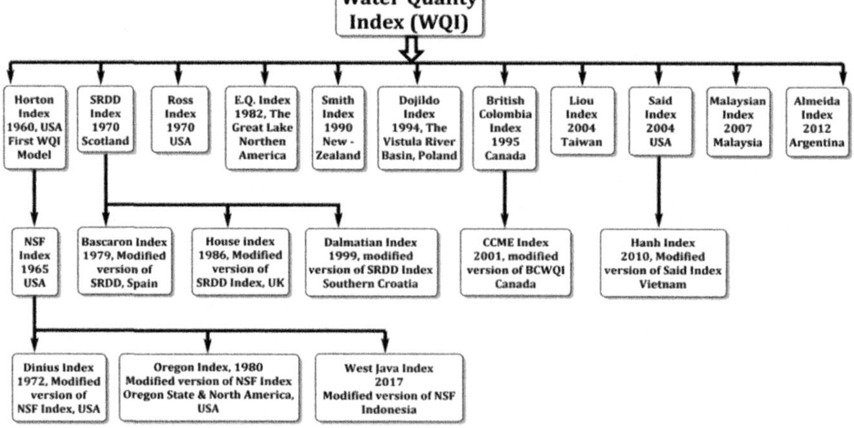

Fig. 1 Historical development of the WQI model

conservation and the realisation that water quality is essential to this effort, the 1970s and 1980s in particular emerged as crucial decades [10]. As shown in the picture, while certain WQI models gained popularity on a worldwide scale, others resonated more with people in a particular geographic area, echoing the water quality issues that are unique to that area [11].

As we continue to examine Fig. 1, it is fascinating to see the variety of models and their regional preferences. The ongoing development and improvement of WQI over time have resulted in its fusion with cutting-edge computational approaches, enabling a more intricate and accurate assessment of water quality [12]. Modern variations of the WQI seamlessly incorporate not just conventional physicochemical characteristics but also embrace contemporary pollutants and environmental issues, attesting to the flexibility and adaptability of the WQI in the current climate of rapid change [13].

2.2 *Advancements and Transformations in WQI Over Time*

The evolution of the Water Quality Index (WQI) over time is a perfect example of environmental science's never-ending search for accuracy and comprehensive understanding. This path has been marked by important turning points that have been motivated by both scientific advancement and the need of changing environmental issues. The initial versions of the WQI were based on the requirement for a unified metric that could offer an extensive yet understandable assessment of water quality [14]. These models focused on a synthesis that would appeal to both specialists and the general public, and they were based on a few fundamental parameters. These indices have to constantly evolve, though, due to the natural dynamic of ecosystems and the complex interactions between a wide range of physical, chemical, and biological elements.

One of the most significant changes in WQI models has been the inclusion of a wider range of characteristics, which is a reflection of the growing recognition that water quality is a complex concept requiring careful analysis [15]. Newer WQI iterations started including criteria that addressed contaminants such as pharmaceuticals, endocrine disruptors, and sophisticated industrial chemicals as there was growing knowledge of emerging pollutants, notably in the area of anthropogenic activities [16].

Furthermore, WQI models now have exceptional precision and analytical depth thanks to the global upsurge in technological development and computing power in the late twentieth and early twenty-first centuries [17]. Real-time water quality analyses and predictions were made possible by the combination of WQI with cutting-edge computational tools, data analytics, and geospatial approaches [18]. Furthermore, as highlighted in recent research, the orientation of WQI has been further adjusted in response to the social desire for fair and sustainable water management. The imperatives of sustainable development are now taken into account by modern WQI frameworks, ensuring that assessments take into account not only current quality but also potential future threats and their socioeconomic ramifications [19].

The development of WQI from its early phases to its sophisticated state now highlights the ongoing interaction between scientific rigour, technological capability, and societal requirements, weaving a story of adaptation, precision, and all-encompassing comprehension.

3 Core Structure and Functionality of the Water Quality Index

3.1 *The Fundamental Structure of WQI*

The Water Quality Index (WQI) is a crucial instrument for evaluating the health and quality of water bodies because it offers a quantitative measure that combines the intricate information from many different water quality criteria into a single, understandable number. This makes it easier to inform decision-makers, stakeholders, and the general public effectively about the state of water quality [20]. The fundamental concept behind the WQI's construction is the assimilation of numerous water quality characteristics, each of which is given a weight based on how significant it is in determining the overall quality of the water. By using a weighted approach, it is ensured that important factors have a significant impact on the index's final value, making it sensitive to and reflective of the actual water conditions. A typical WQI structure, as shown in Fig. 2, offers a clear visual picture of how many models and regions of usage have been developed and used across time, spanning from the 1960s to the recent decades [21]. The evolutionary development of WQI models highlights the index's adaptability and flexibility in incorporating new criteria and approaches as our understanding of water quality evolves.

Over the years, a number of formulas have been presented to quantify the WQI. The principle of obtaining the index by averaging sub-indices (SI) weighted by their relative significance (W_i) is encapsulated by Eq. (1) [22].

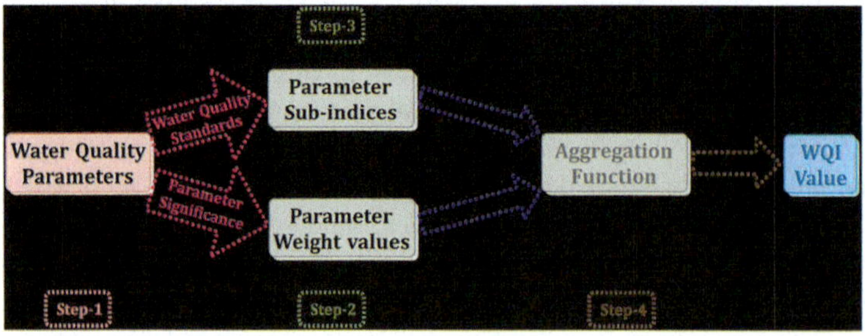

Fig. 2 General structure of WQI model

$$\mathrm{WQI} = \prod_{i=1}^{n} \mathrm{SI}^{W_i} \quad (1)$$

With this multiplicative method, the entire index would be considerably impacted by a negative value in any of the factors, making it a sensitive measure. The ability of the WQI to quantitatively capture the level of water quality underpins both its fundamental structure and functionality. The index continues to be a useful, accurate, and comprehensive measure for monitoring and communicating water quality because of its structure, which is based on a variety of weighted characteristics, and mathematical formulation [23].

3.2 *Key Components and Mechanisms in WQI Formulation*

The Water Quality Index (WQI) combines different quality measures into a single indicative value to provide a cogent depiction of the health of the water. In order to create a solid WQI formulation that will ensure the index's dependability, thoroughness, and relevance, it is necessary to carefully analyse a number of important factors and their underlying mechanisms.

3.2.1 Parameter Selection, Weight Assignments, and Their Mathematical Representations

Building a trustworthy WQI starts with selecting the right water quality criteria. The quality of the water body should be represented by these criteria holistically, taking into account its chemical, biological, and physical features [24, 25]. After being chosen, each parameter needs to be given a weight that accounts for its importance in terms of the overall quality of the water. Weighting makes ensuring that important factors have a significant impact on the final WQI rating. The Water Quality Index (WQI) is mathematically represented to emphasise its robustness. According to the fundamental equation (Eq. 2), the WQI is calculated [26] as follows:

$$\mathrm{WQI} = \sum_{i=1}^{n} W_i Q_i \quad (2)$$

where W_i is the weighing factor, and Q_i = is the rating value of parameter i and n = number of sub-indices. Each parameter's sub-index [22] is given in Eq. (3):

$$I = \sum_{i=1}^{n} I_i W_i \quad (3)$$

where I_i = sub-index of each parameter.

Furthermore, the WQI can also be represented by Eq. (4) [27]:

$$\mathrm{WQI} = \sum_{i=1}^{n} \mathrm{SI}_i W_i \tag{4}$$

where SI_i = sub-index of each parameter, W_i is the weighting factor, and, n is the number of sub-indices. To aggregate sub-index results, the weighted arithmetic mean formula can be enhanced by using the un-weighted harmonic square mean formula.

3.2.2 Deriving Sub-Indices, Their Aggregation, and Associated Calculations

Sub-indices provide a more detailed understanding of each water quality measure. Each sub-index is derived from a particular water quality parameter, essentially converting the parameter's unprocessed data into a standardised score and making it easier to include the parameter in the overall WQI.

These sub-indices are combined to ensure that the WQI provides a comprehensive evaluation of water quality. Strong mathematical formulas are often used in this aggregation. For instance, Eq. (5) [28] employs the harmonic mean of the inverses of the squared sub-indices:

$$\mathrm{WQI} = \sqrt{\frac{n}{\sum_{i=1}^{n} \frac{1}{\mathrm{SI}i^2}}} \tag{5}$$

Additionally, Eq. (6) [29] introduces a method for identifying the lowest sub-index value, ensuring that the most critical water quality parameter is highlighted:

$$I_{\min} = \sum \min\left(I_{\mathrm{sub1}}, I_{\mathrm{sub2}}, \ldots I_{\mathrm{subn}}\right) \tag{6}$$

where $I_{\min}$ represents the lowest sub-index value among the parameters. Furthermore, Eq. (7) [30] proposes a new approach, the British Columbia Water Quality Index (BCWQI), which incorporates specific criteria indicated by F_1, F_2, and F_3 to provide a broadened understanding of water quality:

$$\mathrm{BCWQI} = 100 - \left(\sqrt{\frac{F_1^2 + F_2^2 + F_3/3^2}{1.453}}\right) \tag{7}$$

These equations collectively contribute to a more comprehensive and nuanced assessment of water quality, allowing for better decision making in water resource management. The delicate integration of these elements utilising mathematically sound equations, well-selected parameters, and wise weighting are what ultimately determine how complex the WQI formulation is. These components have been thoughtfully incorporated to ensure that WQI remains an essential tool for monitoring and regulating water quality.

4 Incorporation of Multi-Criteria Decision Making (MCDM) in WQI

Evaluation of the Water Quality Index (WQI) frequently necessitates the synthesis of a number of characteristics, each with a distinct importance and impact on water quality. Employing Multi-Criteria Decision Making (MCDM) techniques can improve the accuracy, dependability, and comprehensiveness of the WQI due to the multi-dimensional character of this evaluation.

4.1 Integration and Advantages of MCDM in WQI

Multi-Criteria Decision Making (MCDM) provides a structured approach for analysing complex decision making problems that involve multiple aspects. It enables a comprehensive, fair, and well-informed examination. Incorporating Multi-Criteria Decision Making (MCDM) methodologies into the calculation of the Water Quality Index (WQI) can provide several benefits. Equation (8) [30] emphasises the importance of weighted sub-indices (s_i) which indicate the relevance and contribution of each parameter to the overall water quality. This equation exemplifies the significance of Multi-Criteria Decision Making (MCDM) by facilitating the systematic incorporation of different water quality metrics according to their assigned weights. The Water Quality Index (WQI) is calculated by summing the products of the weights (w) and scores (s) for each parameter, and then dividing by the total number of parameters (n).

$$\mathrm{WQI} = \left[\frac{w_1 s_1 + w_2 s_2 + w_3 s_3 + \ldots + w_n s_n}{w_1 + w_2 + w_3 + \cdots + w_n} \right] m_1 m_2 \tag{8}$$

Equation (9) [31] demonstrates the process of integrating the weighted sub-indices to create a full quality score, which involves summation. This process, which is essential in MCDM approaches, enables the index to incorporate several elements without reducing the significance of each component individually.

$$\text{Brown WQI} = \sum_{i=1}^{n} w_i s_i \tag{9}$$

The primary advantages of Multi-Criteria Decision Making (MCDM) in Water Quality Index (WQI) are:

Precision: By taking into account the relative relevance of each criterion, MCDM's structured evaluation ensures a more accurate portrayal of water quality.

Comprehensiveness: MCDM ensures a holistic assessment, capturing the entire quality spectrum, by taking into account several criteria.

Flexibility: Because MCDM is inherently adaptable, it is possible to adjust to different settings and requirements.

Engagement of Stakeholders: MCDM techniques frequently promote multi-stakeholder participation, ensuring that various points of view are taken into account and the WQI is more robust and palatable.

Sensitivity Analysis: MCDM supports sensitivity analyses, which make it simple to assess how changing one or more parameters may affect the WQI [32].

Essentially, incorporating MCDM approaches into WQI development improves the evaluation process and ensures that the index remains dynamic, flexible, and representative of diverse water quality conditions and stakeholder priorities.

4.2 Real-World Application, Utility of MCDM, and Its Mathematical Insight

It is advantageous both practically and theoretically to use Multi-Criteria Decision Making (MCDM) approaches when determining Water Quality Indices (WQIs). The scope, accuracy, and adaptability of WQIs are improved by the real-world applications of MCDM, which also better align them with local conditions and particular water management issues.

Applications in the Real World: Water bodies suffer numerous pollution-related problems on a global scale. Each water body presents a different matrix of toxins, from industrial effluents to agricultural runoff. Authorities and academics can weigh the relative effects of various pollutants by using MCDM approaches, ensuring that WQI computations accurately reflect the state of water quality. For instance, MCDM can be used to provide bigger weights to industrial pollutants in urban environments where they may be prevalent, ensuring the resulting WQI is as high as possible.

Figure 3 exemplifies the comprehensive application of MCDM in diverse settings, highlighting its adaptability and relevance across different environmental matrices [33].

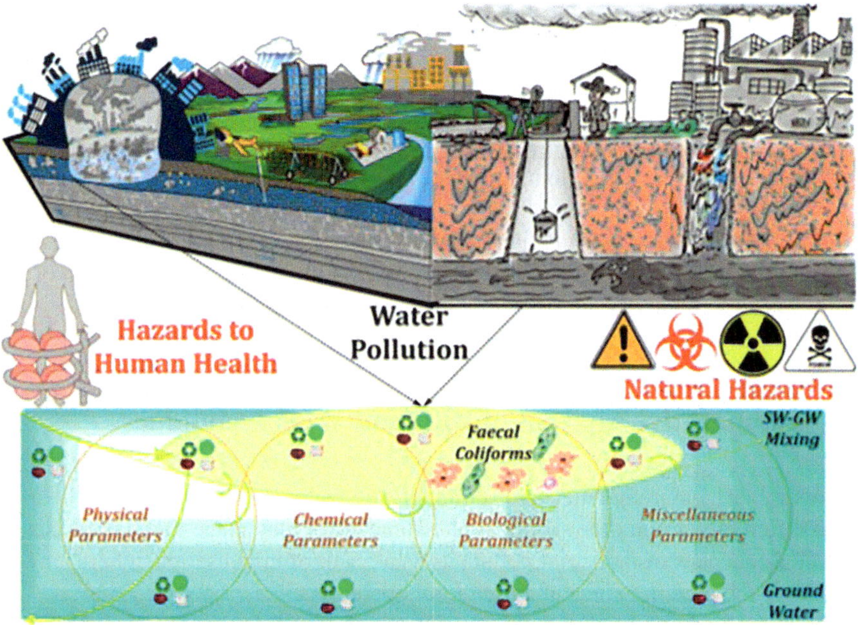

Fig. 3 Common structure of the multi-criteria decision making (MCDM) method: total goal (water quality assessment) based on multiple criteria (drinking, irrigation, commercial, domestic) and alternatives (physical, chemical, biological parameters)

4.2.1 Mathematical Insight into MCDM in WQI

Equation (10) [34] calculates the squared weighted sum of the sub-indices. This equation utilises the squared deviation method, which guarantees that the impact of each sub-index is squared, hence assigning more significance to significant departures from the average.

$$\text{SRDD} - \text{WQI} = \frac{1}{100}\left(\sum_{i=1}^{n} S_i W_i\right)^2 \quad (10)$$

Equation (11) [35] employs the harmonic mean method to combine many failure parameters, guaranteeing that every element of the water quality system that fails to satisfy the set standards has a substantial impact on the overall index.

$$\text{WQI} = 100 - \left[\frac{\sqrt{F_1^2 + F_2^2 + F_3^2}}{1.732}\right] \quad (11)$$

Equation (12) [36] provides a ratio-based assessment that measures the percentage of failed parameters out of the total number, scaled to 100. This approach guarantees that the Water Quality Index (WQI) is immediately influenced by the frequency of failures, hence enhancing the index's responsiveness and rendering it a more accurate reflection of real-life circumstances.

$$F_1 = \left[\frac{\text{number of failed parameters}}{\text{total number of parameters}} \right] \times 100 \tag{12}$$

The utilisation of these equations collectively improves the resilience and responsiveness of the Water Quality Index, guaranteeing its precise representation of the influence of notable deviations and deficiencies in water quality parameters.

5 Comparative Analysis of Different WQI Techniques

Geographical, environmental, and administrative factors influence WQI methodologies differently around the world. While some areas may place more emphasis on the consequences of industrial pollutants as a result of urbanisation, others may do so due to the predominance of agricultural operations.

Each method offers a distinct viewpoint, benefits, and drawbacks. Understanding the advantages and disadvantages of various approaches through comparative analysis helps decision-makers and researchers select or create WQIs that are most appropriate for their particular problems. Such comparison analyses also draw attention to commonalities and best practices, encouraging global partnerships and shared learning. Stakeholders can strive towards a more comprehensive, flexible, and accurate water quality evaluation system by knowing the workings, results, and field efficacies of various WQIs.

5.1 Variations and Similarities in Global WQI Models: A Mathematical Overview

Different areas and countries have different Water Quality Indices (WQIs), adapted to the particular environmental and anthropogenic challenges they confront. But underlying these differences, mathematical structures reveal startling parallels that serve as the basis for these indices.

Variations: Every geographic region on Earth faces unique water-related difficulties. For example, in metropolitan locations where there is industrial activity, heavy metals can provide a substantial danger, while in rural places, the main issue may be nutrient runoffs. As a result, Water Quality Indices (WQIs) include different attributes to precisely address local concerns. The heterogeneity in these indices is illustrated by the subsequent piecewise function, which depicts a fuzzy membership function for a specific water quality metric. This function, expressed in Eq. (13) [37], incorporates many intervals to determine the values of membership.

$$\mu_A(x) = \begin{cases} 0 & \text{if } x \leq a \text{ or } x > d \\ \dfrac{x-a}{b-a} & \text{if } a < x \leq b \\ 1 & \text{if } b < x \leq c \\ \dfrac{d-x}{d-c} & \text{if } c < x \leq d \end{cases} \tag{13}$$

In this context, $\mu_A(x)$ represents the value of the membership function for the parameter x, which varies from 0 to 1. The parameters a, b, c, and d specify the intervals in which the membership value varies. a represents the lower bound, d represents the upper bound, and b and c indicate the intermediate thresholds when the membership function transitions between distinct values. This equation has the ability to be applied in a variety of scenarios and provides an advanced method for controlling uncertainty through exact intervals.

Similarities: The fundamental mathematical concepts that govern WQIs remain consistent, despite regional variations. A weighted aggregation approach is implemented by a significant number of WQIs, irrespective of the variation in parameters or their weights. The minimum membership function is used to depict the intersection of two fuzzy sets, A and B, as illustrated in Eq. (14) [38].

$$\mu_{A \cap B}(x) = \min\left(\mu_A(x), \mu_B(x)\right) \tag{14}$$

The membership function value for the intersection of sets A and B at point x is represented by $\mu_{A \cap B}(x)$. The function $\min(\mu_A(x), \mu_B(x))$ calculates the minimum of the membership values of x in both sets, indicating the extent to which x is a member of the intersection of A and B. In the same way, the maximum membership function can be used to represent the union of two fuzzy sets A and B, as shown in Eq. (15) [39].

$$\mu_{A \cup B}(x) = \max\left(\mu_A(x), \mu_B(x)\right) \tag{15}$$

The membership function value for the union of sets A and B at point x is denoted by $\mu_{A \cup B}(x)$. The function $\max(\mu_A(x), \mu_B(x))$ calculates the maximum of the membership values of x in both sets, indicating the extent to which x is a member of the union of A and B. The uniform implementation of fundamental mathematical principles in various WQI systems, irrespective of regional water quality issues, is illustrated by these equations.

5.2 *Evaluating Eclipsing and Uncertainties in WQI Metrics with Mathematical Modelling*

Eclipsing happens when one parameter dominates another in the WQI computation because of its large value, potentially producing false findings. On the other hand, uncertainties are a result of a variety of factors, such as measurement errors, seasonal changes, or arbitrary weighting. The parameters and their uncertainties are displayed in Table 1, which makes it easier to see the magnitude and possible causes of these fluctuations.

Table 1 Comparative analysis of the sources of eclipsing and uncertainty in WQI models

Index name	Issue/ source of uncertainty	Ref.
Oregon index	The eclipsing problem in water quality indices (WQIs) is a result of a complex parameter selection process, and the aggregation function further contributes to uncertainty	[40, 41]
Horton index	Nutrient elements were not included in the parameter group of the model, and the aggregation function is a significant source of uncertainty due to the poorly defined coefficient factors	[42]
House index	The aggregation function is a primary source of instability, and the eclipsing problem is a result of the parameter selection method employed in the model	[43]
NSF index	The parameter selection phase facilitates model eclipsing, which results in the introduction of a new multiplicative aggregation function as a result of the original function's lack of sensitivity	[44, 45]
CCME index	The model does not specify the water quality parameters or necessitate weighting values, both of which contribute to eclipsing problems. Furthermore, the CCME model implements intricate aggregation functions that may result in ambiguity in the final index ranking	[43, 45]
Smith index	The model lacks a specified number of required water quality parameters, uses complex sub-index equations leading to eclipsing issues, and does not require parameter weight values	[46]
Malaysian index	The model employed only common water quality parameters and excluded toxic and biological indicators	[47]
Said index	The parameter selection process was based on literature, and parameter standardisation was not required	[46]
Dojildo index	The model did not specify the number of required water quality parameters, and the parameter selection processes were based on the intent of the end user	[48]
Almeida index	The closed parameter selection techniques do not allow for the inclusion of other important water quality indicators, and the specific concentration values of water quality are directly employed as a sub-index	[41]
Liou index	The eclipsing problem emerges due to a selection procedure that fixes parameters and eliminates other relevant parameters	[41, 43, 48]
WQI for west java	The fundamental cause of eclipsing is the model's parameter selection process, which relies on the availability of monitoring data	[31]
FIS	The process of selecting parameters can worsen the problem of data eclipse in the model	[15]
Hanh index	The model employs complex parameter selection procedures and assigns equal weight to all selected parameters	[43]

The uncertainty in WQI can be quantified mathematically. This is expressed by the Eq. (16) [49], which captures the idea of inverse proportionality:

$$\mu_{\bar{A}}(x) = 1 - \mu_A(x) \tag{16}$$

In Eq. (16), $\mu_A(x)$ represents the membership function of parameter x, and $\mu_{\bar{A}}(x)$ represents the uncertainty or inverse membership. This equation indicates that if the certainty (or membership) of a parameter is known, its uncertainty can be calculated as the complement of the membership function. Additionally, uncertainty in WQIs can be evaluated using a weighted average algorithm, which accounts for both the parameter's value and the degree of membership [50–53], as represented in Eq. (17):

$$Z = \frac{\int \mu(z) z dz}{\int \mu(z) dz} \tag{17}$$

In this equation, Z is the weighted average, $\mu(z)$ is the membership function of parameter z, and the integrals in the numerator and denominator calculate the weighted sum and the total membership, respectively. To model the effect of these uncertainties in a more comprehensive manner, the following matrix equation (Eq. 18) is used [54–56].

$$B = W.R = [w_1 w_2 \ldots w_i] . \begin{bmatrix} \mu_{11}\ \mu_{12} \cdots \mu_{1j} \\ \mu_{21}\ \mu_{22} \cdots \mu_{2j} \\ \cdots \quad \cdots \quad \cdots \quad \cdots \\ \mu_{i1}\ \mu_{i2} \cdots \mu_{ij} \end{bmatrix} = [b_1 \ \ b_2 \ldots b_i] \tag{18}$$

Here, B represents the output vector reflecting the overall water quality issue, W is the vector of weights assigned to each parameter, and R is the matrix of membership values for each parameter across different conditions. Each element w_i is the weight for parameter i, and μ_{ij} represents the membership value of parameter i under condition j. The resulting vector B captures the combined effect of various parameters and their uncertainties.

6 WQI in Different Environmental Contexts

Water Quality Indices (WQIs) are crucial instruments that can be used in a variety of aquatic situations rather than just one type of water body. These indices may play a crucial role in guiding public health recommendations, conservation activities, and policy decisions.

6.1 Use of WQI in Diverse Water Bodies: Analytical Perspectives

Regardless of their size or origin, all bodies of water, whether vast oceans, serene lakes, meandering rivers, or man-made reservoirs, provide distinct challenges. It is crucial to design and utilise different Water Quality Indices (WQIs) that are customised to the specific characteristics of each water body. This is because both human activities and natural processes have a substantial influence on water quality. The analytical model given in Eq. (19) [54] is utilised to demonstrate the dependency and influence of several parameters on water quality.

$$Z_{ij} = a_{i1}x_{j1} + a_{i2}x_{j2} + a_{i3}x_{j3} + \ldots + a_{im}m_{jm} \tag{19}$$

Equation (19) defines Z_{ij} as the aggregate impact of multiple parameters on water quality. The coefficients a_{ik} indicate the individual influence of each parameter, whereas x_{jk} represents the values of these parameters. This equation represents the interplay between many attributes and their combined impact on the overall quality of water. An alternative viewpoint is presented by the standardisation equation, as shown in Eq. (20) [1]:

$$Z_{ij} = \frac{x_{ij} - \overline{x_j}}{S_j} \tag{20}$$

In this context, Z_{ij} represents the standardised value of parameter i at instance j, x_{ij} is the actual value of the parameter, $\bar{x}_j$ is the mean value, and S_j is the standard deviation. This formula calculates the difference between each parameter and the average value, providing information about which variables depart from the normal range. This allows for specific actions to be implemented. Equation (21) highlights the significance of fluctuations in water quality metrics [1] throughout time and space.

$$E^2_{ij} = \sum_{j=1}^{n} \left(Z_{ij} - Z_{kj}\right)^2 \tag{21}$$

In this equation, E_{ij}^2 represents the squared differences between specific characteristics at various times or locations, which help in predicting future trends and understanding how water quality may be changing. This comprehensive approach not only provides a better understanding of current conditions but also paves the way for targeted interventions to protect and optimise the use of this invaluable resource.

6.2 Regional WQI Applications, Practices, and Their Mathematical Formulations

The Water Quality Index (WQI) has regional as well as worldwide applicability. The creation and application of specialised WQI models are required due to the many locations' disparate environmental, geological, and anthropogenic characteristics. This allows these indices to accurately catch the subtleties and offer a more realistic picture of the water quality.

6.2.1 WQI Model Regional Variations and Similarities

We may infer the differences and similarities in WQI applications across various geographical regions by using Fig. 4 (prepared based on discussion presented in [33]). This visual illustration makes it easier to comprehend how various conditions,

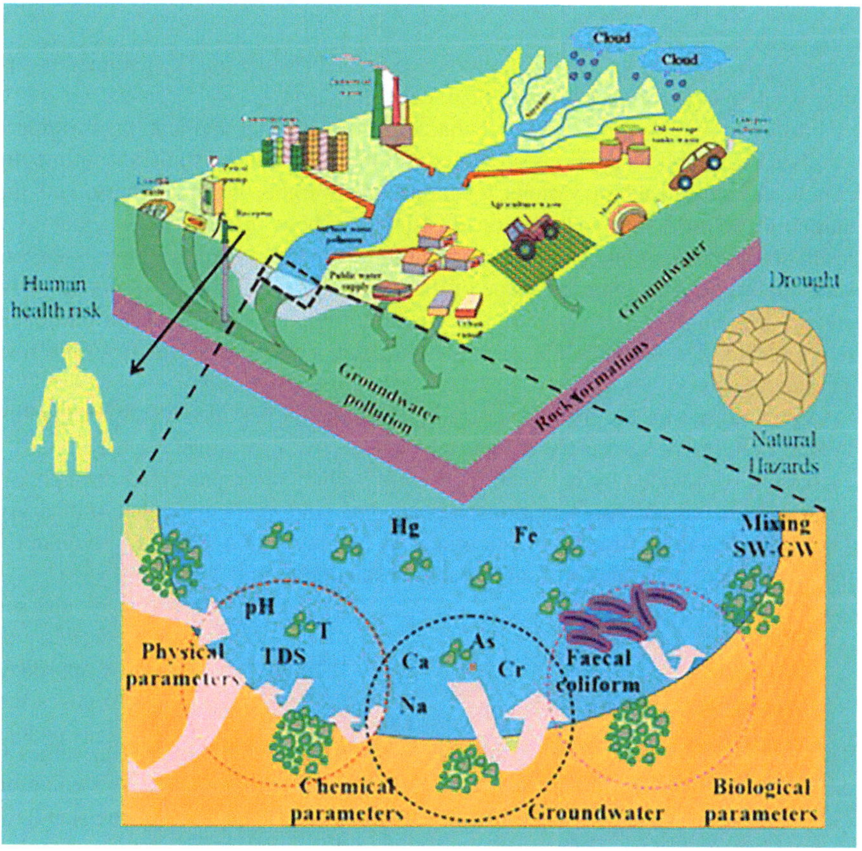

Fig. 4 Degradation of surface water and groundwater quality by anthropogenic and natural processes from various sources and its impact on human health

Table 2 Overview of WQI model applications (total and by study area) in literature

WQI model	Number of uses	Marine/sea/coastal study area	Lake study area	River study area
Almeida	1	–	–	1
Bascaron	4	1	–	3
BCWQI	1	–	–	1
CCME	36	3	5	28
Di	–	–	–	–
Dalmatian	1	1	–	–
Dojildo	1	–	–	1
EQI	2	–	1	1
FIS	12	1	1	10
Horton	7	1	–	6
MWQI	8	1	1	6
NSF	18	–	1	17
Oregon	2	–	–	2
Smith	2	–	–	2
SRDD	6	–	–	6

from dry to tropical locations, can produce a range of water quality parameters and necessitate the use of various analytical techniques.

Further highlighting the regional characteristics in WQI applications is Table 2 (prepared based on discussions presented in [46]). The tabular statistics outline the parameters prioritised in various locations, the weights given to them, and the unique difficulties that each region deals with. These difficulties, whether they result from commercial operations, agricultural runoffs, or natural processes, are very important in shaping the shape of the regional WQI model. In terms of mathematics, regional WQIs frequently use special formulations to meet their particular requirements. The formula offers a way to measure a parameter‘s fluctuation across a certain range.

A generalised metric for evaluating water quality is provided by Water Quality Indices (WQIs). Nevertheless, a nuanced and effective instrument for water quality evaluation is provided by regional specificities, which are represented through tailored models, visual representations, tabular data, and mathematical formulations. The sub-index value SI can be determined by employing Eq. (22) [57] to quantify the deviation of a parameter from a specified standard.

$$\mathrm{SI} = \left[(S_1 - S_2) \times \left(\frac{x_i - x_1}{x_2 - x_1} \right) \right] - S_1 \tag{22}$$

The sub-index scores S_1 and S_2 in this formula correspond to the parameter values x_1 and x_2, respectively, while x_i is the observed value of the parameter. This equation assists in determining the extent to which a parameter deviates from the standard. Equation (2) provides an alternative formulation that expresses the same concept in a slightly different manner [57].

$$\mathrm{SI} = \left[(S_1 - S_2) \times \left(\frac{x_1 - x_i}{x_1 - x_2} \right) \right] - S_1 \tag{23}$$

Here, SI denotes the sub-index value, while S_1, S_2, x_1, x_2, and x_i have the same meanings as previously defined. This variation provides another method to assess the departure of a parameter from its standard. A more straightforward model that offers insights into how a particular parameter performs relative to the maximum observed value is the ratio-based method, expressed as Eq. (24) [57]:

$$\mathrm{SI} = \frac{x_i}{x_{\max}} \tag{24}$$

Parameter x_i, the observed value, SI is the sub-index value, and $x_{\max}$ is the maximum observed value in this equation. If you want an easy way to see how well a parameter is doing, go no further than this ratio. By adapting these models and mathematical formulas to the unique requirements of each area, we can conduct precise water quality assessments.

7 Innovation and Future of WQI

As science and technology continue to progress, future Water Quality Indices (WQIs) should be increasingly precise and thorough. We expect water quality indicators (WQIs) to be understood and used in a whole new light once big data analytics, AI, and the IoT are integrated into water quality assessment. With the use of previous data, machine learning algorithms can forecast changes in water quality, and IoT devices may gather data in real time to guarantee fast actions. As molecular biology and genomics continue to make great strides, WQIs may soon include molecular biological factors, giving an even more comprehensive picture of water quality. Despite the exciting potential of these innovations, they also bring new difficulties. Challenges include ensuring data security, fostering interdisciplinary collaboration, and improving the accuracy of AI forecasts. However, if lawmakers, scientists, and IT professionals work together, WQIs have a bright, accurate, and comprehensive future.

7.1 Prospective Technological Integrations in WQI and the Role of Advanced Mathematical Equations

The relationship between technology and the Water Quality Index (WQI) is growing as advancements in technology have the potential to redefine the future of water quality assessment. The incorporation of intricate technological systems enables more profound analysis, enhanced data collection techniques, and advanced forecasting models, all enhanced by sophisticated mathematical algorithms. The weighted formula illustrates the significance of weighted components in determining the ultimate WQI. Equation (25) [29].

$$\mathrm{WQI} = \frac{\left[W_1 S_{1+} W_2 S_{2\ldots} W_n S_n\right]}{W_1 W_2 \ldots W_n} M_1 M_2 \tag{25}$$

where W_i denotes the weight allocated to each parameter, and S_i indicates the sub-index value for each parameter. This technique highlights the importance of allocating varying weights to different criteria based on the area context, ensuring that more important aspects have a stronger impact on the overall Water Quality Index (WQI). Furthermore, a thorough assessment of water quality can be achieved by utilising a cumulative equation, as demonstrated in Eq. (26) [58]:

$$\mathrm{WQI} = \sum_{i=1}^{n} S_i \tag{26}$$

In this formula, S_i represents the sub-index values for each parameter, and the summation encompasses the overall water quality. This highlights the value of summative evaluations and points to potential technological advancements that could enable sensors to quickly compute such summations. The significance of weights in WQI computations is further emphasised by another formula, as represented in Eq. (27) [42]:

$$\mathrm{WQI} = \sum_{i=1}^{n} W_i \times S_i \tag{27}$$

Here, W_i denotes the weight assigned to each parameter, and S_i represents the sub-index value for each parameter. By incorporating AI-driven weight assignment tools that adapt their behaviour in response to changing environmental conditions, the accuracy of this method could be dramatically increased. These advanced mathematical formulations enable a detailed and accurate assessment of water quality, tailored to specific regional needs and conditions.

7.2 Directions for Future Research in WQI: Potential Areas of Mathematical Enhancements

Robust research endeavours are crucial to stay abreast of the continuous development of Water Quality Indices (WQIs), particularly in terms of enhancing their mathematical underpinnings. Improving these models would not only enhance the accuracy of evaluation but also take into consideration different environmental variables. Equation (28) [4] emphasises the correlation between parameter values and their corresponding priority scores.

$$\text{WQI} = \frac{\sum_{i=1}^{n} P_i \times C_i}{\sum_{i=1}^{n} P_i} \tag{28}$$

The formula assigns a priority score, represented by P_i, to each parameter i, while C_i reflects the concentration of parameter i. This equation exhibits substantial research potential, especially in scenarios where there is a scarcity or variability of data for particular parameters. Equation (29) [59] introduces a quadratic change to the sub-indices.

$$\text{WQI} = \frac{1}{100} \sum_{i=1}^{n} (S_i)^2 \tag{29}$$

S_i represents the value of the sub-index for parameter ii. Subsequent research can investigate the impacts of these quadratic alterations and their ability to adjust to various water quality conditions. Equation (30) [59] introduces the concept of a weighted quadratic transformation.

$$\text{WQI} = \frac{1}{100} \sum_{i=1}^{n} (W_i \times S_i)^2 \tag{30}$$

The equation uses the notation W_i to represent the weight applied to parameter i, and S_i to represent the sub-index value for parameter i. This introduces an additional level of intricacy and could be pivotal in regions where the factors that determine water quality differ significantly. The outlook for WQI research is optimistic, particularly due to the use of state-of-the-art technologies and advanced mathematical models. Together, these developments can provide a thorough, accurate, and adaptable structure for analysing water quality [60].

8 Conclusion

The Water Quality Index (WQI) continues to be an instrument that is crucial when it comes to assessing and communicating the current state of health of bodies of water. This paper highlights the significant evolution of WQI models, which has been driven by recent breakthroughs in environmental science and technology. These developments have been the driving force behind the considerable evolution of WQI models. There has been a considerable improvement in the resilience and accuracy of WQIs as a result of the adoption of Multi-Criteria Decision Making (MCDM) approaches. Additionally, these WQIs are now more adaptable to a number of environmental circumstances. Despite the fact that there are variations based on location, comparative studies have shown that the underlying mathematical principles that regulate WQIs are consistent worldwide. This is the case even though there are variations dependent on location. WQI research appears to have a bright future ahead of it thanks to the integration of cutting-edge technology such as artificial intelligence and the internet of things. These technologies are able to give evaluations that are both more accurate and more up-to-date. It is essential to engage in ongoing research and to work across disciplines in order to further improve these models and address the growing concerns regarding the environment. Therefore, this will guarantee that water quality indicators (WQIs) continue to be useful tools for the management of water resources, which is necessary in order to guarantee the sustainability of these resources.

References

1. Daughney, C. J., Raiber, M., Moreau-Fournier, M., Morgenstern, U., & Raaij, R. V. D. (2012). Use of hierarchical cluster analysis to assess the representativeness of a baseline groundwater quality monitoring network: Comparison of New Zealand's national and regional groundwater monitoring programs. *Hydrogeology Journal, 20*, 185–200.
2. Trabelsi, R., & Zouari, K. (2019). Coupled geochemical modeling and multivariate statistical analysis approach for the assessment of groundwater quality in irrigated areas: A study from north eastern of Tunisia. *Groundwater for Sustainable Development, 8*, 413–427.
3. Ponsadailakshmi, S., Sankari, S. G., Prasanna, S. M., & Madhurambal, G. (2018). Evaluation of water quality suitability for drinking using drinking water quality index in Nagapattinam District, Tamil Nadu in southern India. *Groundwater for Sustainable Development, 6*, 43–49.
4. Pathak, S., Gupta, S., & Ojha, C. S. P. (2021). Assessment of groundwater vulnerability to contamination with modified DRASTIC index: A case study in Haridwar, Uttarakhand, India. *ASCE Journal of Hazardous, Toxic, and Radioactive Waste, 25*(2), 04020081.
5. Almeida, C., González, S. O., Mallea, M., & González, P. (2012). A recreational water quality index using chemical, physical and microbiological parameters. *Environmental Science and Pollution Research, 19*, 3400–3411.
6. Gani, A., Singh, M., Pathak, S., & Hussain, A. (2024). Groundwater quality index development using ANN model of Delhi Metropolitan City. *India. Environmental Science and Pollution Research*, 1–16.
7. Singh, K. P., Malik, A., Mohan, D., & Sinha, S. (2011). Water quality assessment and its relation to human health: A review. *Stochastic Environmental Research and Risk Assessment, 25*, 761–779.
8. Shukla, B. K., Bashir, M., Sharma, P. K., & Abhimanyu. (2021). An analytical investigation of surface water quality and pollution status in Srinagar, Jammu and Kashmir, India. *Journal of Green Engineering, 11*(1), 952–962.
9. Desa, E., Bhatia, S., Iftikhar, H., & Sajjad, H. (2020). Historical development of the WQI model and its implications on global water quality assessment. *Journal of Hydrologic Engineering, 25*(2), 04019049.
10. Sarkar, C., & Abbasi, S. A. (2006). Qualitative assessment of several surface water and groundwater sources and a tap water source for drinking and irrigation purposes. *Environmental Monitoring and Assessment, 118*, 393–413.
11. Keeley, M., Koburger, A., Dolowitz, D. P., Medearis, D., Nickel, D., & Shuster, W. (2013). Perspectives on the use of green infrastructure for stormwater management in Cleveland and Milwaukee. *Environmental Management, 51*, 1093–1108.

12. Reza, R., & Singh, G. (2010). Assessment of groundwater quality status by using water quality index method in Orissa, India. *World Applied Sciences Journal, 9*, 1392–1397.
13. Shikuku, V. (Ed.). (2023). *Artificial intelligence applications in water treatment and water resource management*. IGI Global.
14. Sharma, D., & Kansal, A. (2011). Water quality analysis of river Yamuna using water quality index in the national capital territory, India (2000–2009). *Applied Water Science, 1*(3–4), 147–157.
15. Hussain, A., Deshwal, A., Priyadarshi, M., Pathak, S., Sambandam, G., Chand, S., & Shukla, A. K. (2024). Landfill leachate analysis from selected landfill sites and its impact on groundwater quality, New Delhi, India. *Environment, Development and Sustainability*, 1–26.
16. Gani, A., Hussain, A., Pathak, S., & Banerjee, A. (2024). An empirical investigation on the elimination of heavy metals using bioremediation method for selected plant species. *Physics and Chemistry of the Earth, 134*, 103568.
17. Mishra, A., & Singh, V. P. (2010). A review of drought concepts. *Journal of Hydrology, 391*(1–2), 202–216.
18. Mukonza, S. S., & Chiang, J. L. (2023). Meta-analysis of satellite observations for united nations sustainable development goals: Exploring the potential of machine learning for water quality monitoring. *Environments, 10*(10), 170.
19. Swart, R., Robinson, J., & Cohen, S. (2003). Climate change and sustainable development: Expanding the options. *Climate Policy, 3*(Suppl. 1), S19–S40.
20. Mladenović-Ranisavljević, I. I., Takić, L., & Nikolić, Đ. (2018). Water quality assessment based on combined multi-criteria decision-making method with index method. *Water Resources Management, 32*, 2261–2276.
21. Debels, P., Figueroa, R., Urrutia, R., Barra, R., & Niell, X. (2005). Evaluation of water quality in the Chillán River (Central Chile) using physicochemical parameters and a modified water quality index. *Environmental Monitoring and Assessment, 110*(1–3), 301–322.
22. Cude, C. G. (2001). Oregon water quality index: A tool for evaluating water quality management effectiveness. *Journal of the American Water Resources Association, 37*(1), 125–137.
23. Singh, S., Singh, V. P., Yadav, S., Kushwaha, V., Kumar, A., & Shukla, B. K. (2023, September). A compendious study on groundwater quality evaluation and monitoring. In *AIP conference proceedings* (Vol. 2800, No. 1). AIP Publishing.
24. Shukla, B. K., Teeli, M. A., Shukla, S. K., Chandra, R., Bharti, N., & Singh, U. (2024). A comprehensive overview of vital water quality parameters. *Lecture Notes in Civil Engineering, 439*, 1–20.
25. Shukla, B. K., Singh, R. D., Singh, N., Singh, S., Burman, Y. K., & Singh, R. (2023). Application of nanomaterials in water purification: A thematic review. *Lecture Notes in Civil Engineering, 281*, 97–113.
26. Almeida, C., González, S., Mallea, M., & González, P. (2017). A review of the use of the basic parameters of water quality as indicators of the presence of priority substances in European rivers. *Science of the Total Environment, 601–602*, 206–219.
27. Brown, R. M., McCleiland, N. I., Deininger, R. A., & O'Connor, M. F. (1972). A water quality index—Do we dare? *Water & Sewage Works, 119*(10), 339–343.
28. Dinius, S. H. (1987). Design of a surface water quality index: The need for a theoretical framework. *Water Research, 21*(8), 963–972.
29. Horton, R. K. (1965). An index number system for rating water quality. *Journal of Water Pollution Control Federation, 37*.
30. Wang, Y., Morgan, R. P., & Gao, J. (2006). Application of a weight of evidence approach in a comparative assessment of multiple environmental data sources. *Environmental Monitoring and Assessment, 116*(1–3), 63–75.
31. Sutadian, A. D., Muttil, N., Yilmaz, A. G., & Perera, B. J. (2016). Development of a water quality index for rivers in West Java Province, Indonesia. *Ecological Indicators, 61*, 574–586.
32. Hajkowicz, S., & Collins, K. (2007). A review of multiple criteria analysis for water resource planning and management. *Water Resources Management, 21*(9), 1553–1566.

33. Akhtar, N., Ishak, M. I. S., Ahmad, M. I., Umar, K., Md Yusuff, M. S., Anees, M. T., et al. (2021). Modification of the water quality index (WQI) process for simple calculation using the multi-criteria decision-making (MCDM) method: A review. *Water, 13*(7), 905.
34. Dalkey, N., & Helmer, O. (1963). An experimental application of the Delphi method to the use of experts. *Management Science, 9*(3), 458–467.
35. Ramirez, J., & Rogers, G. (2002). History and use of zero-valent iron for groundwater remediation. *Groundwater Remediation Technologies Analysis Center, 4*(1), 1–13.
36. Ulusoy, U. (2023). A review of particle shape effects on material properties for various engineering applications: From macro to nanoscale. *Minerals, 13*(1), 91.
37. Seifi, A., Dehghani, M., & Singh, V. P. (2020). Uncertainty analysis of water quality index (WQI) for groundwater quality evaluation: Application of Monte-Carlo method for weight allocation. *Ecological Indicators, 117*, 106653.
38. Prasad, B., & Kumar, P. (2020). A comparative study of various water quality indices for groundwater suitability. *Environmental Geochemistry and Health, 42*, 1483–1500.
39. Shukla, A. K., Ojha, C. S. P., & Garg, R. D. (2017). Application of overall index of pollution (OIP) for the assessment of the surface water quality in the upper ganga river basin, India. In *Development of water resources in India* (pp. 135–149). Springer International Publishing. https://doi.org/10.1007/978-3-319-55125-8_12
40. Gupta, A. K., & Gupta, S. K. (2021). Development of a water quality index (WQI) for river water quality assessment. *Environmental Monitoring and Assessment, 193*(4), 1–13.
41. Swamee, P. K., & Tyagi, A. (2000). Describing water quality with aggregate index. *Journal of Environmental Engineering, 126*(5), 451–455.
42. Brown, R. M., McClelland, N. I., Deininger, R. A., & Tozer, R. G. (1970). A water quality index-do we dare. *Water and Sewage Works, 117*(10), 339–343.
43. Sutadian, A. D., Muttil, N., Yilmaz, A. G., & Perera, B. J. C. (2016). Development of river water quality indices—A review. *Environmental Monitoring and Assessment, 188*, 1–29.
44. Brown, R. M., McClelland, N. I., Deininger, R. A., & Landwehr, J. M. (1973, January). Validating the WQI. In *Proceedings of the National Meeting of American Society of Civil Engineers on water resources engineering, Washington, DC, USA, 29 January–2 February 1973* (Vol. 29).
45. Shukla, A. K., Ojha, C. S. P., Mijic, A., Buytaert, W., Pathak, S., Garg, R. D., & Shukla, S. (2018). Population growth, land use and land cover transformations, and water quality nexus in the upper Ganga River basin. *Hydrology and Earth System Sciences, 22*(9), 4745–4770. https://doi.org/10.5194/hess-22-4745-2018
46. Uddin, M. G., Nash, S., & Olbert, A. I. (2021). A review of water quality index models and their use for assessing surface water quality. *Ecological Indicators, 122*, 107218.
47. Abdul Maulud, K. N., Fitri, A., Wan Mohtar, W. H. M., Wan Mohd Jaafar, W. S., Zuhairi, N. Z., & Kamarudin, M. K. A. (2021). A study of spatial and water quality index during dry and rainy seasons at Kelantan River basin, peninsular Malaysia. *Arabian Journal of Geosciences, 14*, 1–19.
48. Shukla, B. K., Gupta, L., Parashar, B., Sharma, P. K., Sihag, P., & Shukla, A. K. (2025). Integrative assessment of surface water contamination using GIS, WQI, and machine learning in urban–industrial confluence zones surrounding the National Capital Territory of the Republic of India. *Water, 17*(7), 1076. https://doi.org/10.3390/w17071076
49. Dahiya, S., Singh, B., Gaur, S., Garg, V. K., & Kushwaha, H. S. (2007). Analysis of groundwater quality using fuzzy synthetic evaluation. *Journal of Hazardous Materials, 147*(3), 938–946.
50. Ocampo-Duque, W., Ferre-Huguet, N., Domingo, J. L., & Schuhmacher, M. (2006). Assessing water quality in rivers with fuzzy inference systems: A case study. *Environment International, 32*(6), 733–742.
51. Shukla, B. K., Sharma, P. K., & Khan, M. A. (2020, May). Physico-chemical study of some surface water bodies of Punjab. In *Journal of physics: Conference series* (Vol. 1531, No. 1, p. 012121). IOP Publishing.
52. Ross, T. J. (2005). *Fuzzy logic with engineering applications*. John Wiley & Sons.

53. Page, M. J., McKenzie, J. E., Bossuyt, P. M., Boutron, I., Hoffmann, T. C., Mulrow, C. D., Shamseer, L., Tetzlaff, J. M., Akl, E. A., Brennan, S. E., & Moher, D. (2021). The PRISMA 2020 statement: An updated guideline for reporting systematic reviews. *British Medical Journal, 372*.
54. Shukla, A. K., Ojha, C. S. P., Shukla, S., & Garg, R. D. (2021). Water quality challenges in Ganga River basin, India. In *The Ganga River basin: A hydrometeorological approach* (pp. 1–19). https://doi.org/10.1007/978-3-030-60869-9_1
55. Lu, R. S., & Lo, S. L. (2002). Diagnosing reservoir water quality using self-organizing maps and fuzzy theory. *Water Research, 36*(9), 2265–2274.
56. Wu, H. Y., Chen, K. L., Chen, Z. H., Chen, Q. H., Qiu, Y. P., Wu, J. C., & Zhang, J. F. (2012). Evaluation for the ecological quality status of coastal waters in East China Sea using fuzzy integrated assessment method. *Marine Pollution Bulletin, 64*(3), 546–555.
57. Misaghi, F., Delgosha, F., Razzaghmanesh, M., & Myers, B. (2017). Introducing a water quality index for assessing water for irrigation purposes: A case study of the Ghezel Ozan River. *Science of the Total Environment, 589*, 107–116.
58. Deininger, R. A., & Landwehr, J. M. (1971). *A water quality index for public water supplies*. Department of Environment and Industrial Health, University of Michigan.
59. Tyson, J. M., & House, M. A. (1989). The application of a water quality index to river management. In *Urban discharges and receiving water quality impacts* (pp. 175–186). Pergamon.
60. Singh, S., Singh, V. P., Garg, P., & Shukla, B. K. (2023, February). Physico-chemical analysis of groundwater in Noida-Ghaziabad region. In *IOP conference series: Earth and environmental science* (Vol. 1110, No. 1, p. 012028). IOP Publishing.

Climate Change Challenges in Western Rajasthan and the Adaptive Response of Haveli Planning and Materials

Ruma Kalla, Vivek Kalla, and Ravish Kumar

1 Introduction

Rajasthan state of India, which is located in the northwestern part of the country, is known for its arid climate and huge desert landscapes, most notably the Thar Desert. This region faces different challenges arising from its physical and climatic situations, which are worsened by the effects of climate change. Among these concerns, rising temperatures, irregular rainfall patterns, and water scarcity stand out as major risks to both the environment and the livelihoods of the region's habitat. Climate change has significantly altered the climatic conditions in Western Rajasthan over the period of time, leading to a rise in temperatures especially during the last few decades. Extreme summers and chilly winters characterize the climate of this region, with intense heat waves becoming more frequent and prolonged due to climate change. Increased temperatures not only affect human health but also aggravate water scarcity and agricultural challenges, thereby amplifying the vulnerability of communities reliant on these resources for sustenance. Erratic rainfall patterns further compound the challenges faced by Western Rajasthan. The region predominantly relies on the southwest monsoon for its annual precipitation, which is highly variable and often insufficient to meet the water demands of its inhabitants. Climate change has heightened the unpredictability of monsoon rainfall, leading to irregularities in its onset, duration, and intensity. Therefore, prolonged droughts and flash

R. Kalla (✉) · R. Kumar
Department of Architecture and Planning, National Institute of Technology, Patna, Bihar, India
e-mail: rumak.ph21.ar@nitp.ac.in; ravish@nitp.ac.in

V. Kalla
Department of Architecture, Sri Venkateshwara College of Architecture, Hyderabad, Telangana, India

S. Pathak et al. (eds.), *Geo-Data Revolution*, Advances in Geographic Information Science, https://doi.org/10.1007/978-3-031-94078-1_19

floods have become more common, disturbing agricultural activities and intensifying water scarcity in the region of study.

A recurring problem in Western Rajasthan, water scarcity has been made worse by climate change-related causes like decreased surface water availability and diminishing groundwater levels. The reliance on groundwater for irrigation and domestic needs has resulted in overexploitation of aquifers, further diminishing already limited water resources. Furthermore, the water shortage problems have been made worse by the deterioration of crises and the attack of water bodies, which presents serious problems to the region's sustainability and water management arrangements. In response to these issues, traditional architectural practices, particularly those demonstrated in haveli planning, provide unique insights into adaptive solutions for mitigating the effects of climate change and fostering resilience in this part of Rajasthan's built environment. Havelis, or the dwellings with traditional courtyards throughout the region, have developed over ages to tackle the harsh climatic conditions of the deserted lands of Rajasthan.

The architectural design of havelis features various elements aimed at enhancing thermal comfort and water management. Thick mud walls, central courtyards, and jharokhas (overhanging balconies) facilitate natural ventilation and passive cooling, helping to alleviate the effects of extreme heat. Additionally, havelis often incorporate water harvesting and storage systems, such as tankas (underground water storage tanks) and khadins (rainwater harvesting structures), to cope with water scarcity and ensure a reliable water supply throughout the year. Overall, the adaptive response of haveli planning and materials to the climate challenges of Western Rajasthan underscores the importance of integrating traditional knowledge and practices into contemporary approaches to sustainable development. By utilizing the lessons learned from traditional architecture, policymakers, planners, and communities, they can enhance the resilience of the built environment and mitigate the adverse effects of climate change on the region's inhabitants and environments.

2 Challenges of Climate Change

2.1 Main Climatic Challenges of Rajasthan

The climate in the western region of Rajasthan is hot and arid, with high temperatures, frequent droughts, and minimal or no rainfall. One of the world's most populous deserts, the Thar Desert, includes the area. With major effects on water supplies, agriculture, and livelihoods, climate change is making matters worse for residents in Western Rajasthan, thus demanding for sustainable solutions.

2.2 *Irregular Rainfall Patterns and Shortage of Water*

One of the most challenging situations in Rajasthan region is the increasing changes in the rainfall patterns. Rainfall in the area of study is irregular and occasional, with most of it falling between July and September during the monsoon season. However, monsoon patterns are becoming increasingly unpredictable due to change in climate, with some regions seeing extended droughts and others seeing more extreme, unexpected rainfall incidents [1].

The erratic nature of rainfall establishes significant challenges for water availability and agricultural status in the western part of Rajasthan. Rainfed agriculture is the primary source of livelihood for many communities in the region, relying heavily on the monsoon rains for crop cultivation. Erratic rainfall patterns not only affect crop yields but also contribute to water scarcity, as groundwater recharge rates decline during periods of drought especially in summers.

2.3 *Increasing Heat Stress and Rising Temperatures*

The climate of Western Rajasthan is harsh, with hot summers and cold winters. Rising temperatures are causing longer heat waves and more frequent hot days, which is making heat stress in the area worse due to climate change. In addition to having an adverse effect on human health, high temperatures also have an effect on energy use, livestock well-being, and agricultural productivity.

The increasing temperatures in Western Rajasthan are also hastening the progress of desertification. Soil deterioration and vegetation loss result from rising temperatures, which also cause a drop in soil moisture levels. Desertification worsens water scarcity and soil deterioration, making sustainable development in the region very difficult [3, 4].

Figure 1 represents the parameters associated with health impacts due to exposure to extreme heat conditions in the region of study.

2.4 *Haveli Planning and Response to Climate Challenges Through Materials*

Western Rajasthan's ancient courtyard homes, known as havelis, are made to withstand the severe weather of the area. Centuries of environmental adaptation are reflected in the architectural style and building materials of havelis, which offer inhabitants' social cohesiveness, water management, and thermal comfort. The subsequent sections explore the ways in which haveli materials and planning have changed to address Western Rajasthan's climate-related issues [5, 6].

Fig. 1 Heat wave condition and heat stroke symptoms and repercussionReference [2] Impacts of 1.5 °C and 2 °C global warming on regional rainfall and temperature change across India, Environmental Research Communications

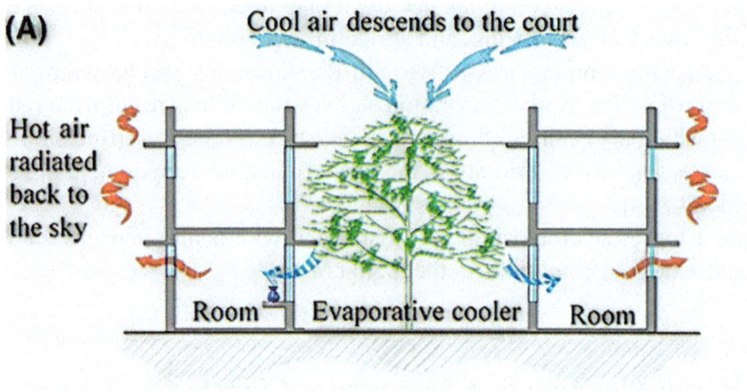

Fig. 2 Section of a courtyard of Haveli

2.5 *Passive Cooling Techniques and Its Application in Havelis*

Havelis maximize natural ventilation and passive cooling by utilizing architectural features like thick mud walls, jharokhas (overhanging balconies), and courtyards. In addition to serving as the haveli's main point, the central courtyard encourages cross-ventilation and airflow throughout the living spaces (Fig. 2). Jharokhas provide shade and sun protection while allowing the air to circulate freely inside the building, supporting patterns of natural ventilation [7].

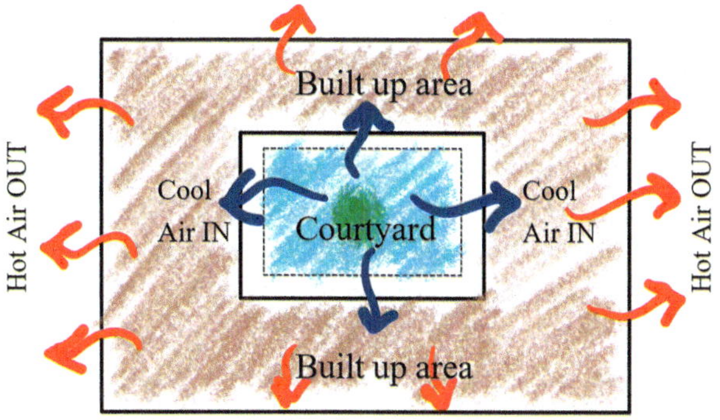

Fig. 3 Plan of a courtyard of Haveli

The thermal mass of walls of haveli, which are made from locally produced materials like mud bricks and lime plaster, helps to manage indoor temperatures (Fig. 3). During the day, the walls absorb heat from the sun and keep the interiors cool. As temperatures drop at night, the walls release the stored heat, allowing indoor temperatures to remain comfortable and pleasant.

2.6 *Water Harvesting in the Region and Its Storage Techniques*

Havelis in Western Rajasthan have always used water collecting and storage methods to deal with water constraint in the region due to its hot and dry climatic conditions. Rooftop catchment systems, also known as tankas or khadins, gather rainfall during the monsoon season and store it underground in tanks or reservoirs. The proposed methods boost local houses to collect and store rainwater for domestic consumption usage, livestock watering purposes, and crop irrigation during the hot and dry times of the year.

Construction of havelis/traditional dwellings with native material resources such as stone, mud, and lime plaster supports sustainability and water conservation in the region. Since the mentioned materials are porous, precipitation can seep into the soil, replenishing groundwater supplies and fostering plant growth. To improve water collection through catchment and reduce runoff of waters, havelis' designs usually have flat roofs with a slight slope toward the gutters that direct rainfall into storage tanks [8].

2.7 *Community Resilience and Social Cohesion*

The planning of havelis promotes and encourages social cohesion and community resilience to oppose climate challenges. Central courtyards serve as gathering spaces for family members and neighbors, fostering social interactions and support networks. During periods of extreme heat, the courtyard provides shade and a cool refuge, allowing residents to mitigate the effects of high temperatures while maintaining social connections [9].

The collective ownership and management of water resources in haveli communities enhance resilience to water scarcity and promote equitable access to water. Community-based water management practices, such as the maintenance of traditional water harvesting structures and the sharing of water resources among households, strengthen social cohesion and adaptive capacity in Western Rajasthan.

3 Dataset Collected and Used

The average rainfall in Western Rajasthan is provided in Table 1.

The pattern of rising temperatures in the area is highlighted in Table 2, which shows a consistent rise in the average temperatures in Western Rajasthan during the previous 20 years.

Table 3 enhances comprehension of the climate change concerns in the region and the adaptive response of haveli planning and materials by providing graphic representations of climatic patterns, water harvesting techniques, and community resilience in Rajasthan [10].

4 Results and Discussion

Findings:

1. **Trends observed in annual rainfall in Western Rajasthan**:
   - The average yearly rainfall in Western Rajasthan has steadily decreased during the past 20 years.
   - The average rainfall decreased gradually from 250 mm to 180 mm between 2000 and 2020.
   - Given the region's increased water scarcity and agricultural concerns, this decline implies a decrease in precipitation levels, thus representing clear scarcity of water.

Table 1 Average annual rainfall in Western Rajasthan

Year	Average rainfall (mm)
2000	250
2010	200
2020	180

Table 2 Table illustrating temperature trends in Western Rajasthan in the past two decades

Year	Average temperature (°C)
2000	30.5
2001	31.2
2002	31.8
2003	32.1
2004	32.5
2005	32.7
2006	33.2
2007	33.6
2008	34.0
2009	34.3
2010	34.8
2011	35.2
2012	35.5
2013	35.9
2014	36.2
2015	36.6
2016	37.0
2017	37.4
2018	37.8
2019	38.1
2020	38.5

Table 3 Traditional water harvesting structures in havelis

Structure	Function
Tanka	Rooftop catchment system for rainwater
Khadin	Earthen embankment for rainwater harvesting
Baori	Stepwell for groundwater recharge

2. **Temperature patterns observed in Western Rajasthan region**:

   – The study of temperature data shows a continual rise in average temperatures across this region of Rajasthan as observed from 2000 to 2020 (Fig. 4).
   – Throughout the duration, the temperatures have progressively increased, with the average temperature hiking from 30.5 °C in 2000 to 38.5 °C in the year 2020; thus, a variation of considerable 8 °C was recorded.
   – This ascending trend in temperatures shows a heating pattern, which carries significant effects for human health, agriculture, and environments in the region of study.

3. **Community resilience index in the region of havelis:**

   – The general level of community resilience in haveli area communities is moderate, according to assessments in the region of study.

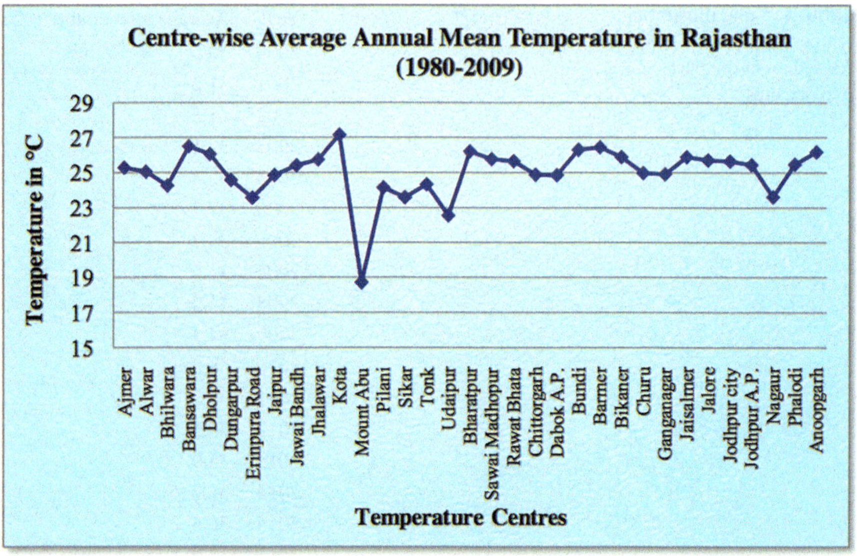

Fig. 4 Temperature variation observation in RajasthanSource: https://www.thefreelibrary.com/Climatic+Change+in+Rajasthan+(1980-2009)%3A+Case+Study+of+Desert...-a0558368309

- The main factors associated with resilience include social unity, water management, and adaptive capacity within the region of havelis.
- There is room for improvement, nevertheless, especially in bolstering community-based adaptation strategies and boosting water conservation initiatives.

5 Discussion

1. **Decreasing rainfall and its impact in Western Rajasthan:**
   - The falling trend in average annual rainfall poses significant problems to water supply and agricultural output in Western Rajasthan (Fig. 5).
   - Reduced precipitation levels can increase the danger of drought, water shortages, and desertification, all of which have a substantial impact on the livelihoods of populations that rely on rainfed agriculture.
   - Implementing effective water conservation practices, such as rainwater harvesting and efficient irrigation systems, is critical to mitigating the negative effects of decreased rainfall on water resources in the region of study with considerable solutions needed [11].

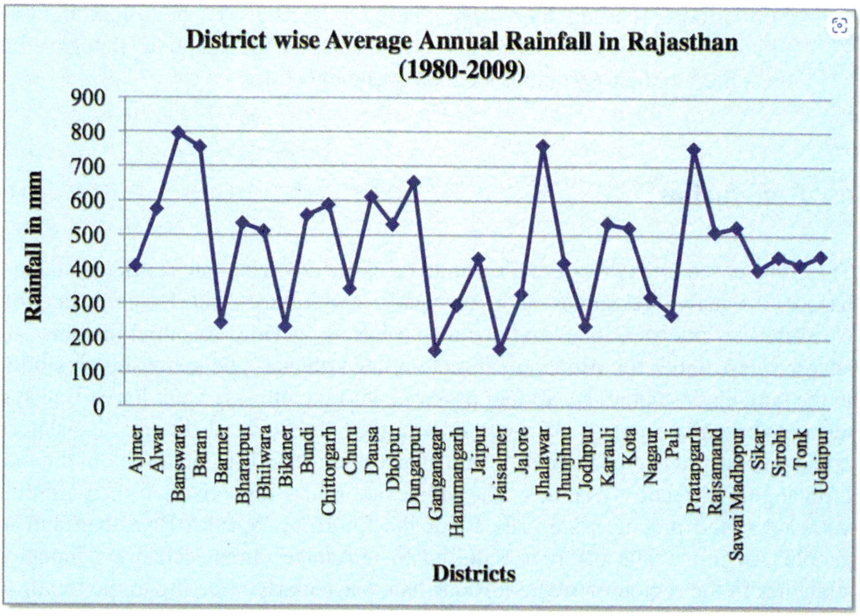

Fig. 5 Average annual rainfall observed in the district of Rajasthan Source: https://www.thefreelibrary.com/Climatic+Change+in+Rajasthan+(1980-2009)%3A + Case+Study+of+Desert...-a0558368309

2. **Consequences of Rising Temperatures:**

   – The persistent rise in temperatures reported in Western Rajasthan has serious consequences for a variety of sectors, including human health, agriculture, and environment [12].
   – Elevated temperatures may worsen heat-related health concerns and increase consumption of energy for cooling.
   – Agriculture is especially vulnerable, as rising temperatures can reduce crop output and cattle health.
   – Developing heat-resistant crop varieties and using sustainable land management techniques are two adaptation strategies that are crucial to addressing these issues and increasing Western Rajasthan's resilience to the increasing climate change [13, 14].

3. **Improving Community Resilience**:

   – The evaluation of community resilience emphasizes how important social cohesiveness and adaptability are to successfully addressing the effects of climate change.
   – Resilience to water scarcity and associated issues can be improved by bolstering community-based water management strategies, such as cooperative water harvesting and storage projects.

– Furthermore, funding knowledge-sharing projects and capacity-building programs can enable communities to create and execute adaptive strategies that are suited to their particular local environment [15].

6 Conclusion

The climate issues that Western Rajasthan faces are deterring, presenting significant hazards to native resources such as water, agriculture, and local occupation. Nevertheless, the traditional architecture of haveli provides essential insights into adaptive techniques for mitigating climate-related impacts and increasing flexibility in the built environment. By adding passive cooling methods, water harvesting systems, and locally sourced materials into current architectural designs, legislators, urban planners, and communities can foster resilience and sustainability in the face of changing climatic dynamics. The preservation and revitalization of traditional knowledge and practices are critical for the future of Western Rajasthan and its people. An immediate effort is required to introduce climate change adaptation strategies in the region of Western Rajasthan, particularly referring to the trends of scanty rainfall and rising temperatures. Highlighting sustainable water management practices and customized adaption techniques is critical to mitigating the adverse effects of climate change and advocating long-term sustainability. As a summary, Western Rajasthan is at a serious juncture from climatic shift, needing coordinated actions to strengthen community resilience and protect perishable vital resources. By strengthening ancient practices and techniques and combining modern solutions, the region is able to overcome the challenges of climate change and lead the path for an environmentally friendly sustainable future.

References

1. Rawat, A., Kumar, D., & Khati, B. S. (2024). A review on climate change impacts, models, and its consequences on different sectors: A systematic approach. *Journal of Water and Climate Change, 15*(1), 104–126. https://doi.org/10.2166/wcc.2023.536
2. Yaduvanshi, A., Zaroug, M., Bendapudi, R., & New, M. (2019). Impacts of 1.5 °C and 2 °C global warming on regional rainfall and temperature change across India. *Environmental Research Communications, 1*(2), 125002.
3. Kaushik, G., & Sharma, K. C. (2015). Climate change and rural livelihoods-adaptation and vulnerability in Rajasthan, India. *Global NEST Journal, 17*(1), 41–49.
4. Bhushan, M. (2014). Accessing the climate change awareness, action planned and suggestions among an internet literate population. *Journal of Global Communication, 7*(1), 55–60.
5. Desai, S., Manapragada, N. V. S. K., Shukla, A. K., & Pignatta, G. (2022). Mould-growth study in building materials exposed to warm and humid climate using heat and mass transfer (HAMT) EnergyPlus simulation method. *Sustainability, 14*(14), 8292. https://doi.org/10.3390/su14148292

6. Midudula, G. R., Shukla, A. K., & Javed, B. (2024). Assessing the wall material for energy efficiency and cost efficiency in warm and humid climate: A case study. In *Sustainable development and geospatial technology: Volume 1: Foundations and innovations* (pp. 171–183). Springer Nature Switzerland. https://doi.org/10.1007/978-3-031-65683-5_9
7. Siva Barathi, A., Manapragada, N. V. S. K., Rai, P. K., Shukla, S., & Shukla, A. K. (2024). Artificial intelligence and machine learning-based building solutions: Pathways to ensure occupant comfort and energy efficiency with climate change. In *Big data, artificial intelligence, and data analytics in climate change research: For sustainable development goals* (pp. 57–79). Springer Nature Singapore. https://doi.org/10.1007/978-981-97-1685-2_4
8. Sharma, C., Shukla, A. K., & Zhang, Y. (2021). Climate change detection and attribution in the ganga-Brahmaputra-Meghna river basins. *Geoscience Frontiers, 12*(5), 101186. https://doi.org/10.1016/j.gsf.2021.101186
9. Gupta, S. K., & Kanga, S. (2024). Optimizing land use for climate mitigation using nature based solution (NBS) strategy: A study on afforestation potential and carbon sequestration in Rajasthan, India. *Case Study, 2*, 36.
10. Kalla, R., & Kumar, R. (2025). Vernacular materials for thermal comfort. *Engineering, Technology & Applied Science Research, 15*(1), 20229–20234. https://doi.org/10.48084/etasr.8524
11. Choudhary, K., & Kumar, R. (2024). Predictions of groundwater variations using regression analysis in Delhi. *SSRG International Journal of Civil Engineering, 12*(1), 36–45. https://doi.org/10.14445/23488352/IJCE-V12I1P105
12. Singh, S., Singh, S. K., Kanga, S., Shrivastava, P. K., Kumar, P., & Meraj, G. (2024). Hydrological dynamics of the Kalisindh and Parbati Rivers: An integrated analysis in the context of the eastern Rajasthan Canal project (ERCP). *Result in Engineering, 21*, 101960.
13. Jomaa, M., Abbes, M., Tadeo, F., & Mami, A. (2019). Greenhouse modeling, validation and climate control based on fuzzy logic. *Engineering, Technology & Applied Science Research, 9*(4), 4405–4410. https://doi.org/10.48084/etasr.2871
14. Minimizing carbon footprint via microalgae as a biological capture—ScienceDirect. Retrieved 2 April 2024, from https://www.sciencedirect.com/science/article/pii/S2772656821000075.
15. Shukla, S., Deeksha Chand, S., Rai, P. K., Shukla, A. K., Misra, R. K., & Singh, U. (2024). Global and Indian perspectives on sustainable development, remote sensing, and climate change: An overview. *Natural Resources Management and Sustainable Livelihoods in the Mountainous Region: Evidence, Gap and Future Strategies*, 73–83. https://doi.org/10.1007/978-981-97-2100-9_6

Estimation of Surface Runoff of Satluj Sub-basin of Bilaspur and Parts of Mandi, Shimla, Solan, Himachal Pradesh, India

Saloni Bauddh, Roshan Mahto, and Neha G. Paswan

1 Introduction

Rainfall, surface runoff, and curve number (CN) method are the fundamental components in the study and management of watersheds. A storm event's rainfall is the primary feature of hydrological cycle, representing the amount of water that falls on a specific area during a storm, encouraging hydrological processes within the watershed, such as soil infiltration, groundwater recharge, and surface runoff.

Surface runoff is the part of rainfall, which flows on land surface towards streams, rivers, and other water bodies, rather than infiltrating into soil. Different land uses (e.g., agricultural lands, built-up land, forestland) affect runoff behavior. Soils with different infiltration rates (e.g., sandy soils vs. clay soils) have different runoff potentials. Generally, steeper slopes contribute to high runoff volumes. Soil moisture condition before rainfall affects the quantity of runoff generated.

The conversion of rainfall into runoff across a catchment area is an intricate hydrological process because it operates in a nonlinear manner, changes over time, and is dispersed across space. Many models have been advanced by researchers to simulate this process over the course of time.

The SCS-CN method is an empirical technique for estimating direct runoff using rainfall events based on antecedent moisture conditions, land use, and soil type. This method is accepted and used widely all across the world for calculating surface

S. Bauddh · R. Mahto (✉)
Department of Civil and Environmental Engineering, Indian Institute of Technology Patna, Bihta, Patna, Bihar, India
e-mail: 2201ce56_saloni@iitp.ac.in; 2201ce53_roshan@iitp.ac.in

N. G. Paswan
Department of Civil Engineering, Indian Institute of Technology Ropar, Rupnagar, Punjab, India
e-mail: neha.23cez0014@iitrpr.ac.in

S. Pathak et al. (eds.), *Geo-Data Revolution*, Advances in Geographic Information Science, https://doi.org/10.1007/978-3-031-94078-1_20

runoff. Curve number has no dimension, and it ranges between 0 and 100 that reflects the runoff potential of an area. AMC_{II} is the standard condition for CN calculations, representing average soil moisture conditions [1].

The model is capable of evaluating the runoff depth for each land-use class exploiting different antecedent moisture condition (AMC) to get runoff depth layer and calculating the highest runoff rate. Modified SCS-CN methodology for Indian condition has been used for estimating surface runoff of the area of interest in this chapter.

For analyzing different types of hydrological processes, temporal and spatial data are required like catchment area calculation, watershed delineation, filling of elevation grid, and determination of flow directions. Integrated use of GIS and remote sensing techniques is common in evaluating hydrological limits.

Watershed, which is also called drainage basin, is a geographic area that drains all precipitation to a common outlet, such as ocean, river, or lake. These are the primary units for hydrological studies, enabling the analysis of water flow, sediment transport, and pollutant dispersion. Watershed management involves the planning and implementation of practices to protect and enhance water resources within the watershed. Watersheds are delineated using topographic data, often derived from digital elevation models (DEMs), to identify drainage divides and flow directions [2].

2 Area of Interest

A specific area of the Satluj River Basin was selected for study, encompassing most of Bilaspur district and including sections of Solan, Mandi, and Shimla districts of Himachal Pradesh (Fig. 1). This area of interest lies between 31.095833°N to 31.521501°N and 76.687500°E to 77.233333°E (Fig. 1). The area of interest has a geographical area of 1098.769 km^2.

The Satluj River rises near Rakshastal, which is fed by Lake Manasarovar at an elevation level (EL) of about 4572 m in the Tibetan Plateau [3]. The river has a total length of around 1450 km, and its basin area (Indian part) is equal to 22,275 km^2 over altitudes ranging in between 500 and 7000 m [4]. It flows through several states of India including Himachal Pradesh, Punjab, and Haryana. In this study, we used rainfall data (daily, monthly) from June 2023 to September 2023 covering up monsoon season of the Satluj Basin in Himachal Pradesh [5].

Satluj River Basin's land use is a mix of forests, agricultural lands, and urban areas, each with distinct runoff characteristics. Deforestation and urban expansion in certain areas have altered the natural land cover, increasing the surface runoff potential and erosion. Soil erosion, a significant issue in this hilly terrain, affects soil infiltration rates and increases surface runoff. It is estimated that the region of interest receives approximately 1106.28 mm of rainfall (about 81.5% during monsoon) annually [6].

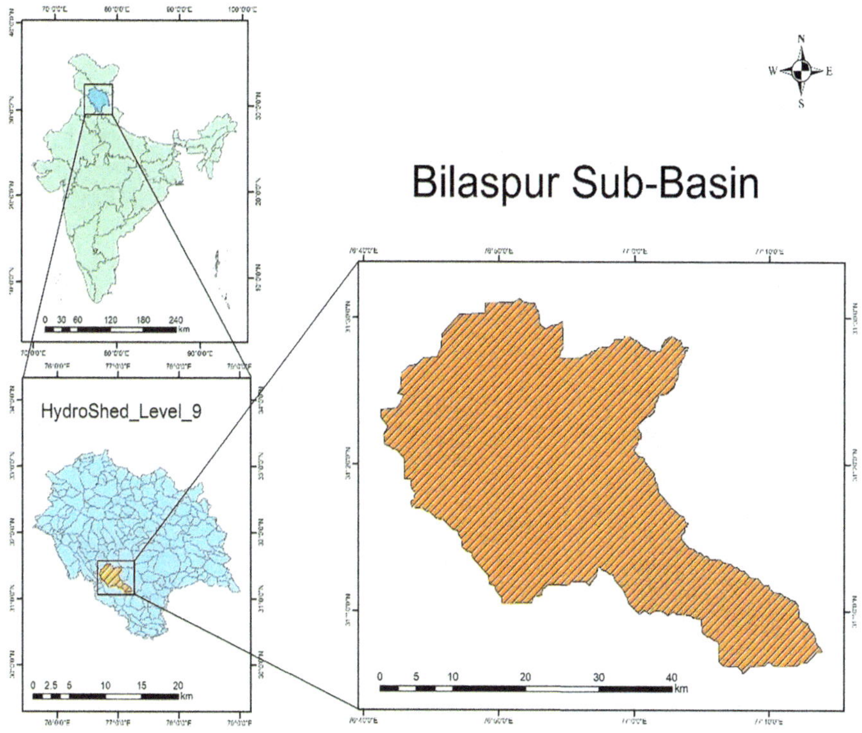

Fig. 1 Location of region of interest

3 Methodology and Dataset

The methodology followed here is highlighted in Fig. 2 that shows flowchart reflecting distinct steps leading towards the final estimation of the surface runoff. USGS website was used to download Landsat-8 satellite images with a 30 m × 30 m resolution for the required area of interest [8]. Then the seven bands of these images were layerstacked, and mosaic of the stacked image was done in ERDAS Imagine 2014. Subsequent clipping and supervised classification resulted in the formation of the final LULC map of area of interest for the year 2023.

Soil data of the region of interest was downloaded by using the Food and Agriculture Organization (FAO) website [9], and rainfall data (daily and monthly) from first June 2023 to 30th September 2023 was collected from India-Water Resources Information System [10]. The hydrobasin map of the world was downloaded from HydroSHEDS website [11], and then georeferencing and clipping of the area of interest were done in ArcGIS 10.7.1 software. The hydrobasin level 9 classification (out of its 12 classification levels) was used for clipping the area of interest.

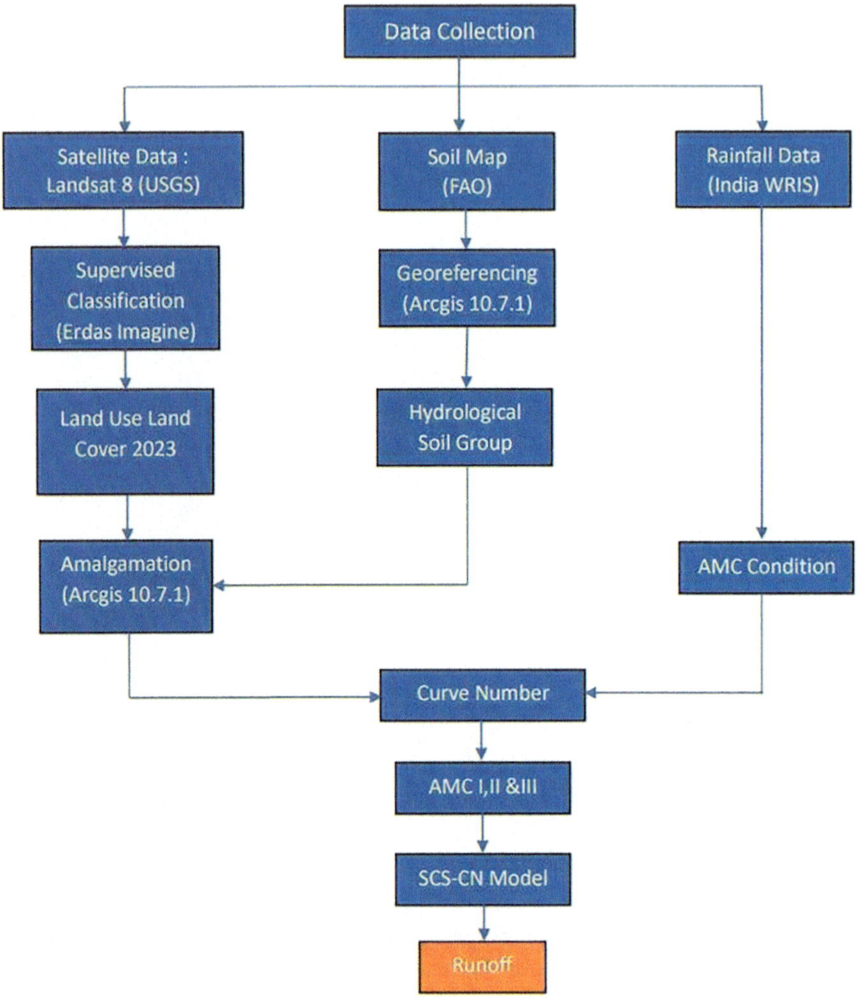

Fig. 2 Method for calculating surface runoff [7]

SRTM data with a 30 m × 30 m resolution was also downloaded from the USGS website [8] and used to prepare a slope map [12]. The downloaded data was used to delineate the watershed. The soil types within the watershed are categorized into hydrologic soil classes C and D on the basis of their infiltrating properties.

The LULC map and obtained HSG maplayer were overlaid in Arcgis10.7.1 and were used as features for the Raster Calculator tool (analysis tool in Arcgis10.7.1) to obtain polygonal shape-area of the soil type (HSG) below individual LULC category. Curve numbers for each derived polygon were established according to SCS-CN method standards. Weighted curve numbers (CN_w) were then generated for the three antecedent moisture conditions (AMC_I, AMC_{II}, AMC_{III}). Using these calculations, the surface runoff was estimated with the SCS-CN method's equation.

4 Assessment of Surface Runoff

4.1 SCS-CN Model

The United States Department of Agriculture (USDA) and the Natural Resources Conservation Service (NRCS) developed the SCS-CN method in the early 1950s. It is a widely accepted and frequently used method for runoff estimation. The methodology comprises using simple empirical formulae, curves, and tables advanced by soil conservation service. The model is also called as the curve number(CN) method. According to the NRSC TR 55, CN values are calculated for various soil textures and land covers. It is being used for the calculation of surface runoff [13]. Water balance equation is the base of this method.

Infiltration, plant interceptions, evaporation, and surface storage that takes place preceding to runoff, all are water losses which contribute to initial abstractions. Generally, its value is considered to be 0.2S. Only those rainfalls are considered for runoff estimation in which the depth of rainfall is more than 0.2S; otherwise, it is not considered for runoff estimation.

This method comprises the below equations [14, 15]:

$$Q = \frac{(P - 0.2S)^2}{(P + 0.8S)}$$

when $P > 0.2S$; otherwise, $Q = 0$.

Here, depth of rainfall in mm is represented by P, depth of surface runoff in mm is represented by Q, and depth of maximum retention in mm is represented by S.

S could be calculated by using the below formula:

$$S = \frac{25{,}400}{\mathrm{CN}} - 254$$

Range of curve number is represented by CN.

5 Result and Discussion

5.1 LULC

The LULC map (Fig. 3) gives a detailed breakdown of types of areas within the study region, including agricultural land, roads, built-up areas, water bodies, settlements, wastelands, and forested areas along with their respective areas in km^2. The area of interest has a geographical area of 1098.769 km^2. Detailed GIS analysis has been done in ArcMap 10.7.1 and ERDAS Imagine 2014 software and has revealed

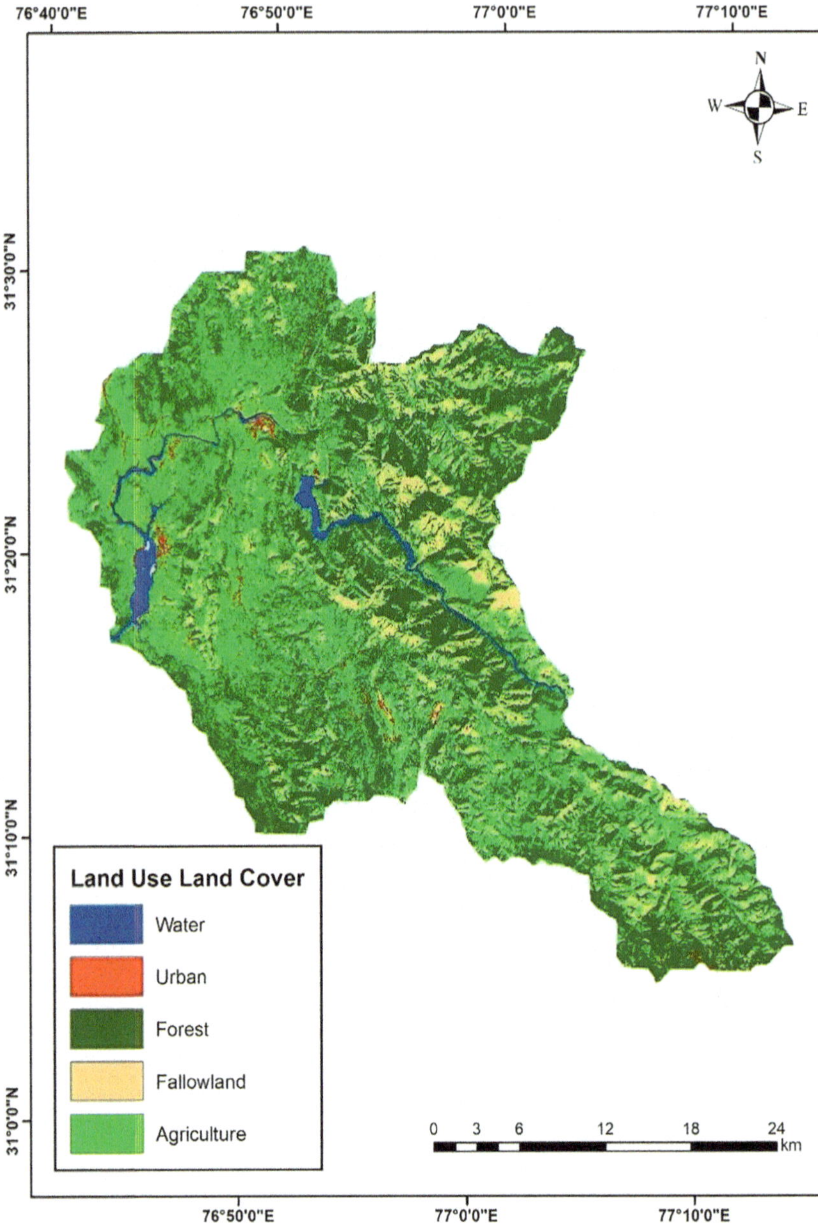

Fig. 3 LULC map

specific areas covered by different categories: forested land spans 536.13 km^2, agricultural land covers 366.88 km^2, fallow land occupies 160.88 km^2, built-up areas extend over 20.21 km^2, and water bodies encompass 23.66 km^2. Weighted curve number will be calculated using this data for the required area of interest.

5.2 Hydrologic Soil Group (HSG)

Hydrologic soil groups (HSGs) are classifications used in hydrology and soil science to categorize soils based on their runoff potential and infiltration characteristics (Fig. 4). These groups are crucial in determining how water moves

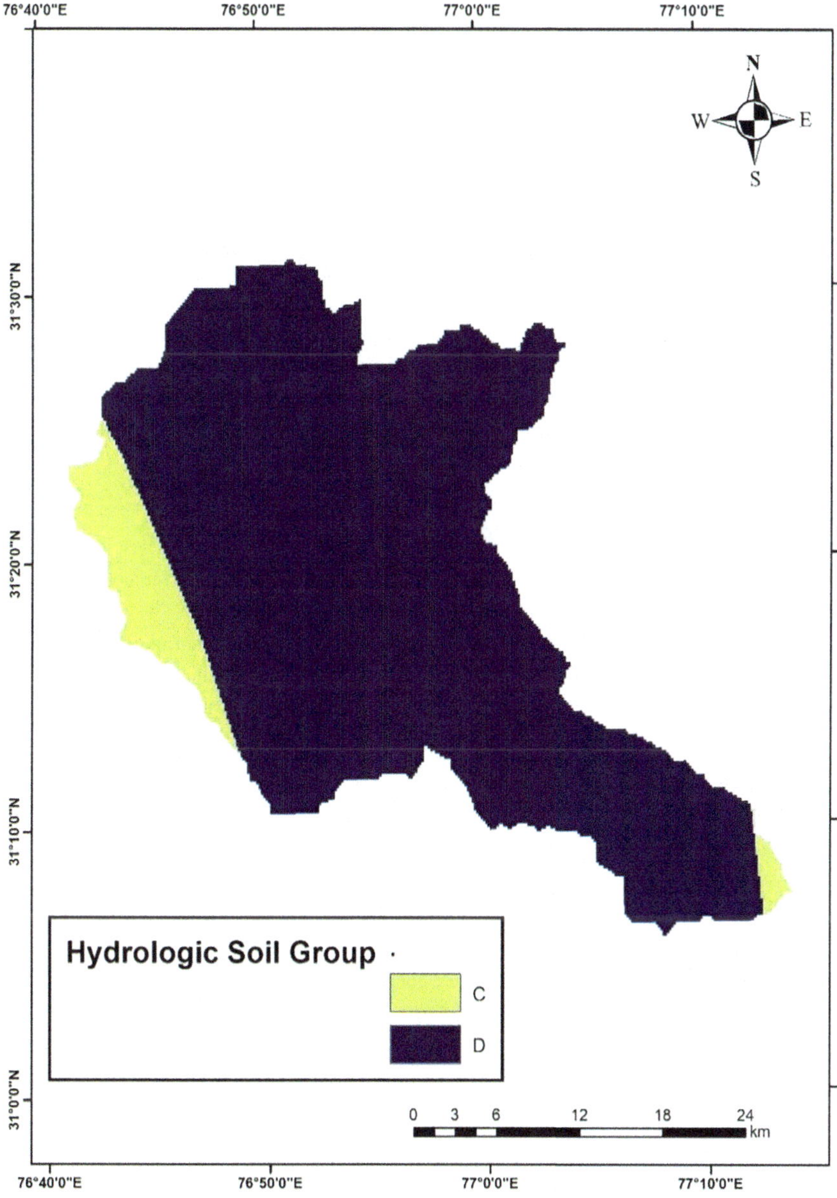

Fig. 4 Hydrological soil group map

through and interacts with the soil, which is vital for various engineering and environmental applications, including stormwater management, land development, and erosion control.

The United States Natural Resources Conservation Service (NRCS) developed a widely used classification criterion that categorizes soils into four main categories: A, B, C, and D on the basis of their properties related to runoff and infiltration (Table 1). Soil texture data (Table 2) sourced from FAO was processed, projected, georeferenced, and digitized using ArcMap 10.7.1 software.

5.3 *Antecedent Moisture Condition (AMC)*

Antecedent moisture condition (AMC) shows the amount of moisture content that exists inside soils at a specific time [17] and is critical for deciding the curve number (CN). The soil's moisture conditions in the sub-basin prior to a storm event are a crucial factor affecting surface runoff. The soil conservation service (SCS) defined three AMC levels (AMC_I, AMC_{II}, AMC_{III}) on the basis of soil moisture and rainfall during dormant and growing seasons [18]. This classification was done on the basis of 5-day antecedent rainfall. Table 3 details these classifications. For this study, we used AMC_{II}, representing avg. soil water content, to calculate curve number for watershed.

Table 1 Hydrologic classification of soil (USDA 1974) [16]

HSG	Soil textures	Runoff potential	Water transmission	Infiltration
A	Sand, sandy loam, loamy sand	Low	High rate	High
B	Loam or silt loam, loam (gravelly) soil	Moderate	Moderate rate	Moderate
C	Loam (gravelly) soil, clayey soil	Moderate	Moderate rate	Slow
D	Clayey soil, rocky outcrops	High	Low rate	Very slow

Table 2 Classification of soil group of area of interest

HSG	Area (km^2)	%
C	84.882712	7.725
D	1013.886311	92.275
Total	1098.769023	100

Table 3 Classification on the basis of moisture content in soil (AMC) [19]

		Five-day antecedent rainfall (mm)	
AMC category	Soil condition	Growing season	Inactive season
I	Dry	<36	<13
II	Average	36–53	13–28
III	Wet	>53	>28

Modification of SCS-CN model has been done for applying it in bigger watersheds (exceeding 15 km^2), and this methodology could be used by calculating a weighted curve number based on the various LULC areas within the watershed. The formula for determining the weighted curve number is as follows [20]:

$$\mathrm{CN}_w = \frac{\left(\mathrm{CN}_1 A_1 + \mathrm{CN}_2 A_2 + \cdots + \mathrm{CN}_n A_n\right)}{\left(A_1 + A_2 + \cdots + A_n\right)}$$

Here, weighted curve number is represented by CN_w, curve number from 0 to 100 is given by CN_i (i is natural number), and A_i (i is natural number) represents area with curve number CN_i.

The given below formulae could be used to calculate AMC_I and $\mathrm{AMC}_\mathrm{III}$ [21]:

$$\mathrm{CN}_\mathrm{I} = \frac{\mathrm{CN}_\mathrm{II}}{2.81 - 0.0128\mathrm{CN}_\mathrm{II}}$$

$$\mathrm{CN}_\mathrm{III} = \frac{\mathrm{CN}_\mathrm{II}}{0.427 + 0.00573\mathrm{CN}_\mathrm{II}}$$

5.4 *Rainfall–Runoff*

The daily rainfall data from June 2023 to September 2023 was collected from India-Water Resource Information System (WRIS) [10]. As shown in Table 4, the rainfall of the region ranges from 93.95 mm to 543.07 mm, while the runoff ranges from 47.57 mm to 477.68 mm. This area experiences about 77.98% of total rainfall received in the whole year. The average daily rainfall is 11.006 mm for the above 4 months. The monthly rainfall graph for the above 4 months of area of interest is given in Fig. 5. The total average volume of surface runoff we got is 302,196,119.25 m^3. This volume was calculated by multiplying the area of the watershed with runoff of that month.

Table 4 Monthly runoff calculation (2023) for growing season

Months	Rainfall (mm)	CN_w	S	Runoff (mm)	Area (km^2)	Volume (m^3)
June 2023	163.5	81.026	59.48	108.8847492	1098.769	119,639,187
July 2023	531.25	81.026	59.48	465.9864412	1098.769	512,011,456
August 2023	543.07	81.026	59.48	477.6841297	1098.769	524864513.5
September 2023	93.95	81.026	59.48	47.5707998	1098.769	52269320.12
Total	1331.77			1100.12612		1,208,784,477

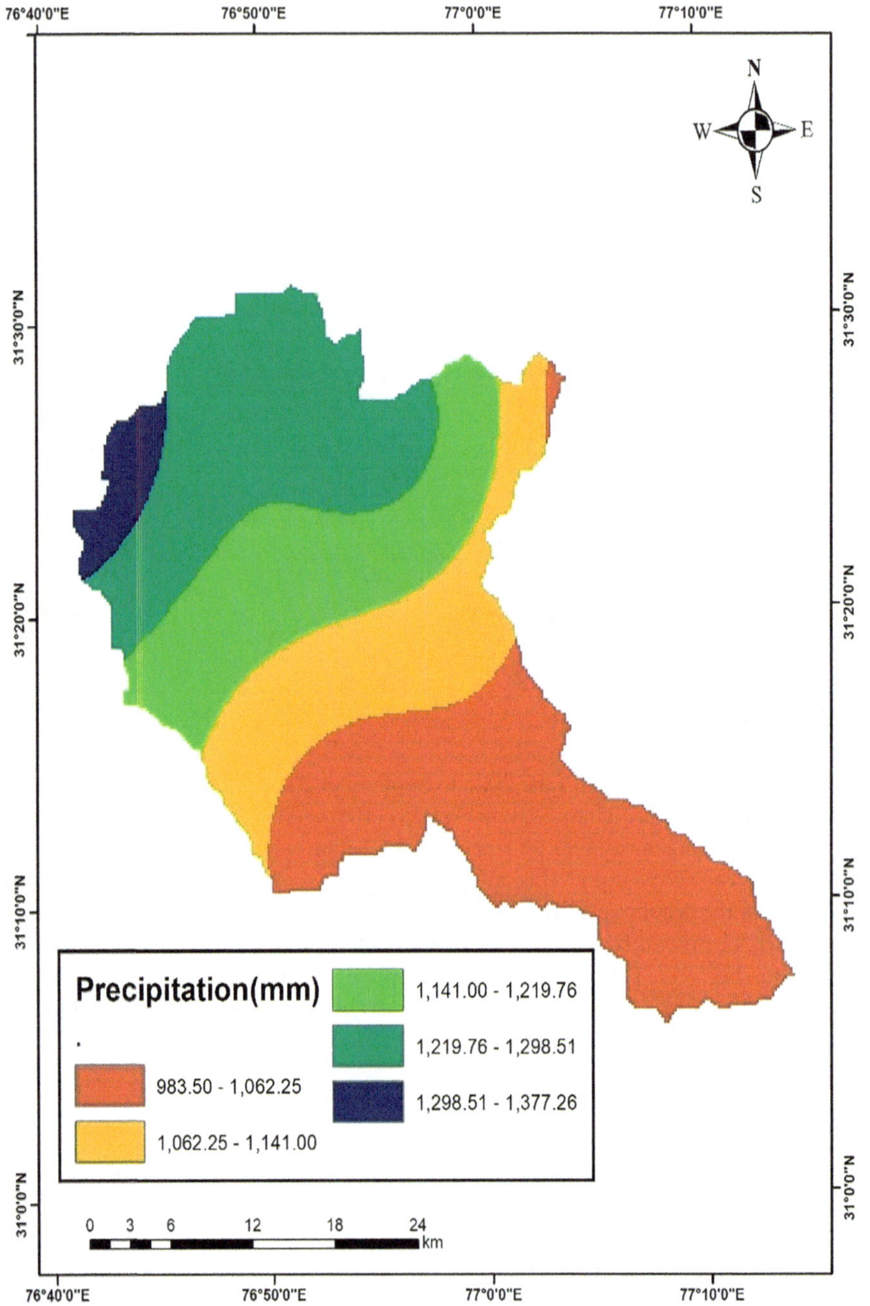

Fig. 5 Precipitation map

Table 5 CN_w calculation

LULC	HSG	Area (km^2)	CN	Area*CN	CN_w
Urban	C	2.30	74	170.2	$AMC_I = 64.203$
	D	15.83	89	1408.87	$AMC_{II} = 81.026$
Forest	C	30.61	70	2142.7	$AMC_{III} = 90.759$
	D	513.11	77	39,509.47	
Fallow land	C	3.39	90	305.1	
	D	149.64	94	14,066.16	
Agriculture	C	39.14	78	3052.92	
	D	320.53	81	25,962.93	
Water	–	23.66	100	2366	

The area of interest is in the outer foothills of the Himalayas. There are no major mountains in this area, i.e., the elevation is not very high, but the terrain is quite hilly. Flat land is limited to narrow strips along rivers and streams [22]. This indicates that the runoff potential of this area is higher. As shown in Table 5, the calculated curve numbers (Fig. 6) in wet, average, and heavy rainfall condition are 64.203, 81.026, and 90.759, respectively. The slope map is represented in Fig. 7, and runoff depth map is represented in Fig. 8.

6 Conclusion

From the above study, it can be concluded that surface runoff majorly depends on soil texture and depth of rainfall in the area of interest (Fig. 9). The combined use of geographic information system (GIS), remote sensing (RS), and SCS-CN method has proven to be an effective approach in estimating surface runoff. Accurate calculation of surface runoff is crucial for managing and conserving water resources, especially in areas where droughts are frequent, to ensure rational managing of water resources. The model used is very sensitive to the curve number values and is not a clear approach on how to fluctuate antecedent moisture conditions. The result of this study showed the curve number in different conditions (i.e., AMCI, AMCII, and AMCIII are 64.203, 81.026, and 90.759, respectively) and an average runoff from June 2023 to September 2023 of the area of interest to be 275.031 mm.

This method has been used for surface runoff assessment in this area because of its stability, predictability, and simplicity. It effectively responds to key runoff giving watershed properties like soil types, LULC, surface condition, and AMC. Combining geographic information system (GIS), remote sensing (RS), and SCS-CN model enhances both the accuracy and speed of runoff estimation.

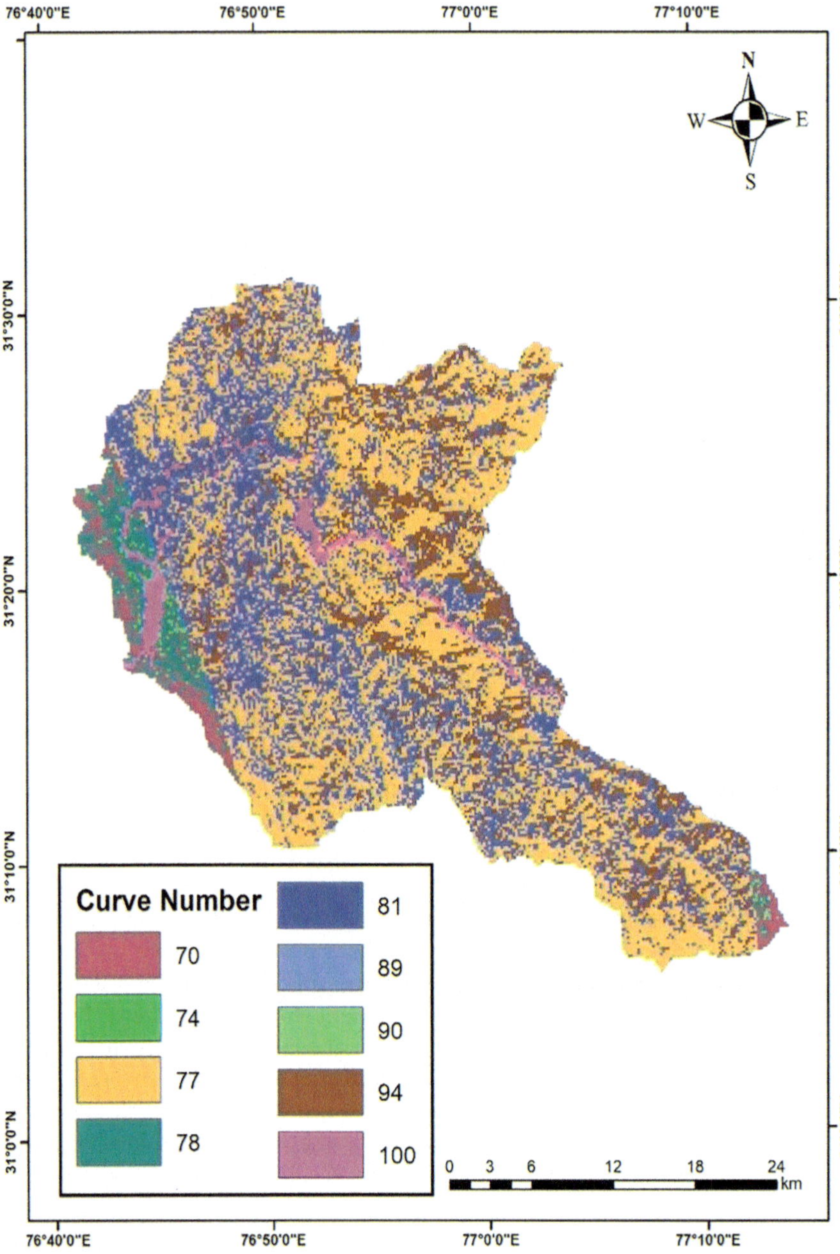

Fig. 6 Curve number map

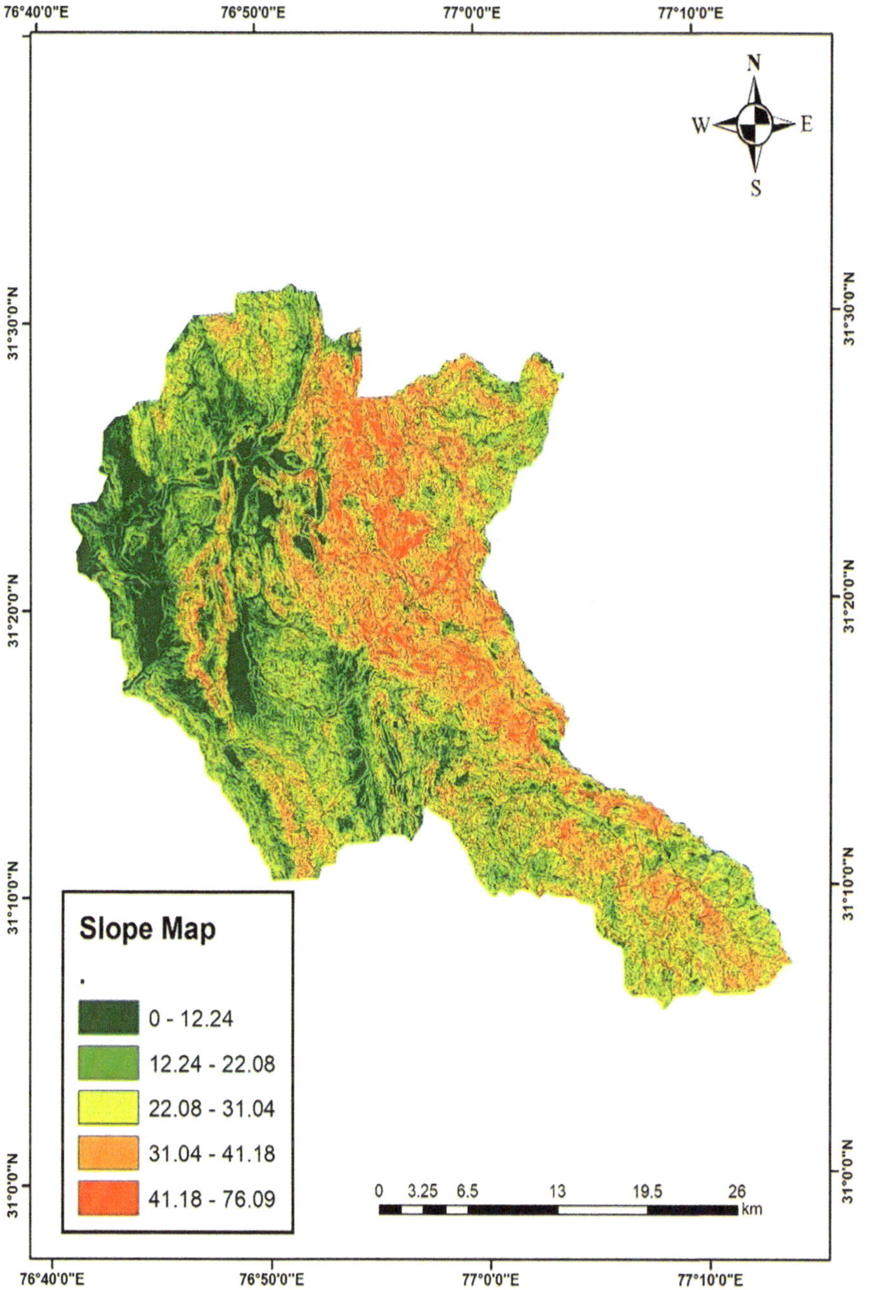

Fig. 7 Slope map

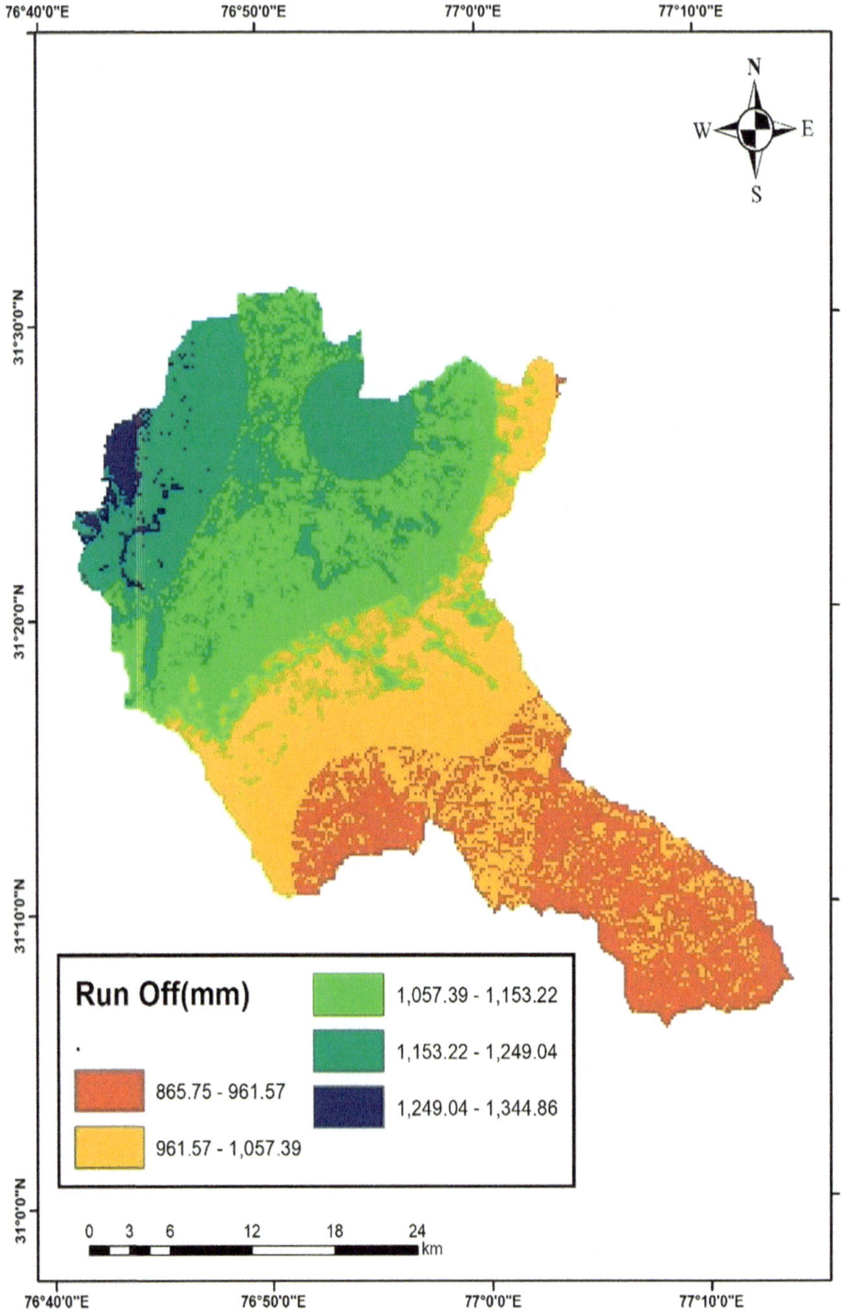

Fig. 8 Runoff map

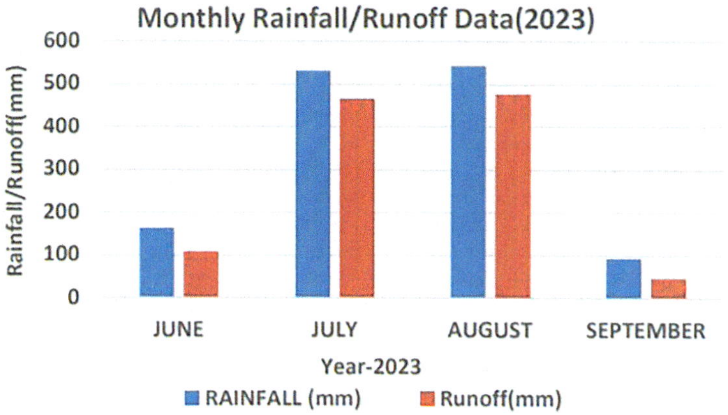

Fig. 9 Monthly rainfall/runoff data of area of interest

References

1. Kumar, A., Kanga, S., Taloor, A. K., Singh, S. K., & Đurin, B. (2021). Surface runoff estimation of Sind river basin using integrated SCS-CN and GIS techniques. *HydroResearch, 4*, 61–74. https://doi.org/10.1016/j.hydres.2021.08.001
2. Edwards, P. J., Williard, K. W. J., & Schoonover, J. E. (2015). Fundamentals of watershed hydrology. *Journal of Contemporary Water Research & Education, 154*(1), 3–20.
3. Anonymous Contributors. "Satluj," India WRIS Wiki. Retrieved from http://indiawris.gov.in/wiki/doku.php?id=satluj&rev=1629789147
4. Singh, J., & Yadav, R. R. (2013). Tree-ring-based seven century long flow records of Satluj River, western Himalaya, India. *Quaternary International, 304*, 156–162. https://doi.org/10.1016/j.quaint.2013.03.024
5. Mishra, P., Jena, D., Thakur, R. R., Chand, S., Javed, B., & Shukla, A. K. (2024). Peri-urban floodscapes: Identifying and analyzing flood risk areas in North Bhubaneswar in Eastern India. *Water, 16*(21), 3019. https://doi.org/10.3390/w16213019
6. Government of India Ministry of Water Resources Central Ground Water Board.
7. M. Sayeed, U. Hasan, & S. K. Singh (2018). *Estimation of surface runoff using geospatial model: A case study of Gandheshwari watershed.* Retrieved from http://www.sedindia.org
8. U.S. Geological Survey, USGS Website. Retrieved from https://earthexplorer.usgs.gov/
9. Food and Agriculture Organization, FAO website, FAO. Retrieved from https://www.fao.org/home/en
10. "India WRIS ," INDIA WRIS WIKI. Retrieved from https://indiawris.gov.in/wris/#/
11. "HydroSHEDS," HydroBASINS. Retrieved from https://www.hydrosheds.org/products/hydrobasins#downloads
12. Siddi Raju, R., Sudarsana Raju, G., & Rajasekhar, M. (2018). Estimation of rainfall runoff using SCS-CN method with RS and GIS Techniques for Mandavi Basin in YSR Kadapa District of Andhra Pradesh, India. *Hydrospatial Analysis, 2*(1), 1–15. https://doi.org/10.21523/gcj3.18020101
13. Shukla, A. K., Ojha, C. S. P., Garg, R. D., Shukla, S., & Pal, L. (2020). Influence of spatial urbanization on hydrological components of the Upper Ganga River Basin, India. *Journal of Hazardous, Toxic, and Radioactive Waste, 24*(4), 04020028. https://doi.org/10.1061/(ASCE)HZ.2153-5515.0000508

14. Shukla, A. K., Pathak, S., Pal, L., Ojha, C. S. P., Mijic, A., & Garg, R. D. (2018). Spatio-temporal assessment of annual water balance models for upper Ganga Basin. *Hydrology and Earth System Sciences, 22*(10), 5357–5371. https://doi.org/10.5194/hess-22-5357-2018
15. HEC-HMS Technical Reference Manual-1. Retrieved from www.hec.usace.army.mil
16. Soil Hydrologic Groups. Retrieved from https://efotg.sc.egov.usda.gov/references/Delete/2017-11-11/hydrogroups.htm
17. Mishra, S. K., Jain, M. K., & Singh, V. P. (2004). *Evaluation of the SCS-CN-based model incorporating antecedent moisture*. Kluwer Academic Publishers.
18. Matomela, N., Tianxin, L., Morahanye, L., Bishoge, O. K., & Ikhumhen, H. O. (2019). Rainfall-runoff estimation of Bojiang lake watershed using SCS-CN model coupled with GIS for watershed management. *Journal of Applied and Advanced Research*, 16–24. https://doi.org/10.21839/jaar.2019.v4i1.263
19. M. Singh & D. P. Satapathy. *Rainfall-runoff estimation using SCS-CN and GIS approach in the Kuakhai watershed of the Mahanadi Basin of Bhubaneswar Odisha*. Retrieved from www.ermt.net
20. C. S. Bhura, N. P. Singh, P. R. Mori, I. Prakash, & K. Mehmood (2015). *Estimation of surface runoff for ahmedabad urban area using SCS-CN method and GIS*. Retrieved from www.ijste.org
21. Moid Mohammed, A., Lakshmi Thatiparthi, V., Rao Pyla, K., & Maryada, A. (2020). Estimation of surface runoff in an ungauged basin using SCS-CN method, a case study of Manair River Basin in Telangana, India. *Applied Ecology and Environmental Sciences, 8*(6), 340–350. https://doi.org/10.12691/aees-8-6-3
22. Arora, N., Kumar, A., & Singal, S. K. (2024). Spatial variation of suspended sediment properties in the Sutlej river basin of Indian Himalayas. *Acta Geophysica*. https://doi.org/10.1007/s11600-024-01322-6

Flood Risk Assessment and Hydrological Response Modelling in the Bhimgoda Region, Upper Ganga Basin, India

Kajal and Shray Pathak

1 Introduction

Disaster is a term which is used very commonly nowadays and can be caused by many kinds of hazards, like floods, landslides, earthquakes, wildfires, and droughts. Hazards are of natural or anthropogenic origin [1]. Worldwide records show flood as the most frequent natural disasters to occur. Global warming and the resulting shifts in climate patterns are key drivers behind this increase. Severe rains combined with a lack of proper infrastructure in urban areas can lead to flood-like situations. Thereby accurate flood predictions and modelling are essential for efficient flood control in such environments. Providing accurate hydrological outputs, such as rainfall-runoff volume over watersheds, is a crucial part of flood forecasting. The accuracy of these outputs has a direct bearing on how reliable flood predictions are.

Precise characterization of hydrological dynamics over time is a requirement for improved flood early warning systems. Many river basins have changed hydrologically as a result of global warming, which leads to an increase or decrease in local and regional temperatures from the AMC [2]. Elevated temperatures cause shorter periods of strong and frequent rainfall, which in turn modifies flood patterns, which may lead to flooding in urban regions due to impervious surface [3]. As a result, flood modelling and forecasting become more vital in tackling the growing threat that changes in climate are posing to the hydrological systems.

The main cause of runoff is rainfall, but it also depends on its interaction with other elements of nature, such as soil type, evaporation, storage, and land use [4]. It is important to study this behaviour as it affects how the rainfall becomes runoff. Using the water balance equation and important runoff variables, HEC-HMS of the

Kajal (✉) · S. Pathak
Department of Civil Engineering, Indian Institute of Technology Ropar, Rupnagar, Punjab, India
e-mail: 2023cem1008@iitrpr.ac.in

S. Pathak et al. (eds.), *Geo-Data Revolution*, Advances in Geographic Information Science, https://doi.org/10.1007/978-3-031-94078-1_21

Hydrologic Engineering Centre simulates precipitation-runoff events and is used for hydrological modelling [5]. HEC-RAS of the Hydrologic Engineering Centre, on the other hand, models water flow along river channels while accounting for channel morphology, provides information about places that have been inundated, and is used for hydraulic modelling [6]. These two models can be used together to evaluate the flooded areas for a particular storm occurrence, as one will give the discharge and the other will provide the flood inundation map. Additionally, the calibrated model can forecast future floodplains based on anticipated precipitation and land use data, which helps with flood risk assessment and management. In this study, the generated model was used to analyse flood risks by mapping floodplain extent, velocity, and shear stress for the return periods of 02, 05, 10, 25, 50, and 100 years.

In the HEC-HMS modelling part, basin model, meteorological model, and control specification model are created and these models provide user with wide range of methods to choose from for the calculation of loss, transformation, and river routing methods [7]. HEC-HMS provides user with the discharge value which is further used in HEC-RAS as input for further calculation for the different map generation of velocity, shear stress, floodplain delineation, and water surface generation(depth). To mitigate environmental and human impacts resulting from the severe effects of climate change, flood maps serve as a crucial tool for early warning systems, helping to minimize the adverse consequences of extreme weather events [6].

This study was carried on a watershed whose output point was at Bhimgoda Barrage. The study area includes densely populated regions like Haridwar, Rishikesh which are located near the banks of river Ganga [8], making them highly vulnerable in case of flooding. A barrage upstream of Haridwar plays a crucial role in flood control, but its overflow or failure due to heavy rainfall is a major concern for the nearby areas. Therefore, accurately calculating runoff is essential, as intense rainfall can generate high runoff volumes, leading to potential flood risks. The region's vulnerability, coupled with its critical infrastructure, makes it an important area for detailed flood mapping and management efforts.

2 Study Area and Methodology

2.1 Study Area

Uttarakhand, a state in India lies in the range of Latitude 28° 44′ & 31° 28' North and longitude 77° 35′ & 81° 01′ East and covers an area of $53{,}483^2$ km [9]. The topography and terrain of the state are mainly hilly with large areas under snow cover and steep slopes. Altitude of the state Uttarakhand changes from 187 m at Sharda Sagar Reservoir to 7186 m, which is at Nanda Devi Peak, the highest land point of Uttarakhand and the second highest mountain in India above Mean Sea

Level (MSL). The state of Uttarakhand shares a common boundary with China and Nepal [10]. The region receives an average annual rainfall of approximately 1200–2500 mm [11], with monsoon accounting for over 70% of the total precipitation. Weather patterns in the region are heavily influenced by its altitude and monsoonal winds, due to which the region is prone to extreme weather events like cloudbursts, flash floods, and landslides [12]. Figure 1 depicts the map of the study area and its digital elevation model (DEM).

2.2 *Methodology*

2.2.1 Curve Number (CN) Grid Map

In this study, CN is calculated by the intersection of soil hydrological group and land use. Raster calculator tool was employed to derive a unique combination of Land Use Land Cover (LULC) categories and Hydrological Soil Group (HSG) classification. This was achieved using the formulation LULC + (100 × Hydrological Soil Group). After CN map generation (Fig. 2), the Digital Elevation Map (DEM) is given as input in HEC-Geo HMS. Basin processing, terrain processing, and

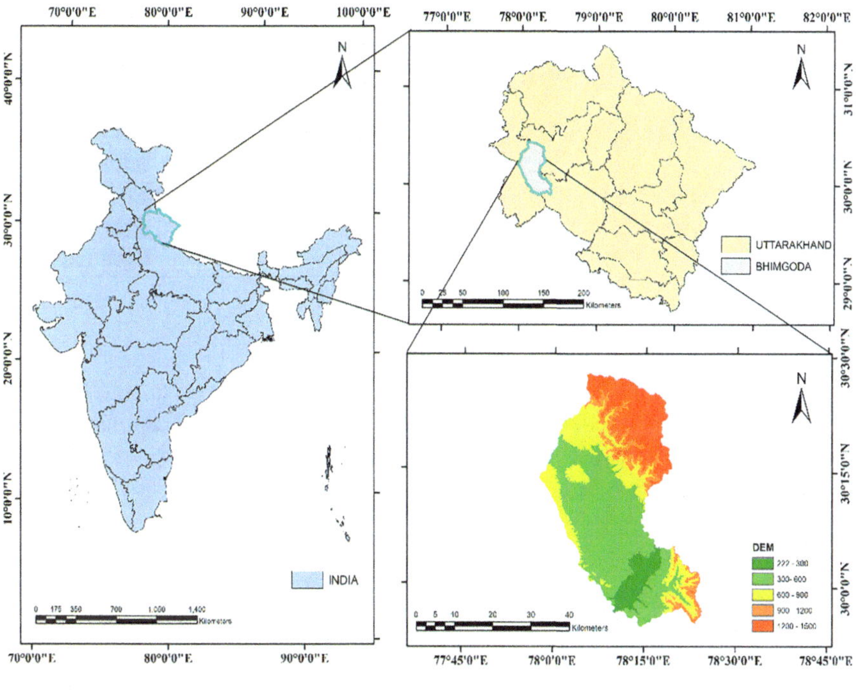

Fig. 1 Map of study area

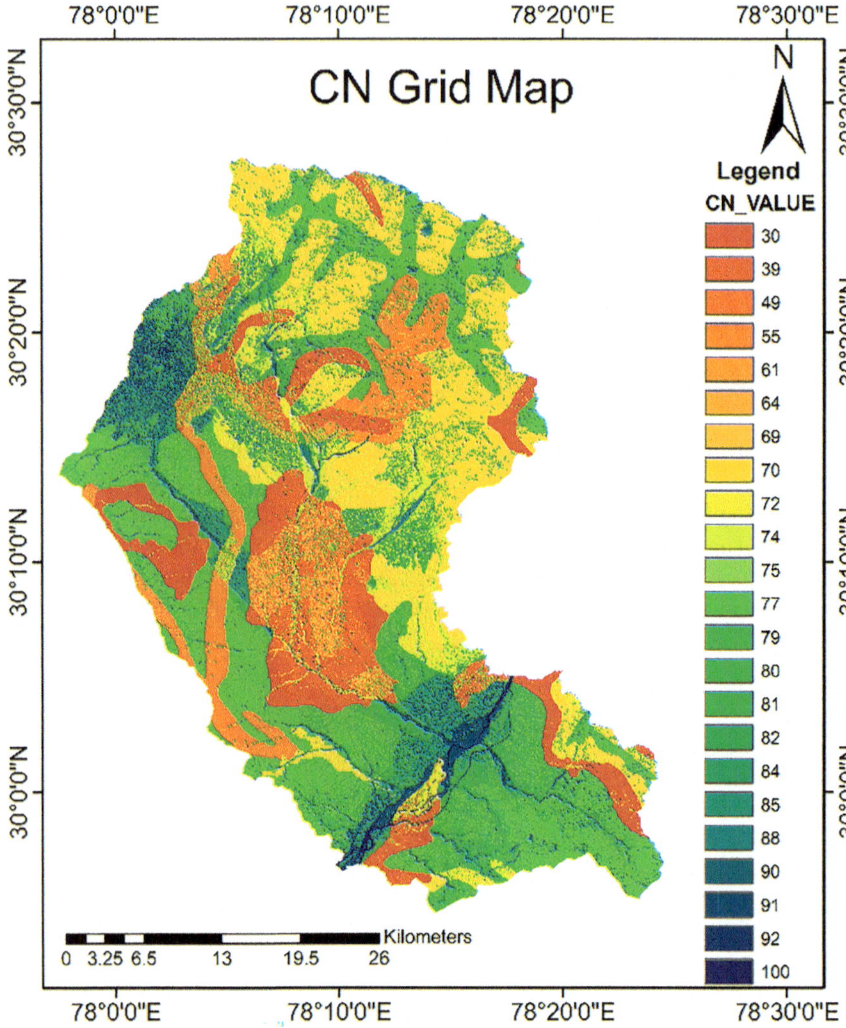

Fig. 2 Curve number map

hydrological parameter computation were done on it which served as foundation for the creation of basin model.

2.2.2 HEC-HMS

The generated basin model was then exported in a format readable by HEC-HMS to model the runoff process for the generation of the meteorological model. Average daily precipitation data from the years 2006 to 2023 were used as input to the

HEC-HMS program using the time series manager tool. In the metrological model, monthly average ET data is incorporated. The control specification function in HEC-HMS helps to specify the time window for which the simulation will be run. With the help of HEC-HMS, the discharge value at each sub-basin and reach is found out whose value is further used in HEC-RAS.

Governing equations used in HEC-HMS are Loss Method—SCS Curve Number, Transform Method—SCS Unit Hydrograph, and River Routing Method—Muskingum Routing Model.

After the model setup, IDF curves (Fig. 3) were constructed for different recurrence intervals by using Gumbel's method. In this method, precipitation depth is computed for a given return period where K_T value denotes the frequency factor for return period T.

$$X_T = \bar{X} + K_T S$$

$\bar{X}$ = mean, S = standard deviation.

2.2.3 HEC-GeoRAS

HEC-GeoRAS prepares the geometric data so that it can be further utilized in HEC-RAS. A digital terrain model of the watershed is created by using the 3D analysis tool (Fig. 4). For hydraulic modelling in HEC-RAS, RAS layers like Stream Centerline, Bank Line, and Flow path Centerline are created. After that river and the reach name are assigned. XS cut lines are then created by using the RAS GEOMETRY tool. All this geometric data is then exported to HEC-RAS for further calculation.

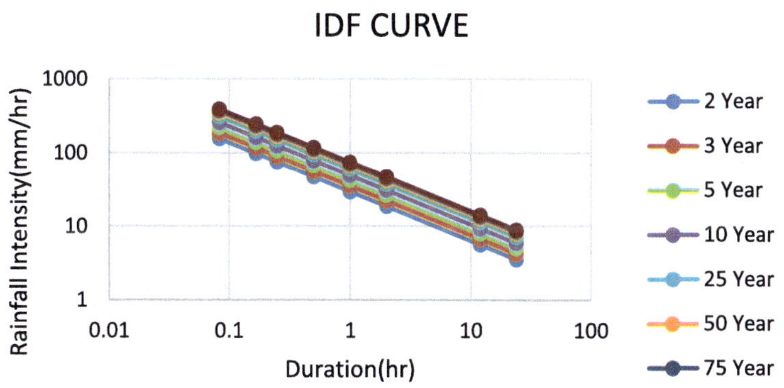

Fig. 3 Intensity–Duration–Frequency (IDF) curve for different return periods

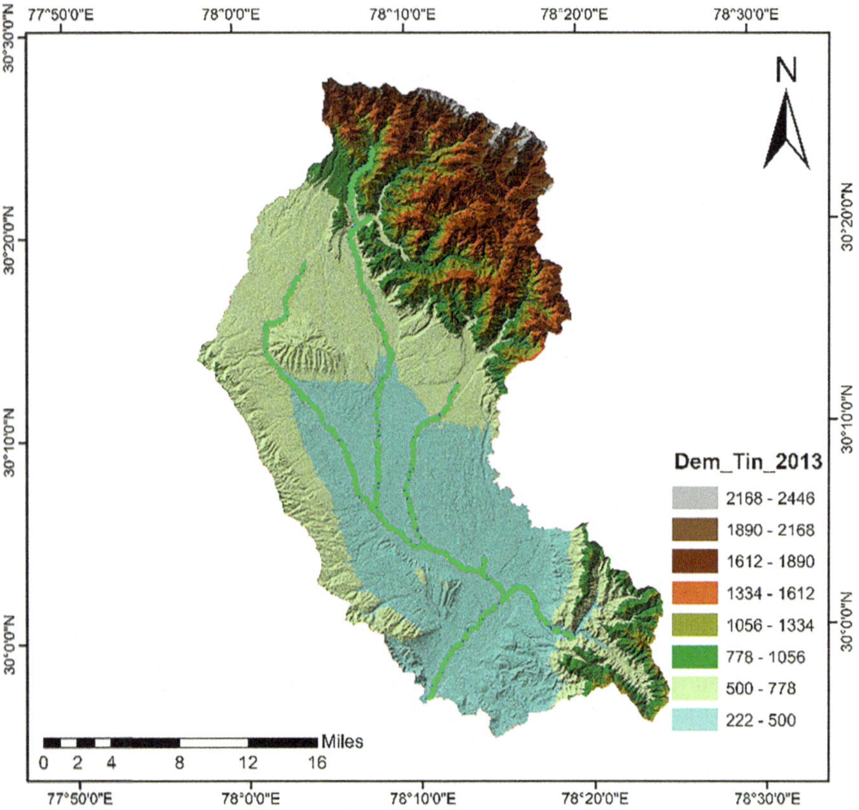

Fig. 4 Triangulated Irregular Network (TIN) model

2.2.4 HEC-RAS

All the necessary files are loaded in the software. Manning's '*n*' value is determined by previous research papers. The geometric data assigning primarily comprises the river centre line's input from US to DS. After the cross-sectioning, the boundary condition has to be assigned in the model. The boundary condition involves steady state analysis of the model. The steady flow data tool in HEC-RAS is used to set up flow rates. The discharge value for each return period is calculated and assigned corresponding to that return period. Discharge values of all reaches, i.e. upstream, downstream, obtained from HEC-HMS were used for the analysis of steady flow. HEC-RAS file is then exported to ArcMap for further calculation. RAS data was imported in ARCMap, and maps for flood inundation, velocity, and shear stress interpolation were generated. After the model setup in HEC-HMS, flood frequency analysis was carried out for different return periods like 2, 5, 10, 25, 50, and 100 years.

Table 1 Satellite imagery applications in landscape mapping

S. No	Title	Data description	Source
1	Satellite image	REScene having pixel resolution of 6.5 m (2013)	Planet scope
2	DEM (Digital Elevation Model)	ALOS PALSAR RTC DEM 12.5 m resolution	NASA earth observation data
3	Daily precipitation	Daily gridded high spatial resolution rainfall data (0.25 × 0.25)	IMD
4	Soil map	Digital maps of soil	NBSSLUP
5	ET (Evapotranspiration) data	Monthly actual evapotranspiration data	USGS-CHIRPS
6	Discharge data	Daily discharge data in cumec	Northern division of Ganga Canal, Roorkee

3 Dataset Used

As illustrated in Table 1, various datasets were utilized to develop an accurate and comprehensive hydrological model using HEC-HMS. Satellite imagery and soil maps were employed to generate the Curve Number (CN) grid map, which plays a vital role in estimating surface runoff by accounting for land use, soil type, and hydrologic conditions. The Digital Elevation Model (DEM) was used to extract key watershed characteristics such as slope, drainage patterns, stream networks, and sub-basin boundaries, all of which are essential for understanding the topography and flow dynamics of the study area. Furthermore, evapotranspiration data was integrated to calculate water losses due to evaporation and plant transpiration, which are critical components of the water balance. Daily precipitation and discharge data were also incorporated into the model setup to simulate rainfall-runoff processes.

4 Results and Discussion

After calibration and validation, for a return period of 2, 5, 10, 25, 50, and 100 years, modelling of the basin was done in HEC-HMS. Peak discharge values corresponding to the reach of the sub-basin are noted to be used later in the study in HEC-RAS for the generation of flood inundation modelling.

Below is a comparative summary highlighting the significant difference in flood inundation areas between a 2-year and 100-year return period (Fig. 5).

The comparison between the 2-year and 100-year return periods reveals a striking difference in the extent of flood inundation within the study area, underscoring the dramatic impact of extreme hydrological events. For a 2-year return period, the inundated area is relatively limited, covering approximately 0.29 km^2 out of the

Fig. 5 Flood inundation map (**a**) 2 years RP (**b**) 100 years RP

total 14 km^2 study area equivalent to just about 2% of the region. This level of inundation typically represents more frequent, low-intensity flood events with limited spatial impact. However, in stark contrast, the scenario for a 100-year return period presents a much more severe outcome. The inundation area expands significantly to 4.42 km^2, accounting for nearly 31% of the total study area. This represents more than a 15-fold increase in the flooded area compared to the 2-year event. Such a substantial rise in inundation highlights the vulnerability of the landscape to high-magnitude flood events and underscores the growing risk posed by climate variability and extreme weather patterns. It also illustrates the critical need for robust risk assessment and the implementation of effective mitigation strategies to safeguard lives, infrastructure, and ecosystems. As return periods increase, so too do the associated hazards, emphasizing the importance of long-term planning and adaptive management in flood-prone regions.

5 Conclusion

This study highlights the vulnerability of the Upper Ganga Basin to catastrophic flood events. Using hydrological and hydraulic modelling tools like HEC-HMS and HEC-RAS, the research successfully mapped flood inundation patterns. The generated flood inundation maps and scenario analyses provide valuable insights into high-risk zones, offering a basis for targeted mitigation strategies.

Flood inundation map for 100 years had the largest extent area-wise when compared to 2, 5, 10, 25, 50 years return period floods. The findings underscore the importance of integrating advanced flood modelling techniques with historical data to enhance disaster preparedness and resilience. A study of this kind is the first one for this watershed and emphasizes the need for sustainable land use planning and effective early warning systems to reduce flood-related risks in this vulnerable region.

References

1. Peker, İ. B., Gülbaz, S., Demir, V., Orhan, O., & Beden, N. (2024). Integration of HEC-RAS and HEC-HMS with GIS in flood modeling and flood hazard mapping. *Sustainability, 16*(3), 1226.
2. Nijssen, B., O'Donnell, G. M., Hamlet, A. F., & Lettenmaier, D. P. (2001). Hydrologic sensitivity of global rivers to climate change. *Climatic Change, 50*, 143–175.
3. Fowler, H. J., Lenderink, G., Prein, A. F., Westra, S., Allan, R. P., Ban, N., et al. (2021). Anthropogenic intensification of short-duration rainfall extremes. *Nature Reviews Earth & Environment, 2*(2), 107–122.
4. Zhang, J., Zhang, H., Xiao, H., Fang, H., Han, Y., & Yu, L. (2021). Effects of rainfall and runoff-yield conditions on runoff. *Ain Shams Engineering Journal, 12*(2), 2111–2116.
5. Shekar, P. R. (2021). Rainfall-runoff modelling of a River Basin using HEC HMS: A review study. *International Journal for Research in Applied Science and Engineering Technology, 9*, 506–508.
6. Al-Hussein, A. A., Khan, S., Ncibi, K., Hamdi, N., & Hamed, Y. (2022). Flood analysis using HEC-RAS and HEC-HMS: A case study of Khazir River (Middle East—Northern Iraq). *Water, 14*(22), 3779.
7. Sahu, M. K., Shwetha, H. R., & Dwarakish, G. S. (2023). State-of-the-art hydrological models and application of the HEC-HMS model: A review. *Modeling Earth Systems and Environment, 9*(3), 3029–3051.
8. Sati, V. P. (2021). Geography of the Ganga Basin. In *The Ganges: Cultural, economic and environmental significance* (pp. 27–42). Springer International Publishing.
9. Arumuga, T. (2024). Study of mountain ecosystem accounting in lower Himalaya range in Uttarkhand, India using geospatial technology. In *Study of mountain ecosystem accounting in lower Himalaya range in Uttarkhand, India using geospatial technology*.
10. Dhapola, D. S., & Sharma, S. (2024). Migration from Uttarakhand's border districts and its strategic implications: A study of district Chamoli. *Journal of Polity and Society, 16*(1).
11. Sharma, U. C., Datta, M., & Sharma, V. (2023). Climate and biodiversity. In *Soils in the Hindu Kush Himalayas: Management for agricultural land use* (pp. 27–57). Springer International Publishing.
12. Kumar, D., & Bhattacharjya, R. K. (2021). Change in rainfall patterns in the hilly region of Uttarakhand due to the impact of climate change. *Applied Environmental Research, 43*(1), 1–13.

Index

S. Pathak et al. (eds.), *Geo-Data Revolution*, Advances in Geographic Information Science, https://doi.org/10.1007/978-3-031-94078-1

The manufacturer's authorised representative in the EU is Springer Nature Customer Service Centre GmbH, Europaplatz 3, 69115 Heidelberg, Germany. If you have any concerns regarding our products, please contact ProductSafety@springernature.com

Printed and bound by CPI Group (UK) Ltd, Croydon, CR0 4YY
30/09/2026
02225571-0013